Character Theory of Finite Groups

Pure and Applied Mathematics

A Series of Monographs and Textbooks

Editors **Samuel Eilenberg and Hyman Bass**

Columbia University, New York

RECENT TITLES

JAMES W. VICK. Homology Theory: An Introduction to Algebraic Topology

E. R. KOLCHIN. Differential Algebra and Algebraic Groups

GERALD J. JANUSZ. Algebraic Number Fields

A. S. B. HOLLAND. Introduction to the Theory of Entire Functions

WAYNE ROBERTS AND DALE VARBERG. Convex Functions

A. M. OSTROWSKI. Solution of Equations in Euclidean and Banach Spaces, Third Edition of Solution of Equations and Systems of Equations

H. M. EDWARDS. Riemann's Zeta Function

SAMUEL EILENBERG. Automata, Languages, and Machines: Volume A and Volume B

MORRIS HIRSCH AND STEPHEN SMALE. Differential Equations, Dynamical Systems, and Linear Algebra

WILHELM MAGNUS. Noneuclidean Tesselations and Their Groups

FRANÇOIS TREVES. Basic Linear Partial Differential Equations

WILLIAM M. BOOTHBY. An Introduction to Differentiable Manifolds and Riemannian Geometry

BRAYTON GRAY. Homotopy Theory: An Introduction to Algebraic Topology

ROBERT A. ADAMS. Sobolev Spaces

JOHN J. BENEDETTO. Spectral Synthesis

D. V. WIDDER. The Heat Equation

IRVING EZRA SEGAL. Mathematical Cosmology and Extragalactic Astronomy

J. DIEUDONNÉ. Treatise on Analysis: Volume II, enlarged and corrected printing; Volume IV

WERNER GREUB, STEPHEN HALPERIN, AND RAY VANSTONE. Connections, Curvature, and Cohomology: Volume III, Cohomology of Principal Bundles and Homogeneous Spaces

I. MARTIN ISAACS. Character Theory of Finite Groups

In preparation

JAMES R. BROWN. Ergodic Theory and Topological Dynamics

CLIFFORD A. TRUESDELL. A First Course in Rational Continuum Mechanics: Volume 1, General Concepts

K. D. STROYAN AND W. A. J. LUXEMBURG. Introduction to the Theory of Infinitesimals

MELVYN BERGER. Nonlinearity and Functional Analysis: Lectures on Nonlinear Problems in Mathematical Analysis

B. M. PUTTASWAMAIAH AND JOHN D. DIXON. Modular Representations of Finite Groups

CHARACTER THEORY OF FINITE GROUPS

I. MARTIN ISAACS

Department of Mathematics
University of Wisconsin
Madison, Wisconsin

AP 1976

ACADEMIC PRESS New York San Francisco London

A Subsidiary of Harcourt Brace Jovanovich, Publishers

ACADEMIC PRESS, INC.
111 Fifth Avenue, New York, New York 10003

United Kingdom Edition published by
ACADEMIC PRESS, INC. (LONDON) LTD.
24/28 Oval Road, London NW1

Library of Congress Cataloging in Publication Data

Isaacs, Irving Martin, (date)
Character theory of finite groups.

(Pure and applied mathematics, a series of monographs and textbooks; 69)
Bibliography: p.
1. Finite groups. 2. Characters of groups. I. Title. II. Series. QA3.P8 [QA171] 510'.8s [512.24]
ISBN 0-12-374550-0 75-19652

AMS (MOS) 1970 Subject Classifications: 20C15, 20C05, 20C25, and 20C20

PRINTED IN THE UNITED STATES OF AMERICA

Contents

Preface

Character theory provides a powerful tool for proving theorems about finite groups. In fact, there are some important results, such as Frobenius' theorem, for which no proof without characters is known. (Until fairly recently, Burnside's $p^a q^b$ theorem was another outstanding example of this.)

Although a significant part of this book deals with techniques for applying characters to "pure" group theory, an even larger part is devoted to the properties of characters themselves and how these properties reflect and are reflected in the structure of the group.

The reader will need to know some basic finite group theory: the Sylow theorems and how to use them and some elementary properties of permutation groups and solvable and nilpotent groups. A knowledge of additional topics such as transfer and the Schur–Zassenhaus theorem would be helpful at a few points but is not essential. The other prerequisites are Galois theory and some familiarity with rings. In summary, the content of a first-year graduate algebra course should provide sufficient preparation.

Chapter 1 consists of ring theoretic preliminaries, and Chapters 2–6 and 8 contain the basic material of character theory. Chapter 7 is concerned with one of the more important techniques for the application of characters to group theory.

The emphasis in all chapters except 1, 9, 10, and 15 is on characters over the complex numbers rather than on modules and representations over other fields. In Chapter 9, irreducible representations over arbitrary fields are considered; and in Chapter 10, this is specialized to subfields of the complex numbers. Chapter 15 is an introduction (and only that) to Brauer's theory of blocks and "modular characters" and so is concerned with connections between complex characters and characteristic p-representations.

The remaining chapters concern more specialized topics. Chapter 12 deals with the connections between the set of degrees of the irreducible characters and the structure of a group. Chapter 13 is essentially an exposition of a paper of Glauberman, and Chapter 14 contains some "classical" results on complex linear groups and a small sampling of more recent developments.

Following each chapter there is a selection of problems. In addition to some routine exercises, these include examples, further results, and extensions and variations of theorems in the text. It is hoped that the reader will find that these problems enhance his understanding (and enjoyment) of character theory.

Acknowledgments

At various times during the preparation of this book I was supported by research grants from the National Science Foundation and a research fellowship from the Sloan Foundation. These organizations deserve (and hereby receive) my sincere thanks. Finally, I thank my teacher, Richard Brauer, both for introducing me to character theory and for his large part in developing the subject which I find so fascinating.

Notation

$M_n(F)$	algebra of $n \times n$ matrices over F
x_V	linear transformation of the A-module V, induced by $x \in A$
A_V	$\{x_V \mid x \in A\}$
A°	regular A-module
$\dot{+}, \sum\cdot$	internal direct sum
$M(V)$	see Definition 1.12
$n_M(V)$	see Lemma 1.13
$J(A)$	the Jacobson radical, Problem 1.4
$\mathrm{Cl}(g)$	conjugacy class of g
Σ_n	symmetric group of degree n
X^T	transpose of the matrix X
$o(g)$	the order of g
$\mathbf{Z}(\chi)$	see Definition 2.26
χ_H	restriction of χ to H
$\det \chi$	see Problem 2.3
ω_χ	the algebra homomorphism $\mathbf{Z}(\mathbb{C}[G]) \to \mathbb{C}$ induced by χ
$[x, y]$	$x^{-1}y^{-1}xy$, the commutator
$v_n(\chi)$	see Lemma 4.4 and the discussion preceding it
$\varphi \times \vartheta$	see Definition 4.20
φ°	see Definition 5.1
ϑ^G	induced character
G_α	stabilizer of α in permutation representation
$I_G(\vartheta)$	inertia group, Definition 6.10

$o(\chi)$	determinantal order of χ, equals the order of det χ in group of linear characters
$\mathbb{Q}_k$	the field $\mathbb{Q}(e^{2\pi i/k})$
$\mathbf{O}^p(G)$, $\mathbf{O}^{p'}(G)$	minimal normal subgroups of p-power index and index prime to p
$R[\mathcal{S}]$	ring of R-linear combinations of $\mathcal{S} \subseteq \mathrm{Irr}(G)$
$R[\mathcal{S}]^\circ$	$\{\vartheta \in R[\mathcal{S}] \mid \vartheta(1) = 0\}$
$P(G, \mathcal{H})$	$\mathbb{Z}[\{1_H{}^G \mid H \in \mathcal{H}\}]$
$g_\pi, g_{\pi'}, g_p, g_{p'}$	see the discussion following Lemma 8.18
$\mathfrak{X}^E$	see the discussion at beginning of Chapter 9
$F(\chi)$	the field generated over F by the values of χ
$m_F(\chi)$	the Schur index, Definition 10.1
$Z(G, A)$, $B(G, A)$, $H(G, A)$	see the discussion preceding Theorem 11.7
$M(G)$	the Schur multiplier, Definition 11.12
$\hat{A}$	the group of linear characters of A
$\mathrm{Ch}(G \mid \vartheta)$, $\mathrm{Irr}(G \mid \vartheta)$	see the discussion preceding Definition 11.23
$\mathrm{core}_G(H)$	$\bigcap H^g$ for $g \in G$
c.d.(G)	$\{\chi(1) \mid \chi \in \mathrm{Irr}(G)\}$
$\mathbf{V}(\chi)$	vanishing-off subgroup, $\langle x \in G \mid \chi(x) \neq 0 \rangle$
d.l.(G)	derived length of G
$\mathbf{F}(G)$	Fitting subgroup of G
$b(G)$	max(c.d.(G))
$\mathbf{O}_p(G)$, $\mathbf{O}_{p'}(G)$	maximal normal subgroup of order a power of p, prime to p
$\mathrm{Irr}_S(G)$	the set of S-invariant $\chi \in \mathrm{Irr}(G)$
$\Phi(G)$	Frattini subgroup of G
$\mathrm{IBr}(G)$	the set of irreducible Brauer characters of G
$d_{\chi\varphi}$	decomposition numbers, Definition 15.9
Φ_φ	the projective character associated with $\varphi \in \mathrm{IBr}(G)$
$\mathrm{Bl}(G)$	the set of p-blocks of G
e_B, λ_B	the idempotent and algebra homomorphism of $\mathbf{Z}(F[G])$ associated with $B \in \mathrm{Bl}(G)$
$\delta(\mathcal{K})$	the set of defect groups for the class $\mathcal{K}$
$\hat{\mathcal{K}}$	the sum of the elements of $\mathcal{K}$ in $F[G]$
$a_B(\mathcal{K})$	coefficient of $\hat{\mathcal{K}}$ in e_B
$\delta(B)$	the set of defect groups for $B \in \mathrm{Bl}(G)$
$d(B)$	defect of $B \in \mathrm{Bl}(G)$

1 Algebras, modules, and representations

Character theory provides a means of applying ring theoretic techniques to the study of finite groups. Although much of the theory can be developed in other ways, it seems more natural to approach characters via rings (or more accurately, algebras). The purpose of this chapter is to provide the reader with the ring theoretic prerequisites needed in the rest of the book.

Many of the results in this chapter are true in more general contexts than those considered here. Nevertheless, an effort has been made to avoid excess generality and to prove only that which will be needed later.

(1.1) DEFINITION Let F be a field and let A be an F-vector space which is also a ring with 1. Suppose for all $c \in F$ and $x, y \in A$, that

$$(cx)y = c(xy) = x(cy).$$

Then A is an *F-algebra*.

We mention some examples of algebras over a field F:

(a) $M_n(F)$ is the algebra of $n \times n$ matrices over the field F.

(b) Let V be an F-vector space. Then $\mathrm{End}(V)$, the set of F-linear transformations of V, is an F-algebra under the following conventions. If $x, y \in \mathrm{End}(V)$, then xy is defined by $(v)xy = ((v)x)y$ and if $c \in F$, then cx is defined by $(v)(cx) = (cv)x$. Of course, $(v)(x + y) = (v)x + (v)y$.

This is a good place to digress briefly to discuss some notational conventions which will be used throughout this book. In writing scalar multiplication in a vector space, the scalar may be written on either side of the vector to which it is applied. Similarly, functions are written on whichever side is convenient, but the rule for function composition is always " fg means do f first, then g." Because of this rule, in those situations where function composition is important [as in example (b)], it will usually be convenient to write functions on the right.

We now return to another example of an algebra, the one of primary importance for us: the group algebra.

(c) Let G be a finite group. Then $F[G]$ is the set of "formal" sums $\{\sum_{g \in G} a_g g \mid a_g \in F\}$. The structure of an F-vector space is given to $F[G]$ in the obvious way and the element of $F[G]$ for which $a_g = 1$ and $a_h = 0$ if $h \neq g$ is identified with g. This identification embeds G into $F[G]$ and in fact G is a basis for $F[G]$. One result of this identification is to give a new meaning for $\sum a_g g$. We may now view this expression not only as a formal sum, but also as an actual sum, a linear combination of the basis vectors. Finally, to define multiplication on $F[G]$, we multiply the basis vectors according to their group multiplication and extend linearly to all of $F[G]$. It is routine to check that this defines the structure of an F-algebra on $F[G]$.

The construction of $F[G]$ suggests a general method of constructing algebras which should be mentioned. Let A be a finite dimensional F-algebra with F-basis $v_1, \ldots, v_n$. We have then $v_i v_j = \sum c_{ijk} v_k$, where $c_{ijk} \in F$ are the *multiplication constants* of A with respect to the basis $\{v_i\}$. It is clear that these constants determine the algebra, so that any n-dimensional algebra may be specified by prescribing n^3 constants $c_{ijk} \in F$. Of course, only a small subset of all possible sets of constants define an algebra since most sets of constants define multiplications that turn out to be nonassociative.

From now on, the word "algebra" in this book will mean a finite dimensional algebra. We make a few observations and definitions before going on to prove anything.

Let A be an F-algebra. Then $F \cdot 1 = \{c1 \mid c \in F\}$ is a subalgebra of A contained in the center $\mathbf{Z}(A)$ since $(c1)x = c(1x) = c(x1) = x(c1)$ for $x \in A$. It is sometimes convenient to identify F with $F \cdot 1$ and thus to view F as a subalgebra of A. If I is a left or right ideal of A as a ring and $x \in I$ and $c \in F$, we have $cx = (c1)x = x(c1) \in I$ and I is a subspace. (If we had not required algebras to contain 1, then ideals would not automatically be subspaces.) If I is an ideal (this means two-sided), then A/I has the structure of an F-algebra in a natural manner.

If A and B are F-algebras and φ is a ring homomorphism from A to B with $\varphi(1) = 1$, it is not necessarily true that φ is an F-linear transformation.

(1.2) DEFINITION Let A and B be F-algebras. Suppose that $\varphi\colon A \to B$ satisfies

(a) $\varphi(xy) = \varphi(x)\varphi(y)$ for all $x, y \in A$;
(b) $\varphi(1) = 1$;
(c) φ is an F-linear transformation.

Then φ is an *algebra homomorphism* or an *F-homomorphism.*

(1.3) DEFINITION Let A be an F-algebra and let V be a finite dimensional F-vector space. Suppose for every $v \in V$ and $x \in A$ that a unique $vx \in V$ is defined. Assume for all $x, y \in A$, $v, w \in V$, and $c \in F$ that

(a) $(v + w)x = vx + wx$,
(b) $v(x + y) = vx + vy$,
(c) $(vx)y = v(xy)$,
(d) $(cv)x = c(vx) = v(cx)$,
(e) $v1 = v$.

Then V is an *A-module.*

Let V be an A-module. Each $x \in A$ defines a map x_V: $V \to V$ by $v \mapsto vx$. By (a) and part of (d) of the definition of a module, $x_V \in \text{End}(V)$. By (b), (c), (e), and part of (d), the map $x \mapsto x_V$ is an algebra homomorphism $A \to \text{End}(V)$. Its image is denoted by A_V.

Some important examples of modules are the following. If $A \subseteq \text{End}(V)$, then V is an A-module in a natural way. If $A = M_n(F)$, then the row space of dimension n over F is an A-module under matrix multiplication. If A is any algebra, then A itself is an A-module under right multiplication. This module is called the *regular* A-module and is denoted by A°.

If V is an A-module and $W \subseteq V$ is an A-invariant subspace, then W is a *submodule of* V. Thus the right ideals of A are exactly the submodules of A°. If W is a submodule of V, then the space V/W becomes an A-module in the usual manner. Note that if I is a proper ideal of A, then the objects A/I, A°/I, and $(A/I)^\circ$ are all defined and all different, being respectively an algebra, an A-module, and the regular (A/I)-module. However, A°/I is an A-module which is annihilated by I (i.e., if $v \in A^\circ/I$ and $x \in I$, then $vx = 0$), and so it may be viewed as an (A/I)-module. As such it becomes $(A/I)^\circ$.

If V and W are A-modules, a linear transformation φ: $V \to W$ is an *A-homomorphism* if $\varphi(vx) = \varphi(v)x$ for all $v \in V$ and $x \in A$. An *A-isomorphism* is an A-homomorphism which is one-to-one and onto and, as is the usual situation, if V and W are A-isomorphic, they are exactly the same "as seen by A." For instance, in this situation they are annihilated by exactly the same set of elements of A.

The set $\text{Hom}_A(V, W)$ of A-homomorphisms from V to W has the structure of an F-space by $(c\varphi)(v) = c(\varphi v)$ for $c \in F$ and $(\varphi + \vartheta)(v) = \varphi(v) + \vartheta(v)$. In addition $\text{Hom}_A(V, V)$ is a ring [remember, $\varphi\vartheta$ is defined by $(v)\varphi\vartheta = (v\varphi)\vartheta$], and in fact $\text{Hom}_A(V, V)$ is an F-algebra. It is exactly the centralizer of A_V in End(V) and is denoted $\mathbf{E}_A(V)$. If $E = \mathbf{E}_A(V)$, then V is an E-module and $\mathbf{E}_E(V) \supseteq A_V$. Later in this chapter we shall find a sufficient condition for equality here.

(1.4) DEFINITION Let V be a nonzero A-module. Then V is *irreducible* if its only submodules are 0 and V.

It is obvious that if $\varphi \in \mathrm{Hom}_A(V, W)$ then ker φ and im φ are submodules of V and W, respectively. The following important lemma is now immediate.

(1.5) LEMMA (*Schur*) If V and W are irreducible A-modules, then every nonzero element of $\mathrm{Hom}_A(V, W)$ has an inverse in $\mathrm{Hom}_A(W, V)$.

An immediate consequence of Schur's lemma is that if V is an irreducible A-module, then $\mathbf{E}_A(V)$ is a *division algebra*, i.e., every nonzero element is invertible.

(1.6) COROLLARY Let F be algebraically closed, A an F-algebra, and V an irreducible A-module. Then $\mathbf{E}_A(V) = F \cdot 1$, the set of scalar multiplications on V.

Proof Clearly, $F \cdot 1 \subseteq \mathbf{E}_A(V)$. Let $\vartheta \in \mathbf{E}_A(V)$. Then ϑ is a linear transformation of a finite dimensional vector space V over F and so has an eigenvalue λ. Then $\vartheta - \lambda 1 \in \mathbf{E}_A(V)$ and is not invertible. Thus $\vartheta - \lambda 1 = 0$ and $\vartheta = \lambda 1 \in F \cdot 1$ as claimed. ∎

To justify the study of irreducible modules we remark that for certain algebras, every module is a direct sum of irreducible ones, and thus to know all irreducibles is to know all modules for these algebras. It happens that group algebras over fields of characteristic zero are among these fortunate algebras. This will follow from Maschke's theorem which will be proved shortly.

(1.7) DEFINITION Let V be an A-module. Suppose for every submodule $W \subseteq V$, there exists another submodule $U \subseteq V$ such that $V = W \dot{+} U$ (the dot indicating that the sum is direct). Then V is a *completely reducible* module.

Observe that irreducible modules are completely reducible as are all modules over fields (i.e., vector spaces).

(1.8) DEFINITION An algebra A is *semisimple* if its regular module, A° is completely reducible.

(1.9) THEOREM (*Maschke*) Let G be a finite group and F a field whose characteristic does not divide $|G|$. Then every $F[G]$-module is completely reducible.

Proof Let V be an $F[G]$-module with submodule W. Let U_0 be an F-subspace of V which is complementary to W, i.e., $V = W \dotplus U_0$. Let φ be the projection map of V onto W with respect to U_0. Now φ is a linear transformation which is not necessarily an $F[G]$-homomorphism. The object of the remainder of the proof is to modify φ in order to create an $F[G]$-homomorphism.

Define ϑ: $V \to W$ by

$$\vartheta(v) = \frac{1}{|G|} \sum_{g \in G} \varphi(vg)g^{-1}.$$

Clearly, ϑ is F-linear. To show that ϑ is an $F[G]$-homomorphism, we compute

$$\vartheta(vh) = \frac{1}{|G|} \sum_{g \in G} \varphi(vhg)g^{-1} = \frac{1}{|G|} \sum_{g} \varphi(vhg)(hg)^{-1}h = \vartheta(v)h$$

since hg runs over G as g does for fixed h.

If $w \in W$, we have $wg \in W$ for all $g \in G$ and thus $\varphi(wg) = wg$. It follows that $\vartheta(w) = w$. Now let $U = \ker \vartheta$, an $F[G]$-submodule of V. We have $\vartheta(v) \in W$, so that $\vartheta(\vartheta)v)) = \vartheta(v)$ and $\vartheta(v - \vartheta(v)) = \vartheta(v) - \vartheta(v) = 0$. Thus $v = \vartheta(v) + (v - \vartheta(v)) \in W + U$ and $V = W + U$. Finally, if $w \in W \cap U$, we have $w = \vartheta(w) = 0$, so that $V = W \dotplus U$ and the proof is complete. ∎

A consequence of Maschke's theorem is that $F[G]$ is semisimple if $\operatorname{char}(F) \nmid |G|$. The converse of this statement is also true and the reader is referred to the problems for a proof.

(1.10) THEOREM Let V be an A-module. Then V is completely reducible iff it is a sum of irreducible submodules.

Proof Suppose $V = \sum V_\alpha$ where the V_α are irreducible. Let $W \subseteq V$. By finite dimensionality, choose $U \subseteq V$ maximal such that $W \cap U = 0$. We claim that $W + U = V$. Otherwise, we may choose $V_\alpha \nsubseteq W + U$ and thus $(W + U) \cap V_\alpha = 0$ by the irreducibility of V_α. It follows that $W \cap (U + V_\alpha) = 0$ and this violates the maximality of U. Thus $V = W \dotplus U$ and V is completely reducible.

Conversely, suppose V is completely reducible and let S be the sum of all of the irreducible submodules of V. If $S < V$, we may write $V = S \dotplus T$ with $T \neq 0$. By finite dimensionality, T contains an irreducible submodule which is a contradiction since $T \cap S = 0$. ∎

(1.11) LEMMA Let V be an A-module and suppose $V = \sum V_\alpha$ where the V_α are irreducible submodules. Then V is the direct sum of some of the V_α's.

Proof Choose $W \subseteq V$ maximal with the property that W is the direct sum of some V_α's. If $W < V$, then $V_\alpha \nsubseteq W$ for some α. Since V_α is irreducible, we have $W \cap V_\alpha = 0$ and $W_1 = W \dotplus V_\alpha > W$. This violates the maximality of W and we conclude that $W = V$ as desired. ▮

We see from the previous two results that the completely reducible modules are exactly the direct sums of irreducible modules. It follows that in order to know all modules for a group algebra over a field of characteristic 0, it suffices to know all irreducible modules.

(1.12) DEFINITION Let V be a completely reducible A-module and let M be an irreducible A-module. The *M-homogeneous* part of V, denoted $M(V)$, is the sum of all those submodules of V which are isomorphic to M.

Observe that if $M \cong N$, then $M(V) = N(V)$. As will be shown shortly, if M and N are nonisomorphic irreducible A-modules, then $M(V) \cap N(V) = 0$. If V has no submodules isomorphic to M, then $M(V) = 0$.

(1.13) LEMMA Let $V = \sum \cdot W_i$ be a direct sum of A-modules with W_i irreducible for all i. Let M be any irreducible A-module. Then

(a) $M(V)$ is an $\mathbf{E}_A(V)$-submodule of V;
(b) $M(V) = \sum \{W_i \mid W_i \cong M\}$;
(c) The number $n_M(V)$ of W_i which are isomorphic to M is an invariant of V, independent of the given direct sum decomposition.

Proof (a) Let $\vartheta \in \mathbf{E}_A(V)$. We need to show $M(V)\vartheta \subseteq M(V)$. It suffices to show that if $W \subseteq V$ and $W \cong M$, then $W\vartheta \subseteq M(V)$. This is sufficient because $M(V)$ is the sum of such W. If $W\vartheta = 0$, there is nothing to prove; if $W\vartheta \neq 0$, then since $W\vartheta$ is a homomorphic image of the irreducible module W, we have $W\vartheta \cong W \cong M$. Thus $W\vartheta \subseteq M(V)$.

(b) Clearly, $\sum \{W_i \mid W_i \cong M\} \subseteq M(V)$. Let π_i be the projection map of V onto W_i. Now let $W \subseteq V$, $W \cong M$. If $\pi_j(W) \neq 0$, then $\pi_j(W) = W_j$ and $W \cong W_j$. Thus $\pi_j(W) \subseteq \sum \{W_i \mid W_i \cong M\}$ for all j. However, $W \subseteq \sum_j \pi_j(W)$ and hence $W \subseteq \sum \{W_i \mid W_i \cong M\}$. It follows that $M(V) \subseteq \sum \{W_i \mid W_i \cong M\}$.

(c) By (b), we have $\dim M(V) = n_M(V) \dim M$ and it is immediate that $n_M(V)$ is an invariant. ▮

By part (b) of Lemma 1.13 it follows that a completely reducible module is the direct sum of its M-homogeneous components for distinct M. In particular, $M(V) \cap N(V) = 0$ if $M \ncong N$.

We wish to discuss "all irreducible A-modules." This poses some difficulties. First, we really mean "all isomorphism classes of irreducible A-modules." It would be convenient to have a representative set of A-modules. By this we mean a set $\mathcal{M}(A)$ of irreducible A-modules with the property that every irreducible A-module is isomorphic to exactly one element of $\mathcal{M}(A)$. Since modules are external to the algebra, it is not clear how one can find a representative set for a given algebra A. To do this we need to produce a set of A-modules large enough to contain copies of all irreducibles. The next lemma shows that the set of homomorphic images of A° suffices.

(1.14) LEMMA Let A be an F-algebra. Then every irreducible A-module is isomorphic to a factor module of A°. If A is semisimple, then every irreducible A-module is isomorphic to a submodule of A°.

Proof Let V be an irreducible A-module, choose $0 \neq v \in V$, and define $\vartheta\colon A \to V$ by $\vartheta(x) = vx$. Then ϑ is clearly F-linear and $\vartheta(xy) = vxy = \vartheta(x)y$ so that $\vartheta \in \mathrm{Hom}_A(A^\circ, V)$. Now, $v \in \mathrm{im}\ \vartheta \subseteq V$ and hence $\mathrm{im}\ \vartheta = V$ since V is irreducible. Let $W = \ker \vartheta$. Then $V \cong A^\circ/W$. If A is semisimple, then $A^\circ = W \dotplus U$ and $A^\circ/W \cong U$. The proof is complete. ∎

Now fix a representative set $\mathcal{M}(A)$ of irreducible A-modules. By part (b) of Lemma 1.13, we have $V = \sum \cdot_{M \in \mathcal{M}(A)} M(V)$ for every completely reducible A-module V. Suppose A is a semisimple algebra. We can then apply the above to A° and write $A^\circ = \sum \cdot M(A^\circ)$. It turns out that $M(A^\circ)$ is actually a two-sided ideal of A. For notational convenience, we write $M(A)$ for $M(A^\circ)$ in what follows.

(1.15) THEOREM (*Wedderburn*) Let A be a semisimple algebra and let M be an irreducible A-module. Then

(a) $M(A)$ is a minimal ideal of A;
(b) if W is irreducible, then it is annihilated by $M(A)$ unless $W \cong M$,
(c) the map $x \mapsto x_M$ is one-to-one from $M(A)$ onto $A_M \subseteq \mathrm{End}(M)$;
(d) $\mathcal{M}(A)$ is a finite set.

Proof If $x \in A$, the map $\vartheta_x\colon y \mapsto xy$ satisfies $\vartheta_x \in \mathbf{E}_A(A^\circ)$. Therefore, by Lemma 1.13(a), $xM(A) = M(A)\vartheta_x \subseteq M(A)$ and $M(A)$ is a left ideal. Since $M(A)$ is a submodule of A°, it follows that it is an ideal of A. Minimality will follow after (c) is proved.

If W is an irreducible A-module with $W \ncong M$, then $W(A) \cap M(A) = 0$ by Lemma 1.13(b). Since $W(A)$ and $M(A)$ are ideals, we have $W(A)M(A) = 0$. By Lemma 1.14, A° has a submodule $W_0 \cong W$ and $W_0 \subseteq W(A)$, so that $M(A)$ annihilates W_0. Since $W \cong W_0$, they have the same annihilator in A and (b) follows.

By (b), it follows that $x_W = 0$ if $x \in M(A)$ and $M \ncong W$. It now follows from the direct sum decomposition $A = \sum \cdot_{M \in \mathcal{M}(A)} M(A)$ that for $y \in A$, we have $y_M = x_M$, where x is the component of y in $M(A)$. We conclude that the map $x \mapsto x_M$ maps $M(A)$ onto A_M. If $x \in M(A)$ and $x_M = 0$, then it follows from (b) that x annihilates every irreducible, and hence every completely reducible A-module. Thus $x = 1x \in A^\circ x = 0$ and (c) is proved.

To show the minimality of $M(A)$, let $I < M(A)$ be an ideal of A. Now $M(A)$ is a sum of submodules isomorphic to M and thus there exists $M_0 \subseteq M(A)$, $M_0 \cong M$, $M_0 \nsubseteq I$. As $M_0 \cap I < M_0$ and $M_0 \cong M$ is irreducible, we have $M_0 \cap I = 0$. Thus $M_0 I \subseteq M_0 \cap I = 0$ and I annihilates M_0 and hence also M. Therefore, if $x \in I$, we have $x_M = 0$. However, $x \mapsto x_M$ is one-to-one for $x \in M(A)$ and thus $x = 0$. We conclude that $I = 0$ and the minimality of $M(A)$ is proved.

Finally, $M(A) \neq 0$ for every M by Lemma 1.14, and yet $A = \sum \cdot_{M \in \mathcal{M}(A)} M(A)$ is finite dimensional. It follows that $|\mathcal{M}(A)|$ is finite and the proof is complete. ▮

Observe that each $M(A)$ is actually an algebra, its unit element being the component of 1 in $M(A)$ under the decomposition $A = \sum \cdot M(A)$. Since the map $x \mapsto x_M$ is an algebra homomorphism from A to A_M, it follows from (c) of the theorem that the restriction of this map to $M(A)$ is an algebra isomorphism from $M(A)$ onto A_M.

Since $M_1(A)M_2(A) = 0$ for $M_1 \ncong M_2$, it follows that every ideal of the algebra $M(A)$ is in fact an ideal of A. We conclude from the minimality of $M(A)$ as an ideal of A, that $M(A)$ is a *simple* algebra, i.e., it has no nontrivial proper ideals. Thus the preceding theorem asserts (among other things) that a semisimple algebra is a direct sum of simple algebras. This is a somewhat more usual statement of Wedderburn's theorem. (It is also true that every simple algebra is semisimple. This follows from Problem 1.5.)

To review the situation now: group algebras over fields of characteristic zero are semisimple; semisimple algebras are direct sums of ideals $M(A)$ and $M(A)$ is naturally isomorphic to A_M. What remains is to study the simple algebras A_M. This is the purpose of the "double centralizer" theorem which follows. The proof we give here is based on an idea of M. Rieffel.

It should be remarked that the hypothesis that A is semisimple is actually superfluous since the theorem is really about A_M which is automatically semisimple when M is completely reducible. See Problem 1.6.

(1.16) THEOREM (*Double Centralizer*) Let A be a semisimple algebra and let M be an irreducible A-module. Let $D = \mathbf{E}_A(M)$. Then $\mathbf{E}_D(M) = A_M$.

Proof It is no loss to replace M by an isomorphic module, and so by Lemma 1.14, we may assume that $M \subseteq A^\circ$. Let $I = M(A)$ so that $M \subseteq I$.

It is clear that $A_M \subseteq \mathbf{E}_D(M)$ so we prove the reverse inclusion. Let $\vartheta \in \mathbf{E}_D(M)$ so that $(m\alpha)\vartheta = (m\vartheta)\alpha$ for $\alpha \in D$. If $m \in M$, define α_m: $M \to A$ by $(x)\alpha_m = mx$. Since $m \in M$ is a right ideal of A, we have $mx \in M$ and α_m: $M \to M$. If $a \in A$ and $x \in M$, we have $(xa)\alpha_m = m(xa) = (mx)a = (x\alpha_m)a$. It follows that $\alpha_m \in \mathbf{E}_A(M) = D$. Thus for $m, n \in M$, we have

$$(*) \qquad (mn)\vartheta = (n\alpha_m)\vartheta = (n\vartheta)\alpha_m = m(n\vartheta).$$

Now fix $n \in M$ with $n \neq 0$ and let e be the unit element of I. We have $AnA \subseteq I$ and by the minimality of the ideal I, we have $e \in I = AnA$ and $e = \sum a_i n b_i$ for suitable $a_i, b_i \in A$. If $m \in M$, we have

$$m = me = m \sum a_i n b_i = \sum (ma_i)(nb_i).$$

Since $ma_i \in M$ and $nb_i \in M$, Equation $(*)$ yields for all $m \in M$ that

$$m\vartheta = \sum ((ma_i)(nb_i))\vartheta = \sum (ma_i)((nb_i)\vartheta) = m \sum a_i((nb_i)\vartheta).$$

Thus $\vartheta = u_M \in A_M$ where $u = \sum a_i((nb_i)\vartheta)$. The proof is complete. ∎

(1.17) COROLLARY Let A be a semisimple algebra over an algebraically closed field F and let M be an irreducible A-module. Then

(a) $A_M = \mathrm{End}(M)$;
(b) $\dim(A_M) = \dim(M(A)) = \dim(M)^2$;
(c) $n_M(A^\circ) = \dim(M)$.

Furthermore, if $\mathcal{M}(A)$ is a representative set of irreducible A-modules, then

(d) $\dim(A) = \sum_{M \in \mathcal{M}(A)} \dim(M)^2$,
(e) $\dim(\mathbf{Z}(A)) = |\mathcal{M}(A)|$.

Proof By Corollary 1.6, $\mathbf{E}_A(M) = F \cdot 1$ for irreducible M and thus $A_M = \mathbf{E}_{F \cdot 1}(M) = \mathrm{End}(M)$. By elementary linear algebra, $\mathrm{End}(M) \cong M_d(F)$, the algebra of $d \times d$ matrices, where $d = \dim M$. Thus $\dim A_M = \dim(\mathrm{End}(M)) = d^2$. Since $M(A) \cong A_M$, we have (b). Now $M(A)$ is the direct sum of $n_M(A^\circ)$ isomorphic copies of M, thus $d^2 = \dim M(A) = n_M(A^\circ) \dim M = d n_M(A^\circ)$, and (c) follows. Since $A = \sum^{\cdot}_{M \in \mathcal{M}(A)} M(A)$, (d) is immediate from (b).

Finally, let $Z^M = \mathbf{Z}(M(A))$. Now $\mathbf{Z}(A_M) = A_M \cap \mathbf{E}_A(M) = A_M \cap F \cdot 1 = F \cdot 1$. Thus $\dim Z^M = \dim(\mathbf{Z}(A_M)) = 1$. Clearly, $\sum^{\cdot}_{M \in \mathcal{M}(A)} Z^M \subseteq \mathbf{Z}(A)$ and $\dim(\sum Z^M) = |\mathcal{M}(A)|$. However, if $z \in \mathbf{Z}(A)$, write $z = \sum a^M$, $a^M \in M(A)$. If $u \in M(A)$, then $ua^M = uz = zu = a^M u$ since the distinct $M(A)$'s annihilate each other. Thus $a^M \in Z^M$ and $\sum Z^M = \mathbf{Z}(A)$. The proof is complete. ∎

Although the word "representations" constitutes one third of the title of this chapter, so far nothing has been said about them. In fact, that isn't really true, because as we shall see, representations are just a different way of looking at modules.

(1.18) DEFINITION Let A be an F-algebra. A *representation* of A is an algebra homomorphism $\mathfrak{X}: A \to M_n(F)$. The integer n is the *degree* of $\mathfrak{X}$. Two representations $\mathfrak{X}$, $\mathfrak{Y}$ of degree n are *similar* if there exists a nonsingular $n \times n$ matrix P, such that $\mathfrak{X}(a) = P^{-1}\mathfrak{Y}(a)P$ for all $a \in A$.

Clearly, similarity is an equivalence relation among representations. Also, if $\mathfrak{Y}$ is a representation of degree n and P is any nonsingular $n \times n$ matrix, then the formula $\mathfrak{X}(a) = P^{-1}\mathfrak{Y}(a)P$ defines a new representation $\mathfrak{X}$.

It is easy to build modules from representations and representations from modules. If $\mathfrak{X}$ is a representation of degree n of the F-algebra A, let V be the n-dimensional row vector space over F. If $v \in V$ and X is any $n \times n$ matrix over F, then $vX \in V$. Define $va = v\mathfrak{X}(a)$ for $a \in A$. It is routine to check that this gives the structure of an A-module to V.

Conversely, if M is an A-module, choose an F-basis for M and let $\mathfrak{X}(a)$ be the matrix of a_M with respect to this basis. It is now easy to check that $\mathfrak{X}$ is a representation. Note that a different choice of basis might give a different representation (and usually does).

Starting with a representation $\mathfrak{X}$, constructing the module V as above, and then choosing the appropriate basis for V and constructing the corresponding representation will result in the original representation $\mathfrak{X}$.

Suppose V and W are A-modules and $\vartheta \in \mathrm{Hom}_A(V, W)$. How does this situation look from the representation point of view? Choose bases $\mathcal{B}_V$ and $\mathcal{B}_W$ for V and W. Since ϑ is a linear transformation from V to W, it has a matrix P with respect to the given bases. Note that P is $m \times n$ where $m = \dim V$ and $n = \dim W$. Now, the fact that $(va)\vartheta = (v\vartheta)a$ for all $v \in V$ and $a \in A$ yields $\mathfrak{X}(a)P = P\mathfrak{Y}(a)$, where $\mathfrak{X}$ and $\mathfrak{Y}$ are the representations given by V and W with respect to the bases $\mathcal{B}_V$ and $\mathcal{B}_W$. If ϑ is a module isomorphism, then P is nonsingular and the above matrix equation yields that $\mathfrak{X}$ and $\mathfrak{Y}$ are similar representations. In particular, the different representations arising from a given module with different choices of basis are all similar.

The above reasoning may be reversed to show that if the representations arising from V and W with respect to bases $\mathcal{B}_V$ and $\mathcal{B}_W$ are similar, then V and W are in fact isomorphic modules. It follows that there is a natural one-to-one correspondence between isomorphism classes of A-modules and similarity classes of representations of A.

If V is an A-module and $W < V$ is a proper nonzero submodule, choose a basis $\mathcal{B}_W$ for W and extend this to $\mathcal{B}_V$, a basis for V. Number $\mathcal{B}_V$ so that the last m vectors are $\mathcal{B}_W$, where $m = \dim W$. Let $\mathfrak{X}$ be the representation of A corresponding to V with respect to the basis $\mathcal{B}_V$ and let $\mathfrak{Y}$ be the representation corresponding to W with respect to the basis $\mathcal{B}_W$. It is then easy to see for $a \in A$ that $\mathfrak{X}(a)$ has the form

$$\mathfrak{X}(a) = \begin{pmatrix} \mathfrak{Z}(a) & \mathfrak{U}(a) \\ 0 & \mathfrak{Y}(a) \end{pmatrix}.$$

Furthermore, $\mathfrak{Z}$ is a representation corresponding to V/W. Note that $\mathfrak{U}$ is a function from A into $(n - m) \times m$ matrices (where $n = \dim V$), but $\mathfrak{U}$ is *not* a representation.

The representation $\mathfrak{X}$ is said to be in *reduced* form and one similar to $\mathfrak{X}$ is *reducible*. Thus the irreducible representations correspond to the irreducible modules.

If there exists a submodule $U \subseteq V$ in the above situation, with $V = W \dotplus U$, then the basis for W may be extended to V by adjoining to it a basis for U. When this is done, the result is that $\mathfrak{U}(a) = 0$ for all $a \in A$. It follows from this discussion that if $\mathfrak{X}$ is any representation corresponding to a completely reducible module, then $\mathfrak{X}$ is similar to a representation in block diagonal form, where each of the blocks is an irreducible representation.

Problems

(1.1) Let V be an A-module. Show that V is completely reducible iff the intersection of all of the maximal submodules of V is trivial.

Hint To prove "if," embed V into a sum of irreducible modules. Recall that our definition of "module" requires finite dimensionality.

Note This problem is "dual" to Theorem 1.10 which implies that V is completely reducible iff it is the sum of its minimal submodules. Theorem 1.10 is true much more generally than we have proved it. It holds for arbitrary modules over rings. Problem 1.1, however, is not valid in this greater generality. A counterexample is the regular module of the ring of integers $\mathbb{Z}$.

(1.2) Let $\mathfrak{X}$ and $\mathfrak{Y}$ be representations of an F-algebra, A. A nonzero matrix P is said to *intertwine* $\mathfrak{X}$ and $\mathfrak{Y}$ if $P\mathfrak{X}(a) = \mathfrak{Y}(a)P$ for all $a \in A$. Assume $\mathfrak{X}$ and $\mathfrak{Y}$ are irreducible.

(a) If P intertwines $\mathfrak{X}$ and $\mathfrak{Y}$, show that P is square and nonsingular.

(b) Assume that F is algebraically closed and that P and Q both intertwine $\mathfrak{X}$ and $\mathfrak{Y}$. Show that $Q = \lambda P$ for some $\lambda \in F$.

(1.3) Show that an algebra A is semisimple iff every A-module is completely reducible.

(1.4) Let A be an algebra. For A-module V, let $\mathscr{A}(V) = \{a \in A \mid Va = 0\}$. Let $J(A) = \bigcap_{M \in \mathscr{M}(A)} \mathscr{A}(M)$, where $\mathscr{M}(A)$ is a representative set of irreducible A-modules. Show

(a) $\mathscr{A}(V)$ is an ideal of A for all V.
(b) $VJ(A) < V$ for every nonzero A-module V.
(c) $J(A)^n = 0$ for some integer n.
(d) If I is a right ideal of A and $I^m = 0$ for some m, then $I \subseteq J(A)$.

Note The ideal $J(A)$ is called the *Jacobson radical* of A; $\mathscr{A}(V)$ is the *annihilator* of V and a right ideal I with $I^n = 0$ is said to be *nilpotent*.

(1.5) Prove that the following are equivalent for the algebra A.

(a) $J(A) = 0$.
(b) A has no nonzero nilpotent right ideals.
(c) A has no nonzero nilpotent ideals.
(d) A is semisimple.

Hint If V is irreducible, then $\mathscr{A}(V)$ is an intersection of maximal right ideals of A.

(1.6) Let A be an algebra and V a completely reducible A-module. Show that the algebra A_V is semisimple.

Note A consequence of Problem 1.6 is that the hypothesis that A is semisimple in the Double Centralizer Theorem 1.16 may be dropped since this theorem is really about A_M for an irreducible module M.

(1.7) Let A be an algebra and V an irreducible A-module. Show that $|\mathscr{M}(A_V)| = 1$.

(1.8) Let G be a group, $H \subseteq G$ a subgroup and F a field with characteristic prime to $|G : H|$. Let V be an $F[G]$-module with submodule W. Suppose that there exists $U_0 \subseteq V$, an $F[H]$-submodule such that $V = W \dotplus U_0$. Show that there exists an $F[G]$-submodule $U \subseteq V$ with $V = W \dotplus U$.

Note This generalization of Maschke's Theorem 1.9 is due to D. G. Higman.

(1.9) Let G be a group and F a field of characteristic p. Suppose $p \mid |G|$ and show that $F[G]$ is not semisimple.

Hint $(\sum_{g \in G} g)^2 = 0$.

(1.10) Let M be an A-module. Show that M is completely reducible iff $MJ(A) = 0$.

Hint $A/J(A)$ is semisimple.

2 Group representations and characters

Let G be a finite group and let F be a field. Suppose $\mathfrak{X}$ is a representation of $F[G]$ with degree n. Since $\mathfrak{X}$ is an algebra homomorphism, $\mathfrak{X}(1) = I$, the identity matrix. It follows for $g \in G$ that $\mathfrak{X}(g)$ is nonsingular and $\mathfrak{X}(g)^{-1} = \mathfrak{X}(g^{-1})$. If we restrict the function $\mathfrak{X}$ to $G \subseteq F[G]$, we obtain a group homomorphism from G into the *general linear group* $\mathrm{GL}(n, F)$, that is, the multiplicative group of nonsingular $n \times n$ matrices over F.

(2.1) DEFINITION Let F be a field and G a group. Then an *F-representation* of G is a homomorphism $\mathfrak{X}: G \to \mathrm{GL}(n, F)$ for some integer n.

We have seen that a representation of $F[G]$ determines an F-representation of G by restriction. Conversely, an F-representation $\mathfrak{X}_0$ of G determines a representation $\mathfrak{X}$ of $F[G]$ by linear extension. That is,

$$\mathfrak{X}(\textstyle\sum a_g g) = \sum a_g \mathfrak{X}_0(g).$$

We shall usually use the same symbol to denote both an F-representation of G and the corresponding representation of $F[G]$. Also, the adjectives "similar" and "irreducible" will be applied to F-representations of G as if they were the corresponding representations of $F[G]$. Some caution is necessary here since if $F \subseteq E$, a larger field, and $\mathfrak{X}$ is an F-representation of G, then $\mathfrak{X}$ is automatically an E-representation. It is entirely possible, however, that $\mathfrak{X}$ is irreducible as an F-representation, and yet is reducible as an E-representation. We will explore this situation in some depth in Chapter 9.

One further triviality which should be mentioned now is the following. If $N \lhd G$ and $\mathfrak{X}$ is an F-representation of G with $N \subseteq \ker \mathfrak{X}$, then there is a unique F-representation $\bar{\mathfrak{X}}$ of G/N defined by $\bar{\mathfrak{X}}(Ng) = \mathfrak{X}(g)$. This formula can also be used to define the representation $\mathfrak{X}$ if $\bar{\mathfrak{X}}$ is given. Note that $\mathfrak{X}$ is irreducible iff $\bar{\mathfrak{X}}$ is. We shall often fail to distinguish between $\mathfrak{X}$ and $\bar{\mathfrak{X}}$.

The trouble with representations is that they contain too much information. If $\mathfrak{X}$ is an F-representation of G of degree n, then for each element of G we have n^2 entries in $\mathfrak{X}(g)$. Some of this data is clearly redundant because it distinguishes between similar representations. The idea behind character theory is to throw away most of the information and to save just enough to be useful. This is done by calculating the traces (that is, the sums of the diagonal entries) of the matrices in question. Recall that if A and B are any two $n \times n$ matrices over a field, then $\mathrm{tr}(AB) = \mathrm{tr}(BA)$.

(2.2) DEFINITION Let $\mathfrak{X}$ be an F-representation of G. Then the *F-character* χ of G *afforded* by $\mathfrak{X}$ is the function given by $\chi(g) = \mathrm{tr}\ \mathfrak{X}(g)$.

As is the case with F-representations of G, we may view F-characters as functions on all of $F[G]$. Note that if the characteristic $\mathrm{char}(F) \neq 0$, then the constant function 0 is an F-character. On the other hand, if $\mathrm{char}(F) = 0$, then 0 is definitely not an F-character because $\chi(1) = \deg \mathfrak{X}$, where $\mathfrak{X}$ is an F-representation of G which affords χ. In this case, we say that $\chi(1)$ is the *degree* of χ.

Most F-characters of a group G are not homomorphisms of any kind. However, if λ is a homomorphism from G into the multiplicative group of F, then $\mathfrak{X}(g) = (\lambda(g))$ is an F-representation of G of degree 1 which affords λ as its character. Characters of degree 1 are called *linear characters*. In particular, the function 1_G with constant value 1 on G is a linear F-character. It is called the *principal* F-character.

(2.3) LEMMA (a) Similar F-representations of G afford equal characters.
(b) Characters are constant on the conjugacy classes of a group.

Proof If P is nonsingular then $\mathrm{tr}(P^{-1}A \cdot P) = \mathrm{tr}(P \cdot P^{-1}A) = \mathrm{tr}(A)$. Both (a) and (b) follow from this observation. To see (b), observe that $\mathfrak{X}(h^{-1}gh) = \mathfrak{X}(h)^{-1}\mathfrak{X}(g)\mathfrak{X}(h)$ if $\mathfrak{X}$ is a representation of G, and hence $\mathrm{tr}(\mathfrak{X}(h^{-1}gh)) = \mathrm{tr}(\mathfrak{X}(g))$. ∎

We make one further general observation. If $\mathfrak{X}$ and $\mathfrak{Y}$ are F-representations of G, then

$$\mathfrak{Z}(g) = \begin{bmatrix} \mathfrak{X}(g) & 0 \\ 0 & \mathfrak{Y}(g) \end{bmatrix}$$

is also an F-representation. Since $\mathrm{tr}\ \mathfrak{Z}(g) = \mathrm{tr}\ \mathfrak{X}(g) + \mathrm{tr}\ \mathfrak{Y}(g)$, it follows that the set of F-characters of G is closed under addition.

We now restrict our attention to the special case that the field $F = \mathbb{C}$, the complex numbers. We emphasize, however, that the subfield consisting of the algebraic elements in $\mathbb{C}$ would work exactly as well, and in fact with only minor modifications, most of what follows works for any algebraically closed field of characteristic not dividing $|G|$.

Let us establish some notation. Fix a finite group G and choose a representative set of irreducible $\mathbb{C}[G]$-modules, $\mathcal{M}(\mathbb{C}[G]) = \{M_1, \ldots, M_k\}$. Choose a basis in each M_i and let $\mathfrak{X}_i$ be the resulting representation of $\mathbb{C}[G]$. Let χ_i be the character afforded by $\mathfrak{X}_i$. It follows that the set $\mathrm{Irr}(G) = \{\chi_1, \ldots, \chi_k\}$ is the set of all irreducible $\mathbb{C}$-characters of G (that is, characters afforded by irreducible representations). Henceforth, the word "character" will mean $\mathbb{C}$-character unless otherwise stated.

Since sums of characters are characters, it follows that $\chi = \sum_{i=1}^{k} n_i\chi_i$ is a character whenever the n_i are nonnegative integers which are not all zero. Conversely, if χ is any character of G afforded by a representation $\mathfrak{X}$ corresponding to a module V, we can decompose V into a direct sum of irreducible modules. It follows that χ is the sum of the corresponding irreducible characters. We have, in fact, $\chi = \sum n_{M_i}(V)\chi_i$.

Corollary 1.17(d) asserts that $\dim(\mathbb{C}[G]) = \sum_{i=1}^{k}(\dim M_i)^2$. Since $\dim \mathbb{C}[G] = |G|$ and $\dim M_i = \deg \mathfrak{X}_i = \chi_i(1)$, we obtain the fundamental formula

$$|G| = \sum_{i=1}^{k} \chi_i(1)^2.$$

It seems natural at this point to ask how we can determine the integer k purely group theoretically, without looking at representations. By Corollary 1.17(e), we have $k = |\mathcal{M}(\mathbb{C}[G])| = \dim \mathbf{Z}(\mathbb{C}[G])$.

(2.4) THEOREM Let $\mathcal{K}_1, \mathcal{K}_2, \ldots, \mathcal{K}_r$ be the conjugacy classes of a group G. Let $K_i = \sum_{x \in \mathcal{K}_i} x \in \mathbb{C}[G]$. Then the K_i form a basis for $\mathbf{Z}(\mathbb{C}[G])$ and if $K_iK_j = \sum a_{ij\nu}K_\nu$, then the multiplication constants $a_{ij\nu}$ are nonnegative integers.

Proof It is clear that the K_i lie in $\mathbf{Z}(\mathbb{C}[G])$. Moreover, they are linearly independent because they are sums of disjoint sets of elements. If $z = \sum a_g g \in \mathbf{Z}(\mathbb{C}[G])$ and $h \in G$, we have $z = h^{-1}zh = \sum a_g g^h$. Comparing the coefficients of g^h on both sides, we obtain $a_{g^h} = a_g$. In other words, the coefficients a_g have the constant value a_i for all $g \in \mathcal{K}_i$. It follows that $z = \sum a_i K_i$ and thus the K_i span $\mathbf{Z}(\mathbb{C}[G])$.

To find $a_{ij\nu}$, pick $g \in \mathcal{K}_\nu$. Then $a_{ij\nu}$ is the coefficient of g in K_iK_j. From the definition of multiplication in a group algebra, this is $|\{(x, y) \mid x \in \mathcal{K}_i, y \in \mathcal{K}_j, xy = g\}|$. Since $a_{ij\nu}$ is the cardinality of a set, it is a nonnegative integer. ∎

(2.5) COROLLARY The number k of similarity classes of irreducible representations of G is equal to the number of conjugacy classes of G.

(2.6) COROLLARY The group G is abelian iff every irreducible character is linear.

Proof Let k be the number of classes of G. Then $k = |G|$ iff G is abelian. Now $|G| = \sum_{i=1}^{k} \chi_i(1)^2$ and $\chi_i(1) \geq 1$ for all i. It follows that $k = |G|$ iff $\chi_i(1) = 1$ for all i. The proof is complete. ▮

We have not yet proved that the χ_i are distinct. To see this, we introduce a little more notation. From the results of Chapter 1 we have the direct sum

$$\mathbb{C}[G] = \sum_{i=1}^{k} \cdot M_i(\mathbb{C}[G]).$$

Let $1 = \sum e_i$ with $e_i \in M_i(\mathbb{C}[G])$. Since $M_j(\mathbb{C}[G])$ annihilates the module M_i if $i \neq j$, we have $\mathfrak{X}_i(e_j) = 0$ in this case. It follows that $\mathfrak{X}_i(e_i) = \mathfrak{X}_i(1) = I$. Therefore, $\chi_i(e_j) = 0$ if $i \neq j$ and $\chi_i(e_i) = \chi_i(1) \neq 0$ and we conclude that the χ_i are distinct as functions on $\mathbb{C}[G]$. Thus the χ_i are also distinct as functions on G.

We may now restate two of the earlier results in a slightly more convenient form.

(2.7) COROLLARY Let G be a group. Then $|\mathrm{Irr}(G)|$ equals the number of conjugacy classes of G and

$$\sum_{\chi \in \mathrm{Irr}(G)} \chi(1)^2 = |G|.$$

For certain very small groups, the information contained in Corollary 2.7 is sufficient to determine the irreducible character degrees. For instance, if $G = \Sigma_3$, the symmetric group on three symbols, then G has exactly three conjugacy classes and $|G| = 6$. It follows that the $\chi_i(1)$ are 1,1, and 2.

Actually, the preceding argument with the e_i's yields more than the fact that the χ_i's are distinct. A *class function* on a group G is a function, φ: $G \to \mathbb{C}$ which is constant on conjugacy classes. All characters are class functions.

(2.8) THEOREM Every class function φ of G can be uniquely expressed in the form

$$\varphi = \sum_{\chi \in \mathrm{Irr}(G)} a_\chi \chi,$$

where $a_\chi \in \mathbb{C}$. Furthermore, φ is a character iff all of the a_χ are nonnegative integers and $\varphi \neq 0$.

Proof The set of class functions of G forms a vector space over $\mathbb{C}$ whose dimension is the number of classes of G. We claim that $\mathrm{Irr}(G)$ is a basis for this space. Since $|\mathrm{Irr}(G)| = k$ = number of classes, it suffices to show that if $\sum a_i\chi_i = 0$, then each $a_i = 0$. This is immediate by evaluation at e_i.

The second statement has already been proved. ▮

If $\chi = \sum_{i=1}^{k} n_i\chi_i$ is a character, then those χ_i with $n_i > 0$ are called the irreducible *constituents* of χ. In general, if ψ is a character such that $\chi - \psi$ is also a character or is zero, then ψ is called a *constituent* of χ. An important consequence of Theorem 2.8 is the following.

(2.9) COROLLARY Let $\mathfrak{X}$ and $\mathfrak{Y}$ be $\mathbb{C}$-representations of a group G. Then $\mathfrak{X}$ and $\mathfrak{Y}$ are similar iff they afford equal characters.

Proof We already know that similar representations afford equal characters.

Let V and W be $\mathbb{C}[G]$-modules corresponding to $\mathfrak{X}$ and $\mathfrak{Y}$ respectively. Then $\mathfrak{X}$ affords the character $\sum n_{M_i}(V)\chi_i$ and $\mathfrak{Y}$ affords $\sum n_{M_i}(W)\chi_i$. If these characters are equal, it follows that $n_{M_i}(V) = n_{M_i}(W)$ for all i. Therefore, V and W are isomorphic to identical direct sums of irreducible $\mathbb{C}[G]$-modules and thus are isomorphic and $\mathfrak{X}$ and $\mathfrak{Y}$ are similar as required. ▮

For a particular group G, the irreducible characters are usually presented in a *character table*: a (square) array of complex numbers whose rows correspond to the χ_i and whose columns correspond to the classes $\mathscr{K}_i$. An example of a character table is the accompanying one for the symmetric group Σ_4 on four symbols.

g:	1	(1 2)	(1 2)(3 4)	(1 2 3 4)	(1 2 3)
$\lvert\mathbf{C}(g)\rvert$:	24	4	8	4	3
$\lvert\mathrm{Cl}(g)\rvert$:	1	6	3	6	8
χ_1:	1	1	1	1	1
χ_2:	1	−1	1	−1	1
χ_3:	2	0	2	0	−1
χ_4:	3	1	−1	−1	0
χ_5:	3	−1	−1	1	0

In this table, the classes are denoted by writing a representative element g in its cyclic notation. (The reader is reminded that in a symmetric group, two elements are conjugate iff they have the same cycle structure.) The sizes of the centralizer $\mathbf{C}(g)$ and of the corresponding conjugacy class $\mathrm{Cl}(g)$ are given for convenience, although they are not, properly speaking, a part of the table. Further character tables are given in the Appendix (page 287). It is rather difficult to describe the process by which these tables are constructed. Usually, various combinations of *ad hoc* arguments and general theorems are necessary. As the student learns some of these theorems, he is urged to try to construct some character tables on his own. The important point is that it is very much easier to construct a character table than it is to construct representations.

Theorem 2.8 tells us that every class function is a linear combination of irreducible characters. For example, if $G = \Sigma_4$ and $\varphi(g)$ is defined to be the number of points moved by $g \in G$, then φ is a class function and is a linear combination of $\chi_1, \chi_2, \ldots, \chi_5$. Thus the row (0, 2, 4, 4, 3) is a linear combination of the five rows of the given table. It turns out that there is an easy method for computing the coefficients, practically by inspection from the table. This is done by using the so called "orthogonality relations," which we are about to derive. These relations are also extremely useful in the construction of character tables.

Before we leave Σ_4 we should comment on the fact that all of the character values, which *a priori* are only known to be complex numbers, in fact turn out to be integers. (Of course the $\chi_i(1)$, being the degrees of representations, are always positive integers.) It is not always true that all character values are ordinary integers, although frequently a large fraction of them are. It is true, however, for all symmetric groups. All of this should become clear later.

The key to the orthogonality relations is to compute explicitly the coefficients of the group elements in the e_i's in terms of the characters. To do this we use the character ρ of G afforded by a representation corresponding to the regular module $\mathbb{C}[G]^\circ$. This *regular character* ρ will be computed in two ways.

(2.10) LEMMA If $g \in G$ and $g \neq 1$, then $\rho(g) = 0$. Also $\rho(1) = |G|$.

Proof We must choose a basis for $\mathbb{C}[G]^\circ$ to obtain a corresponding representation. We simply take G, in some ordering, as the basis and let $\mathfrak{R}$ be the corresponding representation. If $\mathfrak{R}(g) = (a_{ij})$, then $a_{ij} = 0$ unless $g_i g = g_j$, in which case $a_{ij} = 1$. Since $\rho(g)$ is the number of g_i satisfying $g_i g = g_i$, the lemma follows. ∎

Since ρ is a character of G, it may be expressed as an integer linear combination of the χ_i. We do this explicitly.

(2.11) LEMMA $\rho = \sum_{i=1}^{k} \chi_i(1)\chi_i$.

Proof If V is any $\mathbb{C}[G]$-module, it may be decomposed as a direct sum of irreducibles. The character afforded by a corresponding representation is $\sum n_{M_i}(V)\chi_i$. Now by Corollary 1.17(c) we know that $n_{M_i}(\mathbb{C}[G]^\circ) = \dim M_i = \chi_i(1)$. The result follows. ▮

It is suggested that the reader use Lemma 2.11 and the character table of Σ_4 to compute ρ explicitly for this group, and check the result against Lemma 2.10.

(2.12) THEOREM $e_i = (1/|G|) \sum_{g \in G} \chi_i(1)\chi_i(g^{-1})g$.

Proof Write $e_i = \sum a_g g$. By Lemma 2.10, we have $\rho(e_i g^{-1}) = a_g|G|$. Lemma 2.11 thus yields

$$a_g|G| = \sum_j \chi_j(1)\chi_j(e_i g^{-1}).$$

Since

$$\mathfrak{X}_j(e_i g^{-1}) = \mathfrak{X}_j(e_i)\mathfrak{X}_j(g^{-1}) = \begin{cases} 0 & \text{if } i \neq j, \\ \mathfrak{X}_i(g^{-1}) & \text{if } i = j, \end{cases}$$

we have $\chi_j(e_i g^{-1}) = \chi_i(g^{-1})\delta_{ij}$, where the Kronecker δ_{ij} is 0 or 1 depending on whether i and j are unequal or equal. We now have

$$a_g|G| = \chi_i(1)\chi_i(g^{-1})$$

and the result follows. ▮

(2.13) THEOREM (*Generalized Orthogonality Relation*) The following holds for every $h \in G$.

$$\frac{1}{|G|}\sum_{g \in G} \chi_i(gh)\chi_j(g^{-1}) = \delta_{ij}\frac{\chi_i(h)}{\chi_i(1)}.$$

Proof The e_i lie in trivially intersecting ideals of $\mathbb{C}[G]$ and thus $e_i e_j = 0$ if $i \neq j$. Since $1 = \sum e_j$, multiplication by e_i yields $e_i^2 = e_i$. We now substitute the formula of Theorem 2.12 into the equation $e_i e_j = \delta_{ij} e_i$ and compare the coefficients of the group elements on both sides.

The coefficient of a fixed $h \in G$ on the right hand side is $(\delta_{ij}/|G|)\chi_i(1)\chi_i(h^{-1})$ and that on the left-hand side is

$$\frac{\chi_i(1)\chi_j(1)}{|G|^2}\sum_{g \in G} \chi_i((hg^{-1})^{-1})\chi_j(g^{-1}).$$

The result now follows by equating these expressions and substituting h for h^{-1}. ▮

By taking $h = 1$ in Theorem 2.13 we obtain

(2.14) COROLLARY (*First Orthogonality Relation*)

$$\frac{1}{|G|}\sum_{g\in G}\chi_i(g)\chi_j(g^{-1}) = \delta_{ij}.$$

Since the expression $\chi(g^{-1})$ has come up several times, we digress for a while to discuss its connection with $\chi(g)$ and some related questions.

(2.15) LEMMA Let $\mathfrak{X}$ be a representation of G affording the character χ and let $g \in G$. Let $n = o(g)$, the order of g. Then

(a) $\mathfrak{X}(g)$ is similar to a diagonal matrix $\mathrm{diag}(\varepsilon_1, \ldots, \varepsilon_f)$;
(b) $\varepsilon_i^{\,n} = 1$;
(c) $\chi(g) = \sum \varepsilon_i$ and $|\chi(g)| \leq \chi(1)$;
(d) $\chi(g^{-1}) = \overline{\chi(g)}$.

Proof The restriction of $\mathfrak{X}$ to the cyclic group $\langle g\rangle$ is a representation of $\langle g\rangle$ and hence it is no loss to assume $G = \langle g\rangle$. By Maschke's theorem and other results of Chapter 1, it follows that $\mathfrak{X}$ is similar to a representation in block diagonal form, with irreducible representations of G appearing as the diagonal blocks. Since $G = \langle g\rangle$ is abelian, Corollary 2.6 asserts that its irreducible representations have degree 1, and thus $\mathfrak{X}$ is similar to a diagonal representation. Now (a) follows, and we may assume that $\mathfrak{X}$ is diagonal. We have $I = \mathfrak{X}(g^n) = \mathfrak{X}(g)^n = \mathrm{diag}(\varepsilon_1^{\,n}, \ldots, \varepsilon_f^{\,n})$. Therefore (b) is proved. It follows that $|\varepsilon_i| = 1$ and $|\sum \varepsilon_i| \leq \sum |\varepsilon_i| = f = \chi(1)$. It is clear that $\chi(g) = \sum \varepsilon_i$ so that (c) follows. Now $\mathfrak{X}(g^{-1}) = \mathfrak{X}(g)^{-1} = \mathrm{diag}(\varepsilon_1^{\,-1}, \ldots, \varepsilon_f^{\,-1})$ so that $\chi(g^{-1}) = \sum \varepsilon_i^{\,-1}$. Since $|\varepsilon_i| = 1$, we have $\varepsilon_i^{\,-1} = \overline{\varepsilon_i}$ and $\chi(g^{-1}) = \overline{\chi(g)}$. The proof is complete. ∎

Combining Corollary 2.14 and Lemma 2.15(d), we obtain

$$\frac{1}{|G|}\sum_{g\in G}\chi_i(g)\overline{\chi_j(g)} = \delta_{ij}.$$

This suggests the following definition.

(2.16) DEFINITION Let φ and ϑ be class functions on a group G. Then

$$[\varphi, \vartheta] = \frac{1}{|G|}\sum_{g\in G}\varphi(g)\overline{\vartheta(g)}$$

is the *inner product* of φ and ϑ.

Some of the obvious properties of this "inner product" are

(a) $[\varphi, \vartheta] = \overline{[\vartheta, \varphi]}$;
(b) $[\varphi, \varphi] > 0$ unless $\varphi = 0$;

(c) $[c_1\varphi_1 + c_2\varphi_2, \vartheta] = c_1[\varphi_1, \vartheta] + c_2[\varphi_2, \vartheta]$;
(d) $[\varphi, c_1\vartheta_1 + c_2\vartheta_2] = \overline{c_1}[\varphi, \vartheta_1] + \overline{c_2}[\varphi, \vartheta_2]$.

Therefore, [,] has all of the properties usually used to define an inner product in linear algebra and analysis. (In fact, this makes the space of class functions into a finite dimensional Hilbert space.)

We know that Irr(G) is a basis for the space of class functions and it is the content of the orthogonality relation that it is, in fact, an orthonormal basis, that is,

$$[\chi_i, \chi_j] = \delta_{ij}.$$

This yields the promised method for expressing an arbitrary class function in terms of the irreducible characters; for if $[\varphi, \chi_i] = c_i$, then $\varphi = \sum c_i\chi_i$.

Another application of the inner product is to determine instantaneously whether or not a given character is irreducible.

(2.17) COROLLARY Let χ and ψ be (not necessarily irreducible) characters of G. Then $[\chi, \psi] = [\psi, \chi]$ is a nonnegative integer. Also χ is irreducible iff $[\chi, \chi] = 1$.

Proof We have $\chi = \sum n_i\chi_i$ and $\psi = \sum m_i\chi_i$, with all n_i and m_i nonnegative integers. Then $[\chi, \psi] = \sum n_i m_i = [\psi, \chi]$ and $[\chi, \chi] = \sum n_i^2$. The result is now immediate, since χ is irreducible exactly when one $n_i = 1$ and all other $n_i = 0$. ∎

The following "second orthogonality relation" is derived from the first and so imposes no new necessary condition for an array of complex numbers to be a character table. Nevertheless, it is often extremely useful in the construction of character tables and in the extraction of information from them.

(2.18) THEOREM (*Second Orthogonality Relation*) Let $g, h \in G$. Then

$$\sum_{\chi \in \mathrm{Irr}(G)} \chi(g)\overline{\chi(h)} = 0$$

if g is not conjugate to h in G. Otherwise, the sum is equal to $|\mathbf{C}(g)|$.

Proof Let $g_1, g_2, \ldots, g_k$ be representatives of the conjugacy classes of G. Let X be the $k \times k$ matrix whose (i, j) entry is $\chi_i(g_j)$. (In other words, X is the character table, viewed as a matrix.) Let D be the diagonal matrix with entries $\delta_{ij}|\mathscr{K}_i|$ where $\mathscr{K}_i$ is the conjugacy class $\mathrm{Cl}(g_i)$. The first orthogonality relation asserts that

$$|G|\delta_{ij} = \sum_{g \in G} \chi_i(g)\overline{\chi_j(g)} = \sum_{\nu=1}^{k} |\mathscr{K}_\nu|\chi_i(g_\nu)\overline{\chi_j(g_\nu)}.$$

This system of k^2 equations may be replaced by the single matrix equation

$$|G|I = XD\overline{X}^{\mathsf{T}},$$

where I is the identity matrix and the superscript denotes transpose.

Since a right inverse for a square matrix is necessarily also a left inverse, this yields

$$|G|I = D\overline{X}^{\mathsf{T}}X.$$

We now write this as a system of equations and obtain

$$|G|\delta_{ij} = \sum_{\nu} |\mathscr{K}_i|\overline{\chi_\nu(g_i)}\chi_\nu(g_j).$$

Since $|G|/|\mathscr{K}_i| = |\mathbf{C}(g_i)|$, this yields

$$\sum_{\chi \in \mathrm{Irr}(G)} \chi(g_j)\overline{\chi(g_i)} = |\mathbf{C}(g_i)|\delta_{ij},$$

which is the desired result. ∎

In the character table for Σ_4 that was given earlier, the size of each conjugacy class was given. We see now that this information is derivable from the body of the table. It was given only for ease of computation of inner products.

As a check on our results so far, observe what happens if we take $h = 1$ in the second orthogonality relation. The reader should recognize what results as a combination of Lemmas 2.10 and 2.11.

Let $E \subseteq \mathbb{C}$ be the field of algebraic numbers. By Lemma 2.15, all of the character values $\chi(g) \in E$ for $\chi \in \mathrm{Irr}(G)$ and $g \in G$. How is $\mathrm{Irr}(G)$, which is a set of functions $G \to E$ related to the set of irreducible E-characters of G, which we denote by $\mathrm{Irr}_E(G)$? In fact these sets are equal.

Suppose $\mathfrak{X}$ is an irreducible E-representation of G which affords $\chi \in \mathrm{Irr}_E(G)$. Now $\mathfrak{X}$ may be viewed as a $\mathbb{C}$-representation and thus χ is a character (that is, a $\mathbb{C}$-character) of G. Since the entire development of character theory up to this point would have been the same over E as over $\mathbb{C}$ and since $\chi \in \mathrm{Irr}_E(G)$, it follows that $[\chi, \chi] = 1$. However χ is a $\mathbb{C}$-character of G and it follows from Corollary 2.17 that $\chi \in \mathrm{Irr}(G)$. Thus $\mathrm{Irr}_E(G) \subseteq \mathrm{Irr}(G)$.

We also have that both $\mathrm{Irr}_E(G)$ and $\mathrm{Irr}(G)$ have the same cardinality, namely that of the set of conjugacy classes of G. Therefore $\mathrm{Irr}_E(G) = \mathrm{Irr}(G)$.

The point of this digression is to suggest that there is something "absolute" about a character table. It is not entirely an artifact of our choice of the particular field $\mathbb{C}$.

Another consequence of this argument which is sometimes useful is that if $\chi \in \mathrm{Irr}(G)$, then χ is afforded by an E-representation of G. This type of consideration will be discussed much more fully in Chapter 9. Until then, we resume our convention that "character" means "$\mathbb{C}$-character."

A great deal of information about a group can be recovered from its character table. In particular, all of the normal subgroups of G can be found. A normal subgroup is a union of conjugacy classes and the word "found" in the preceding sentence means that those sets of conjugacy classes whose unions form subgroups can be listed. In particular, the orders of all normal subgroups and the inclusion relations among them can be determined.

(2.19) LEMMA Let $\mathfrak{X}$ be a $\mathbb{C}$-representation of G which affords the character χ. Then $g \in \ker \mathfrak{X}$ iff $\chi(g) = \chi(1)$.

Proof If $g \in \ker \mathfrak{X}$, then $\mathfrak{X}(g) = I = \mathfrak{X}(1)$ and $\chi(g) = \chi(1)$. Conversely, by Lemma 2.15, $\chi(g) = \varepsilon_1 + \cdots + \varepsilon_f$, where ε_i is a root of unity and $f = \chi(1)$. Since $|\varepsilon_i| = 1$, the equation $\chi(g) = f$ forces $\varepsilon_i = 1$ for all i. Now $\mathfrak{X}(g)$ is similar to $\operatorname{diag}(\varepsilon_1, \ldots, \varepsilon_f) = I$ and therefore $\mathfrak{X}(g) = I$ and the proof is complete. ∎

(2.20) DEFINITION Let χ be a character of G. Then $\ker \chi = \{g \in G \mid \chi(g) = \chi(1)\}$.

(2.21) LEMMA Let χ be a character of G with $\chi = \sum n_i \chi_i$ for $\chi_i \in \operatorname{Irr}(G)$. Then $\ker \chi = \bigcap \{\ker \chi_i \mid n_i > 0\}$. Also $\bigcap \{\ker \chi_i \mid 1 \le i \le k\} = 1$.

Proof Since $|\chi_i(g)| \le \chi_i(1)$ by Lemma 2.15, $\chi(g) = \chi(1)$ forces $\chi_i(g) = \chi_i(1)$ whenever $n_i \ne 0$. The reverse inclusion is trivial and the first assertion follows.

To prove the second statement, consider the regular character ρ. By Lemma 2.10, $\ker \rho = 1$ and the result follows. ∎

The normal subgroups $N_i = \ker \chi_i$ can be found by inspection from the character table of a group G. We claim that every normal subgroup is the intersection of some of the N_i's and thus can be found from the character table. To see this, let $N \lhd G$ and let $\mathfrak{R}$ be the regular representation of the group G/N so that $\ker \mathfrak{R} = N/N$. Now view $\mathfrak{R}$ as a representation of G with kernel N and let χ be the corresponding character of G. Then $N = \ker \mathfrak{R} = \ker \chi = \bigcap \{N_i \mid [\chi, \chi_i] \ne 0\}$.

Given a normal subgroup N of G (where "given" means listing the classes $\mathscr{K}_i$ which it contains), we may calculate $|N|$ from the character table using $|N| = \sum \{|\mathscr{K}_i| \mid \mathscr{K}_i \subseteq N\}$. (Recall that $|\mathrm{Cl}(g)| = |G : \mathbf{C}(g)|$ and $|\mathbf{C}(g)| = \sum_{\chi \in \operatorname{Irr}(G)} |\chi(g)|^2$ so that $|\mathscr{K}_i|$ is determined from the character table.)

It follows from this discussion that G is simple iff $\ker \chi = 1$ for all non-principal $\chi \in \operatorname{Irr}(G)$ and therefore simplicity of a group can be easily determined from its character table.

The group G is solvable iff it has a chain of normal subgroups, $1 = M_0 \subseteq M_1 \subseteq \cdots \subseteq M_n = G$ such that $|M_i : M_{i-1}|$ is a prime power for all i, $1 \leq i \leq n$. Since the M_i can be located and their orders determined from the character table of G, it follows that the table determines solvability or nonsolvability of G.

This may be a good place to remark that the character table of G does not determine G up to isomorphism. In fact if p is any prime, there are two nonisomorphic nonabelian groups of order p^3. These two groups have identical character tables.

Let $N \lhd G$. It seems natural to ask whether the character tables of N and of G/N can be calculated from that of G. The answer is "no" for N but "yes" for G/N.

We have already observed that there is a one-to-one correspondence between representations of G/N and representations of G with kernel containing N. Furthermore, under this correspondence, irreducible representations correspond to irreducible representations. This situation may be interpreted in terms of characters as follows.

(2.22) LEMMA Let $N \lhd G$.

(a) If χ is a character of G and $N \subseteq \ker \chi$, then χ is constant on cosets of N in G and the function $\hat{\chi}$ on G/N defined by $\hat{\chi}(Ng) = \chi(g)$ is a character of G/N.

(b) If $\hat{\chi}$ is a character of G/N, then the function χ defined by $\chi(g) = \hat{\chi}(Ng)$ is a character of G.

(c) In both (a) and (b), $\chi \in \operatorname{Irr}(G)$ iff $\hat{\chi} \in \operatorname{Irr}(G/N)$.

Usually, we shall identify χ and $\hat{\chi}$. Under this identification, we have $\operatorname{Irr}(G/N) = \{\chi \in \operatorname{Irr}(G) \mid N \subseteq \ker \chi\}$. To demonstrate what is happening here, let us consider the example $G = \Sigma_4$ and N the normal subgroup of order 4. The classes of G which are contained in N are the identity and Cl((1 2)(3 4)). Referring to the character table on p. 17, we see that the irreducible characters χ_i of G with $N \subseteq \ker \chi_i$ are χ_1, χ_2, and χ_3 and so under above identification, we have $\operatorname{Irr}(G/N) = \{\chi_1, \chi_2, \chi_3\}$.

In order to write the character table for G/N we need to know its conjugacy classes. If $\mathscr{K}$ is a class of G, then $\overline{\mathscr{K}}$, its image in G/N, is a class of G/N. However, distinct classes of G may have equal images in G/N. This poses no problem since if $g, h \in G$, then $\bar{g}$ and $\bar{h}$ are conjugate in $\bar{G} = G/N$ iff $\chi(\bar{g}) = \chi(\bar{h})$ for all $\chi \in \operatorname{Irr}(G/N)$. (The second orthogonality relation, applied to G/N proves this.)

The part of the character table of $G = \Sigma_4$ corresponding to the characters of G/N is

	$\mathscr{K}_1$	$\mathscr{K}_2$	$\mathscr{K}_3$	$\mathscr{K}_4$	$\mathscr{K}_5$
χ_1:	1	1	1	1	1
χ_2:	1	-1	1	-1	1
χ_3:	2	0	2	0	-1

Observe that columns 1 and 3 are identical, as are columns 2 and 4. Deleting repeats, we obtain the character table

1	1	1
1	-1	1
2	0	-1

for G/N. (In this case $G/N \cong \Sigma_3$.)

The preceding discussion provides a second method for computing $|N|$ from the character table of G. Namely, by using the fact that

$$\begin{aligned}|G:N| &= \sum \{\chi(1)^2 \mid \chi \in \mathrm{Irr}(G/N)\} \\ &= \sum \{\chi(1)^2 \mid \chi \in \mathrm{Irr}(G) \quad \text{and} \quad N \subseteq \ker \chi\}.\end{aligned}$$

By Corollary 2.7, a group is abelian iff all of its irreducible characters are linear. It follows that given $N \lhd G$, the character table of G determines whether or not G/N is abelian. There is no known way to determine from the table whether or not N is abelian.

(2.23) COROLLARY Let G be a group with commutator subgroup G'. Then

(a) $G' = \bigcap\{\ker \lambda \mid \lambda \in \mathrm{Irr}(G), \lambda(1) = 1\}$;
(b) $|G : G'| =$ the number of linear characters of G.

Proof If λ is a linear character of G, then λ is a homomorphism into the abelian multiplicative group of $\mathbb{C}$. It follows that $G' \subseteq \ker \lambda$. Since G/G' is abelian, all $\chi \in \mathrm{Irr}(G/G')$ are linear and thus $\mathrm{Irr}(G/G') = \{\lambda \in \mathrm{Irr}(G) \mid \lambda(1) = 1\}$. (This equality, of course, depends on the identification of characters of G/G' with characters of G.) Finally, for any $N \lhd G$, we have $N = \bigcap\{\ker \chi \mid \chi \in \mathrm{Irr}(G) \text{ and } N \subseteq \ker \chi\}$ and hence (a) follows.

The number of linear characters of G is equal to the total number of irreducible characters of the abelian group G/G' and hence equals $|G/G'|$. The proof is now complete. ∎

The information in Corollary 2.23(b) is useful for finding the set of character degrees of a group G. For instance, if G is a nonabelian group of order 27, then $|G : G'| = 9$ and G has exactly 11 conjugacy classes. By the preceding, G has exactly nine linear characters and two nonlinear irreducible characters χ and ψ. We have

$$27 = |G| = 9 + \chi(1)^2 + \psi(1)^2.$$

Since the only way that $27 - 9 = 18$ can be written as a sum of two squares is $3^2 + 3^2$, it follows that $\chi(1) = 3 = \psi(1)$.

Before leaving the discussion of character tables and factor groups, we mention an amusing result. The following could be proved without characters but it is somewhat tricky to do so.

(2.24) COROLLARY Let $g \in G$ and $N \lhd G$. Then $|\mathbf{C}_{G/N}(Ng)| \leq |\mathbf{C}_G(g)|$.

Proof From the second orthogonality relation, we have

$$\begin{aligned} |\mathbf{C}_{G/N}(Ng)| &= \sum_{\chi \in \mathrm{Irr}(G/N)} |\chi(Ng)|^2 = \sum \{|\chi(g)|^2 \mid \chi \in \mathrm{Irr}(G), N \subseteq \ker \chi\} \\ &\leq \sum_{\chi \in \mathrm{Irr}(G)} |\chi(g)|^2 = |\mathbf{C}_G(g)|. \quad \blacksquare \end{aligned}$$

We now discuss the connections between characters and the center of a group.

(2.25) LEMMA Let $\mathfrak{X}$ be an irreducible $\mathbb{C}$-representation of G of degree n. Suppose A is an $n \times n$ matrix over $\mathbb{C}$ which commutes with $\mathfrak{X}(g)$ for all $g \in G$. Then $A = \alpha I$ for some $\alpha \in \mathbb{C}$.

Proof Let M be the n-dimensional row space over $\mathbb{C}$ so that M is an irreducible $\mathbb{C}[G]$-module via $m \cdot a = m\mathfrak{X}(a)$ for $m \in M$ and $a \in \mathbb{C}[G]$. Let ϑ: $M \to M$ be defined by $m\vartheta = mA$. Then

$$\vartheta \in \mathbf{E}_{\mathbb{C}[G]}(M) = \mathbb{C} \cdot 1$$

and $\vartheta = \alpha \cdot 1$ for some $\alpha \in \mathbb{C}$. The result follows. ∎

(2.26) DEFINITION Let χ be a character of G. Then $\mathbf{Z}(\chi) = \{g \in G \mid |\chi(g)| = \chi(1)\}$.

If $H \subseteq G$ and $\mathfrak{X}$ is a representation of G, then its restriction to H, denoted $\mathfrak{X}_H$, is a representation of H. Similarly, the restriction χ_H of a character χ of G to H is a character of H and we can write

$$\chi_H = \sum_{\psi \in \mathrm{Irr}(H)} n_\psi \psi$$

for suitable integers n_ψ. Note that if $\chi_H \in \mathrm{Irr}(H)$, then $\chi \in \mathrm{Irr}(G)$. Of course, the converse of this statement is false.

(2.27) LEMMA Let χ be a character of G and let $Z = \mathbf{Z}(\chi)$ and $f = \chi(1)$. Let $\mathfrak{X}$ be a representation of G which affords χ. Then

(a) $Z = \{g \in G \mid \mathfrak{X}(g) = \varepsilon I \text{ for some } \varepsilon \in \mathbb{C}\}$;
(b) Z is a subgroup of G;
(c) $\chi_Z = f\lambda$ for some linear character λ of Z;
(d) $Z/\ker \chi$ is cyclic;
(e) $Z/\ker \chi \subseteq \mathbf{Z}(G/\ker \chi)$.

Furthermore, if $\chi \in \operatorname{Irr}(G)$, then

(f) $Z/\ker \chi = \mathbf{Z}(G/\ker \chi)$.

Proof By Lemma 2.15, $\mathfrak{X}(g)$ is similar to $\operatorname{diag}(\varepsilon_1, \ldots, \varepsilon_f)$, with $|\varepsilon_i| = 1$, $1 \leq i \leq f$. Since $\chi(g) = \sum \varepsilon_i$, it follows that $|\chi(g)| = f$ iff all ε_i are equal. Since the only matrix similar to εI is εI itself, conclusion (a) follows.

Define the function $\lambda: Z \to \mathbb{C}$ by $\mathfrak{X}(g) = \lambda(g)I$ for $g \in Z$. It follows for $g, h \in Z$ that $\mathfrak{X}(gh) = \lambda(g)\lambda(h)I$ and hence Z is a subgroup and λ is a homomorphism (linear character) of Z. We have that $\chi(g) = f\lambda(g)$ for $g \in Z$ and (b) and (c) have been proved.

Clearly, $\ker \chi = \ker \lambda$ and thus $Z/\ker \chi$ is isomorphic to the image of λ, a finite multiplicative subgroup of the field $\mathbb{C}$. This subgroup is necessarily cyclic and (d) follows. Also, $\ker \chi = \ker \mathfrak{X}$ and $\mathfrak{X}(Z) \subseteq \mathbf{Z}(\mathfrak{X}(G))$ and (e) is an immediate consequence.

Finally, if $g(\ker \chi) \in \mathbf{Z}(G/\ker \chi)$, then $\mathfrak{X}(g) \in \mathbf{Z}(\mathfrak{X}(G))$. If $\chi \in \operatorname{Irr}(G)$, then by Lemma 2.25, we conclude that $\mathfrak{X}(g) = \varepsilon I$ for some $\varepsilon \in \mathbb{C}$. Now (f)follows from (a) and the proof is complete. ∎

(2.28) COROLLARY Let G be a group. Then

$$\mathbf{Z}(G) = \bigcap\{\mathbf{Z}(\chi) \mid \chi \in \operatorname{Irr}(G)\}.$$

Proof Since $(\mathbf{Z}(G) \ker \chi)/\ker \chi \subseteq \mathbf{Z}(G/\ker \chi)$, it follows from Lemma 2.27(f) that $\mathbf{Z}(G) \subseteq \mathbf{Z}(\chi)$. Conversely, suppose $g \in \mathbf{Z}(\chi)$ for every $\chi \in \operatorname{Irr}(G)$. It follows that $g(\ker(\chi)) \in \mathbf{Z}(G/\ker \chi)$ and thus for any $x \in G$, the commutator

$$[g, x] = g^{-1}x^{-1}gx \in \ker \chi.$$

Thus $[g, x] \in \bigcap\{\ker \chi \mid \chi \in \operatorname{Irr}(G)\} = 1$ and g commutes with x. Since $x \in G$ was arbitrary, we have $g \in \mathbf{Z}(G)$. ∎

It is apparent from Corollary 2.28 that $\mathbf{Z}(G)$ can be located from the character table of G. It follows that it can be determined from the table whether or not G is nilpotent. This is done by finding $\mathbf{Z}(G)$, then finding the character table of $G/\mathbf{Z}(G)$, and iterating this process. The sequence of subgroups of G which results is the upper central series and G is nilpotent iff this sequence reaches G.

Some information about character degrees can be obtained using Lemma 2.27(c). We need a lemma first.

(2.29) LEMMA Let $H \subseteq G$ and let χ be a character of G. Then

$$[\chi_H, \chi_H] \le |G:H|[\chi, \chi]$$

with equality iff $\chi(g) = 0$ for all $g \in G - H$.

Proof We have

$$|H|[\chi_H, \chi_H] = \sum_{h \in H} |\chi(h)|^2 \le \sum_{g \in G} |\chi(g)|^2 = |G|[\chi, \chi]$$

since $|\chi(g)|^2 \ge 0$ for $g \in G - H$. Equality thus holds iff $\chi(g) = 0$ for all $g \in G - H$. The result follows. ∎

(2.30) COROLLARY Let $\chi \in \text{Irr}(G)$. Then $\chi(1)^2 \le |G:\mathbf{Z}(\chi)|$. Equality occurs iff χ vanishes on $G - \mathbf{Z}(\chi)$.

Proof By Lemma 2.27(c), we have $\chi_{\mathbf{Z}(\chi)} = \chi(1)\lambda$ and thus $[\chi_{\mathbf{Z}(\chi)}, \chi_{\mathbf{Z}(\chi)}] = \chi(1)^2[\lambda, \lambda] = \chi(1)^2$. Therefore

$$\chi(1)^2 \le |G:\mathbf{Z}(\chi)|[\chi, \chi] = |G:\mathbf{Z}(\chi)|$$

with equality iff χ vanishes on $G - \mathbf{Z}(\chi)$. ∎

We already knew, of course, that $\chi(1)^2 \le |G|$ for $\chi \in \text{Irr}(G)$. We now have the slight improvement that $\chi(1)^2 \le |G:\mathbf{Z}(G)|$. Equality can occur here, and when it does, $\mathbf{Z}(\chi) = \mathbf{Z}(G)$ and χ vanishes on $G - \mathbf{Z}(G)$. It has been conjectured that only in a solvable group is it possible to have $\chi(1)^2 = |G:\mathbf{Z}(G)|$ with $\chi \in \text{Irr}(G)$. As of this writing, the question is still open. Observe that the nonabelian groups of order 27 which were discussed earlier give examples where equality occurs.

(2.31) THEOREM Suppose that $\chi \in \text{Irr}(G)$ and that $G/\mathbf{Z}(\chi)$ is abelian. Then $|G:\mathbf{Z}(\chi)| = \chi(1)^2$.

Proof It suffices to prove that χ vanishes on $G - \mathbf{Z}(\chi)$. Let $g \in G - \mathbf{Z}(\chi)$. Then by Lemma 2.27(f), we have that there exists $h \in G$ with $g^{-1}h^{-1}gh \notin \ker \chi$. However, since $G/\mathbf{Z}(\chi)$ is abelian, we have $g^{-1}h^{-1}gh = z \in \mathbf{Z}(\chi)$. Now, if $\mathfrak{X}$ is a representation of G which affords χ, then $\mathfrak{X}(z) = \varepsilon I$ and $\varepsilon \ne 1$ since $z \notin \ker \mathfrak{X}$. We have $\mathfrak{X}(gz) = \mathfrak{X}(g)\mathfrak{X}(z) = \varepsilon\mathfrak{X}(g)$ and thus $\chi(gz) = \varepsilon\chi(g)$. However, $gz = h^{-1}gh$ and so $\chi(gz) = \chi(g)$. Since $\varepsilon\chi(g) = \chi(g)$ and $\varepsilon \ne 1$, we have $\chi(g) = 0$ as desired. ∎

A character χ of G is said to be *faithful* if $\ker \chi = 1$. Every group has a faithful character, namely its regular character ρ; but not every group has a faithful irreducible character.

(2.32) THEOREM (a) If G has a faithful irreducible character, then $\mathbf{Z}(G)$ is cyclic.

(b) If G is a p-group and $\mathbf{Z}(G)$ is cyclic, then G has a faithful irreducible character.

Proof (a) Let $\chi \in \mathrm{Irr}(G)$ be faithful. By Lemma 2.27(f), $\mathbf{Z}(G) = \mathbf{Z}(\chi)$ and by part (d) of that lemma, $\mathbf{Z}(\chi)$ is cyclic.

(b) Since G is a p-group, it follows that if $1 \neq N \lhd G$, then $N \cap \mathbf{Z}(G) \neq 1$. Now let Z be the unique subgroup of order p in the cyclic group $\mathbf{Z}(G)$, so that $Z \subseteq N$ for every nontrivial normal subgroup N of G. Since $\bigcap\{\ker \chi \mid \chi \in \mathrm{Irr}(G)\} = 1$, it follows that $Z \nsubseteq \ker \chi$ for some $\chi \in \mathrm{Irr}(G)$. We conclude that $\ker \chi = 1$ and the proof is complete. ∎

Problem 2.19 provides an example to show that the full converse of Theorem 2.32(a) is not true.

Problems

(2.1) (a) Let $\mathfrak{X}$ be an irreducible F-representation of G over an arbitrary field. Show that $\sum_{g \in G} \mathfrak{X}(g) = 0$ unless $\mathfrak{X}$ is the principal representation.

(b) Let $H \subseteq G$ and $g \in G$ be such that all elements of the coset Hg are conjugate in G. Let χ be a $\mathbb{C}$-character of G such that $[\chi_H, 1_H] = 0$. Show that $\chi(g) = 0$.

Hint (b) Compute the trace of $\sum_{h \in H} \mathfrak{X}(hg)$, where $\mathfrak{X}$ affords χ.

In the following, all characters are over $\mathbb{C}$.

(2.2) (a) Let χ be a character of G. Show that χ is afforded by a representation $\mathfrak{X}$ such that all entries of $\mathfrak{X}(g)$ for all $g \in G$ lie in some field $F \subseteq \mathbb{C}$ with $|F : \mathbb{Q}| < \infty$.

(b) Let $\varepsilon = e^{2\pi i/n}$, where $n = |G|$ and let χ be a character of G. (Note that $\chi(g) \in \mathbb{Q}[\varepsilon]$ for all $g \in G$ by Lemma 2.15.) Let σ be an automorphism of the field $\mathbb{Q}[\varepsilon]$ and define χ^σ: $G \to \mathbb{C}$ by $\chi^\sigma(g) = \chi(g)^\sigma$. Show that χ^σ is a character and that $\chi^\sigma \in \mathrm{Irr}(G)$ iff $\chi \in \mathrm{Irr}(G)$.

(2.3) Let χ be a character of G. Define $\det \chi$: $G \to \mathbb{C}$ as follows. Choose $\mathfrak{X}$ affording χ and set

$$(\det \chi)(g) = \det \mathfrak{X}(g).$$

Show that $\det \chi$ is a uniquely defined linear character of G.

(2.4) (a) Let G be a nonabelian group of order 8. Show that G has a unique nonlinear irreducible character χ. Show that $\chi(1) = 2$, $\chi(z) = -2$, and $\chi(x) = 0$, where $z \in G' - \{1\}$ and $x \in G - G'$.

(b) If $G \cong D_8$, show that $\det \chi \neq 1_G$.
(c) If $G \cong Q_8$, show that $\det \chi = 1_G$.

Hint Show that $\ker(\det \chi)$ contains all elements of order 4. Use Lemma 2.15.

Note Although D_8 and Q_8 have identical character tables, the map $\det: \mathrm{Irr}(G) \to \mathrm{Irr}(G)$ is not the same for both groups.

(2.5) (a) Find a real representation of D_8 which affords the character χ of Problem 2.4(a).
(b) Show that this cannot be done for the group Q_8.

(2.6) Let χ, ψ be characters of G. Define $\chi\psi: G \to \mathbb{C}$ by $(\chi\psi)(g) = \chi(g)\psi(g)$.

(a) If $\psi(1) = 1$, show that $\chi\psi$ is a character.
(b) If $\psi(1) = 1$, show that $\chi\psi \in \mathrm{Irr}(G)$ iff $\chi \in \mathrm{Irr}(G)$.
(c) If $\psi = \bar{\chi}$ (that is, $\psi(g) = \overline{\chi(g)}$) and $\chi(1) > 1$, show that $\chi\psi \notin \mathrm{Irr}(G)$.

Note In Chapter 4 we will show that $\chi\psi$ is always a character.

(2.7) Let G be abelian and write $\hat{G} = \mathrm{Irr}(G)$.

(a) Show that $\hat{G}$ is an abelian group under the multiplication of Problem 2.6.
(b) If $H \subseteq G$, let $H^{\perp} = \{\lambda \in \hat{G} \mid H \subseteq \ker \lambda\}$. Show that $\perp$ is a bijection from the set of subgroups of G onto the set of subgroups of $\hat{G}$.
(c) Show that $G \cong \hat{G}$.

Hints There is a natural isomorphism of G onto $\hat{\hat{G}}$. Use this for (b). For (c), use the fundamental theorem of abelian groups.

(2.8) Let χ be a faithful character of G. Show that $H \subseteq G$ is abelian iff every irreducible constituent of χ_H is linear.

(2.9) (a) Let χ be a character of an abelian group A. Show

$$\sum_{x \in A} |\chi(x)|^2 \geq |A|\chi(1).$$

(b) Let $A \subseteq G$ with A abelian and $|G : A| = n$. Show that $\chi(1) \leq n$ for all $\chi \in \mathrm{Irr}(G)$.

(2.10) Suppose $G = \bigcup_{i=1}^{n} A_i$, where the A_i are abelian subgroups of G and $A_i \cap A_j = 1$ if $i \neq j$.

(a) Let $\chi \in \mathrm{Irr}(G)$. Show that if $\chi(1) > 1$, then $\chi(1) \geq |G|/(n - 1)$.
(b) If G is nonabelian, then $|A_i| \leq n - 1$ for each i and $n - 1 \geq (|G|)^{1/2}$.

Hints For (a), bound $\sum_{g \in G} |\chi(g)|^2$ using Problem 2.9(a). For (b), use Problem 2.9(b).

(2.11) Let $g \in G$. Show that g is conjugate to g^{-1} in G iff $\chi(g)$ is real for all characters χ of G.

Note An element of a group which is conjugate to its inverse is called a *real* element. If G has any real elements other than 1, then G must necessarily have even order.

(2.12) Let $|G| = n$ and let $g \in G$. Show that $\chi(g)$ is rational for every character χ of G iff g is conjugate to g^m for every integer m with $(m, n) = 1$.

Hints Let ε be a primitive nth root of 1 in $\mathbb{C}$ and let $E = \mathbb{Q}[\varepsilon]$. Let $\mathcal{G}$ be the Galois group of E over $\mathbb{Q}$. Given $(m, n) = 1$, show that there exists $\sigma \in \mathcal{G}$ with $\chi(g^m) = \chi(g)^\sigma$ for all $g \in G$ and all characters χ. Conversely, for every $\sigma \in \mathcal{G}$, there is an m such that this formula holds.

(2.13) Let $|G'| = p$, a prime. Assume that $G' \subseteq \mathbf{Z}(G)$. Show that

$$\chi(1)^2 = |G : \mathbf{Z}(G)|$$

for every nonlinear $\chi \in \mathrm{Irr}(G)$.

(2.14) Let $H \subseteq G' \cap \mathbf{Z}(G)$ be cyclic of order n and let m be the maximum of the orders of the elements of G/H. Assume that n is a prime power and show that $|G| \geq n^2 m$.

Hints Choose $\chi \in \mathrm{Irr}(G)$ with $H \cap \ker \chi = 1$. Let $\lambda = \det \chi$ (as in Problem 2.3). We have $\chi_H = \chi(1)\mu$ with $\mu \in \mathrm{Irr}(H)$. Using λ and μ, show that $\chi(1) \geq n$. Finish the proof using Problem 2.9(b).

Note The assumption on n can be removed using Corollary 5.4.

(2.15) Let $\chi \in \mathrm{Irr}(G)$ be faithful and suppose $H \subseteq G$ and $\chi_H \in \mathrm{Irr}(H)$. Show that

$$\mathbf{C}_G(H) = \mathbf{Z}(G).$$

(2.16) Let $H \subseteq G$ and let χ be a (possibly reducible) character of G which vanishes on $G - H$. Assume either that $H = 1$ or that G is abelian. Show that $|G : H|$ divides $\chi(1)$.

Hint Let λ be an irreducible constituent of χ_H. Under either hypothesis, find $\mu \in \mathrm{Irr}(G)$ with $\mu_H = \lambda$. Compute $[\chi, \mu]$ and conclude $|G : H| \mid [\chi_H, \lambda]$.

Note A natural common generalization of the situations $H = 1$ and G is abelian is $H \subseteq \mathbf{Z}(G)$. Is the result true under the hypothesis $H \subseteq \mathbf{Z}(G)$?

(2.17) Let $A < G$ be abelian and assume there exists $\chi \in \mathrm{Irr}(G)$ with $\chi(1) = |G : A|$. Show that G has a nontrivial normal abelian subgroup.

Hint Show that χ vanishes on $G - A$.

(2.18) Let $A \lhd G$ and suppose $A = \mathbf{C}_G(a)$ for every $a \neq 1$, $a \in A$. Assume further that G/A is abelian. Show that G has exactly $(|A| - 1)/|G:A|$ nonlinear irreducible characters and that these all have degree equal to $|G:A|$ and vanish on $G - A$.

Hints Let $k = |\mathrm{Irr}(G)|$. By counting classes, show that

$$k \leq 1 + (|A| - 1)/|G:A| + (|G| - |A|)/|A|.$$

Using characters, show that

$$k \geq |G:A| + (|G| - |G:A|)/|G:A|^2.$$

Use Problem 2.9(b).

Note The group G of Problem 2.18 is a special case of a Frobenius group. The character theory of such groups will be discussed more fully later. Observe that although the hypotheses of 2.18 are very special, this situation does arise frequently. Some examples are A_4 and nonabelian groups of order pq where p and q are primes with $p \mid (q - 1)$.

(2.19) Let $E = \langle x_1, x_2, x_3, x_4 \rangle$ be an elementary abelian group of order 16. Let $P = \langle y \rangle$ be cyclic of order 3. Let P act on E by

$$x_1{}^y = x_2, \quad x_2{}^y = x_1x_2, \quad x_3{}^y = x_4, \quad x_4{}^y = x_3x_4.$$

Let G be the semidirect product $E \rtimes P$. Show that $\mathbf{Z}(G) = 1$ but that G does not have a faithful irreducible character.

Hint The smallest possible degree for a faithful character of E is 4.

(2.20) Let $\mathfrak{X}$ and $\mathfrak{Y}$ be irreducible $\mathbb{C}$-representations of G and define the functions $a_{ij}(g)$ and $b_{ij}(g)$ by $\mathfrak{X}(g) = (a_{ij}(g))$ and $\mathfrak{Y}(g) = (b_{ij}(g))$. Write

$$S_{pqrs} = \sum_{g \in G} a_{pq}(g) b_{rs}(g^{-1}).$$

Show that $S_{pqrs} = 0$ if $\mathfrak{X}$ and $\mathfrak{Y}$ are not similar. If $\mathfrak{X} = \mathfrak{Y}$, show that $S_{pqrs} = 0$ unless $p = s$ and $q = r$ in which case $S_{pqqp} = |G|/\deg \mathfrak{X}$.

Hint Let $P_{qr} = \sum_{g \in G} \mathfrak{X}(g) E_{qr} \mathfrak{Y}(g^{-1})$ where E_{qr} is the $(\deg \mathfrak{X}) \times (\deg \mathfrak{Y})$ matrix with all entries zero except the (q, r) entry which equals one. Note that $\mathfrak{X} P_{qr} = P_{qr} \mathfrak{Y}$. Use Schur's lemma. For the last statement use Lemma 2.25 and compute $\mathrm{tr}(P_{qr})$.

Note The results of this problem are called the *Schur relations*. They can be used to give another proof of the orthogonality relations.

3 Characters and integrality

One of the most celebrated applications of character theory to pure group theory is Burnside's theorem which asserts that a group with order divisible by at most two primes is solvable. The proof of this theorem (and much of the rest of character theory) depends on properties of algebraic integers. We begin by establishing some of the most basic of these properties.

(3.1) DEFINITION An *algebraic integer* is a complex number which is a root of a polynomial of the form

$$x^n + a_{n-1}x^{n-1} + \cdots + a_0,$$

where $a_i \in \mathbb{Z}$ for $0 \leq i \leq n - 1$.

(3.2) LEMMA The rational algebraic integers are precisely the elements of $\mathbb{Z}$.

Proof If $a \in \mathbb{Z}$, then a is a root of the polynomial $x - a$ and thus is an algebraic integer. Conversely, let r/s be an algebraic integer with $r, s \in \mathbb{Z}$. We may assume that $(r, s) = 1$. We have

$$(r/s)^n + a_{n-1}(r/s)^{n-1} + \cdots + a_0 = 0.$$

Now multiply by s^n and rearrange terms to obtain

$$r^n = -s(a_{n-1}r^{n-1} + a_{n-2}sr^{n-2} + \cdots + a_0 s^{n-1}).$$

We conclude that $s | r^n$. However, since $(r, s) = 1$, this yields $s = \pm 1$ and $r/s \in \mathbb{Z}$ as desired. ∎

Frequently, the word "integer" is used to mean an algebraic integer, and the elements of $\mathbb{Z}$ are referred to as "rational integers." One of the most important properties of the set of algebraic integers is that it is a ring. In other words, sums and products of integers are integers. This fact seems surprising from the definition, but it is not hard to prove indirectly. This we proceed to do.

(3.3) LEMMA Let $X = \{\alpha_1, \ldots, \alpha_k\}$ be a finite set of algebraic integers. Then there exists a ring S satisfying

(a) $\mathbb{Z} \subseteq S \subseteq \mathbb{C}$;
(b) $X \subseteq S$;
(c) there exists a finite subset, Y of S such that every element of S is a $\mathbb{Z}$-linear combination of elements of Y.

Proof The integer α_i satisfies an equation of the form

$$\alpha_i^{n_i} = f_i(\alpha_i),$$

where f_i is a polynomial of degree $n_i - 1$ with coefficients in $\mathbb{Z}$. Let $Y = \{\alpha_1^{r_1}\alpha_2^{r_2} \cdots \alpha_k^{r_k} \mid 0 \le r_i \le n_i - 1\}$ and let S be the set of all $\mathbb{Z}$-linear combinations of elements of Y.

Using the equation $\alpha_i^{n_i} = f_i(\alpha_i)$, any power of α_i may be written as a $\mathbb{Z}$-linear combination of $1, \alpha_i, \alpha_i^2, \ldots, \alpha_i^{n_i - 1}$. It follows from this that the product of any two elements of Y lies in S and hence S is a ring. All of the properties claimed for S are now clear. ▮

Condition (c) of the above lemma may be paraphrased by saying that S is finitely generated as a $\mathbb{Z}$-module. We now prove a strong converse to Lemma 3.3.

(3.4) THEOREM Let S be a ring with $\mathbb{Z} \subseteq S \subseteq \mathbb{C}$. Suppose that S is finitely generated as a $\mathbb{Z}$-module. Then every element of S is an algebraic integer.

Proof Let $s \in S$ and let $Y = \{y_1, \ldots, y_n\} \subseteq S$ have the property that every element of S is a $\mathbb{Z}$-linear combination of elements of Y. We then have

$$sy_i = \sum_j a_{ij} y_j$$

for all i, with $a_{ij} \in \mathbb{Z}$. Let A be the matrix (a_{ij}) and let v be the column, $\mathrm{col}(y_1, \ldots, y_n)$. Then

$$Av = sv$$

and thus s is a root of the polynomial

$$f(x) = \det(xI - A).$$

It follows that s is an algebraic integer and the proof is complete. ▮

(3.5) COROLLARY Sums and products of algebraic integers are algebraic integers.

Proof Let α and β be algebraic integers. By Lemma 3.3, there exists a ring S with $\mathbb{Z} \subseteq S \subseteq \mathbb{C}$ such that $\alpha, \beta \in S$ and S is finitely generated as a $\mathbb{Z}$-module. Since $\alpha + \beta$ and $\alpha\beta \in S$, it follows from Theorem 3.4 that they are algebraic integers. ∎

(3.6) COROLLARY Let χ be a character of a group G. Then $\chi(g)$ is an algebraic integer for all $g \in G$.

Proof By Lemma 2.15, we know that $\chi(g) = \varepsilon_1 + \cdots + \varepsilon_f$, where the ε_i are roots of a polynomial of the form $x^n - 1$, and therefore are algebraic integers. The result now follows. ∎

We can now see the reason for the assertion made in Chapter 2 that all of the entries in the character table of the symmetric group Σ_n lie in $\mathbb{Z}$. If $g \in \Sigma_n$ and m is relatively prime to $o(g)$, then g^m and g have identical cycle structures and therefore these elements are conjugate in Σ_n. It follows from Problem 2.12 that $\chi(g)$ is rational for all $\chi \in \mathrm{Irr}(\Sigma_n)$. Since $\chi(g)$ is an algebraic integer, Lemma 3.2 yields that $\chi(g) \in \mathbb{Z}$ as claimed.

Let G be a group and $\chi \in \mathrm{Irr}(G)$. We wish to define a function ω depending on χ, from the center of the group algebra $\mathbb{C}[G]$ into $\mathbb{C}$. Let $\mathfrak{X}$ be any representation which affords χ. If $z \in \mathbf{Z}(\mathbb{C}[G])$, then we may conclude from Lemma 2.25 that $\mathfrak{X}(z) = \varepsilon I$ for some $\varepsilon \in \mathbb{C}$. Observe that since the only matrix similar to εI is εI itself, the complex number ε does not depend on the choice of the particular representation affording χ. We now define ω by setting $\omega(z) = \varepsilon$. In other words

$$\mathfrak{X}(z) = \omega(z)I$$

for all $z \in \mathbf{Z}(\mathbb{C}[G])$. We shall often write $\omega = \omega_\chi$ in order to emphasize the dependence of ω on χ.

Since $\mathfrak{X}$ is an algebra homomorphism, it is easy to see that ω is also a homomorphism. In particular, ω is $\mathbb{C}$-linear and hence to determine ω on $\mathbf{Z}(\mathbb{C}[G])$, it suffices to calculate its values on a basis. Such a basis is given by the class sums for the conjugacy classes of G. Let $\mathscr{K}$ be a class with sum $K \in \mathbb{C}[G]$ and let $g \in \mathscr{K}$. Calculation of traces in the equation $\mathfrak{X}(K) = \omega(K)I$ yields

$$\chi(1)\omega(K) = \chi(K) = \sum_{x \in \mathscr{K}} \chi(x) = |\mathscr{K}|\chi(g)$$

and thus

$$\omega_\chi(K) = \frac{\chi(g)|\mathscr{K}|}{\chi(1)}.$$

Note that it follows from this formula that the functions ω_χ are determined by the character table of G.

(3.7) THEOREM Let $\chi \in \operatorname{Irr}(G)$ and let K be a class sum in $\mathbb{C}[G]$. Then $\omega_\chi(K)$ is an algebraic integer.

Proof Let $\mathscr{K}_1, \ldots, \mathscr{K}_k$ be the classes of G, with corresponding class sums $K_1, \ldots, K_k$. By Theorem 2.4, we have $K_i K_j = \sum_\nu a_{ij\nu} K_\nu$ where $a_{ij\nu} \in \mathbb{Z}$. Since $\omega = \omega_\chi$ is an algebra homomorphism from $\mathbf{Z}(\mathbb{C}[G])$ to $\mathbb{C}$, we have

$$\omega(K_i)\omega(K_j) = \sum_\nu a_{ij\nu}\omega(K_\nu).$$

Let S be the set of all $\mathbb{Z}$-linear combinations of the $\omega(K_i)$. It follows that S is closed under multiplication. Since $\omega(1) = 1$, it follows that $\mathbb{Z} \subseteq S \subseteq \mathbb{C}$ and Theorem 3.4 applies. All of the elements of S are therefore algebraic integers and the proof is complete. ▮

It should be emphasized that the fact that $\chi(g)|\mathrm{Cl}(g)|/\chi(1)$ is an algebraic integer does not follow from the fact that $\chi(g)$ is integral since division of an integer by an integer does not usually result in an integer.

We proceed now toward Burnside's solvability theorem. The essence of the argument is contained in the next result.

(3.8) THEOREM (*Burnside*) Let $\chi \in \operatorname{Irr}(G)$ and let $\mathscr{K}$ be a conjugacy class of G with $g \in \mathscr{K}$. Suppose that $(\chi(1), |\mathscr{K}|) = 1$. Then either $g \in \mathbf{Z}(\chi)$ or else $\chi(g) = 0$.

Proof We know that $\chi(g)|\mathscr{K}|/\chi(1)$ is an algebraic integer. Since $(\chi(1), |\mathscr{K}|) = 1$, we may choose rational integers u and v so that $u\chi(1) + v|\mathscr{K}| = 1$. Thus

$$\frac{\chi(g)(1 - u\chi(1))}{\chi(1)} = v\frac{\chi(g)|\mathscr{K}|}{\chi(1)}$$

is an algebraic integer. Since $u\chi(g)$ is also integral, it follows that $\alpha = \chi(g)/\chi(1)$ is an algebraic integer. Suppose that $g \notin \mathbf{Z}(\chi)$, so that $|\chi(g)| < \chi(1)$ and $|\alpha| < 1$.

Now let $n = o(g)$ and let E be the splitting field for the polynomial $x^n - 1$ over $\mathbb{Q}$ in $\mathbb{C}$ so that $\alpha \in E$. Let $\mathscr{G}$ be the Galois group of E over $\mathbb{Q}$. Since $\chi(g)$ is a sum of $\chi(1)$ roots of unity, so is $\chi(g)^\sigma$ for each $\sigma \in \mathscr{G}$. It follows that $|\chi(g)^\sigma| \leq \chi(1)$

and $|\alpha^\sigma| \le 1$ for $\sigma \in \mathscr{G}$. We have then

$$\left|\prod_{\sigma \in \mathscr{G}} \alpha^\sigma\right| < 1.$$

For each $\sigma \in \mathscr{G}$, α^σ satisfies the same rational polynomials that α satisfies and hence is integral. Therefore $\beta = \prod \alpha^\sigma$ is an algebraic integer. However, β is clearly fixed by all $\sigma \in \mathscr{G}$ and therefore $\beta \in \mathbb{Q}$ by elementary Galois theory. It follows from Lemma 3.2 that $\beta \in \mathbb{Z}$. Since $|\beta| < 1$, we have $\beta = 0$ and hence $\alpha^\sigma = 0$ for some σ. Therefore $0 = \alpha = \chi(g)/\chi(1)$ and $\chi(g) = 0$. The proof is complete. ∎

(3.9) THEOREM Let G be a nonabelian simple group. Then $\{1\}$ is the only conjugacy class of G which has prime power size.

Proof Suppose $g \in G$, $|\mathrm{Cl}(g)| = p^a$, and $g \ne 1$. Let $\chi \in \mathrm{Irr}(G)$, $\chi \ne 1_G$. Then $\ker \chi = 1$ since G is simple and $\mathbf{Z}(\chi) = \mathbf{Z}(G) = 1$ since G is nonabelian. Thus if $p \nmid \chi(1)$, then $\chi(g) = 0$ by Theorem 3.8. Now

$$0 = \rho(g) = \sum_{\chi \in \mathrm{Irr}(G)} \chi(1)\chi(g) = 1 + \sum_{\chi \in \mathrm{Irr}(G);\ p \mid \chi(1)} \chi(1)\chi(g).$$

We have $-1 = p\alpha$, where

$$\alpha = \sum \frac{\chi(1)}{p} \chi(g),$$

the sum being taken over $\chi \in \mathrm{Irr}(G)$ where $p \mid \chi(1)$. It follows that $\alpha = -1/p$ is an algebraic integer and this violates Lemma 3.2. ∎

(3.10) THEOREM Let $|G| = p^a q^b$, where p and q are primes. Then G is solvable.

Proof Use induction on $|G|$. We may assume $|G| > 1$ and choose a maximal proper normal subgroup N. If $N > 1$, then by the inductive hypothesis, N and G/N are solvable and thus G is solvable and the result follows.

Suppose then $N = 1$, so that G is simple. Let $P \ne 1$ be a Sylow subgroup of G. We may choose $g \in \mathbf{Z}(P)$, $g \ne 1$. Then $|\mathrm{Cl}(g)| = |G : \mathbf{C}(g)|$ divides $|G : P|$, which is a prime power. It now follows from Theorem 3.9 that the simple group G is abelian and the proof is complete. ∎

We now obtain some strong results about the degrees of the irreducible characters of a group G. The fact is that $\chi(1) \mid |G : \mathbf{Z}(\chi)|$ for $\chi \in \mathrm{Irr}(G)$. We shall first prove the weaker statement that the irreducible character degrees divide the group order. This proof is much less complicated and serves to motivate the stronger proof.

(3.11) THEOREM Let $\chi \in \mathrm{Irr}(G)$. Then $\chi(1) \mid |G|$.

Proof From the first orthogonality relation we have

$$|G| = \sum_{g \in G} \chi(g)\chi(g^{-1}).$$

We wish to rewrite this equation in terms of ω_χ. Let $\mathscr{K}_1, \mathscr{K}_2, \ldots, \mathscr{K}_k$ be the classes of G, with class sums K_i and representative elements g_i. We have then

$$|G| = \sum_{i=1}^{k} |\mathscr{K}_i| \chi(g_i)\chi(g_i^{-1}) = \sum_{i=1}^{k} \chi(1)\omega(K_i)\chi(g_i^{-1}),$$

where $\omega = \omega_\chi$. This yields

$$|G|/\chi(1) = \sum \omega(K_i)\chi(g_i^{-1}),$$

which is an algebraic integer. Since $|G|/\chi(1)$ is rational, it lies in $\mathbb{Z}$ and the result follows. ∎

(3.12) THEOREM Let $\chi \in \mathrm{Irr}(G)$. Then $\chi(1) \mid |G : \mathbf{Z}(\chi)|$.

Proof Since χ may be viewed as a character of $G/\ker \chi$, it is no loss to assume that $\ker \chi = 1$. Under this assumption, $\mathbf{Z}(G) = \mathbf{Z}(\chi)$.

For $x, y \in G$, define $x \equiv y$ if there exists $z \in Z = \mathbf{Z}(G)$ such that x is conjugate to yz. It is easy to check that $\equiv$ is an equivalence relation and thus partitions G into equivalence classes. We claim that $|\chi(x)|$ is constant as x runs over one of these classes. To see this, observe that $\chi_Z = \chi(1)\lambda$, where λ is a faithful linear character of Z, and that $\chi(yz) = \lambda(z)\chi(y)$ for $z \in Z$ and $y \in G$. If $x \equiv y$, then $x^g = yz$ for some z and $\chi(x) = \lambda(z)\chi(y)$. Since $|\lambda(z)| = 1$, the claim follows.

Let $\mathscr{C}_1, \mathscr{C}_2, \ldots, \mathscr{C}_r$ be those $(\equiv)$-classes on which χ does not vanish. We have then

$$|G| = \sum_{g \in G} |\chi(g)|^2 = \sum_{i=1}^{r} |\mathscr{C}_i| |\chi(g_i)|^2,$$

where the g_i are representatives for the $\mathscr{C}_i$. We claim $|\mathscr{C}_i| = |\mathrm{Cl}(g_i)||Z|$. Clearly, every $x \in \mathscr{C}_i$ is of the form yz where $y \in \mathrm{Cl}(g_i)$ and $z \in Z$. It suffices to show that all of these elements yz are distinct. Suppose that $y_1 z_1 = y_2 z_2$, $y_1, y_2 \in \mathrm{Cl}(g_i)$, and $z_1, z_2 \in Z$. Then

$$\chi(y_1)\lambda(z_1) = \chi(y_2)\lambda(z_2)$$

and $\chi(y_1) = \chi(y_2) = \chi(g_i) \neq 0$. Thus $\lambda(z_1) = \lambda(z_2)$ and hence $z_1 = z_2$ since λ is faithful on Z. Thus $y_1 = y_2$ and the claim is established.

We have now

$$\begin{aligned}|G| = \sum |\mathscr{C}_i| |\chi(g_i)|^2 &= \sum |\mathrm{Cl}(g_i)| \chi(g_i)\chi(g_i^{-1})|Z| \\ &= \sum \chi(1)\omega(K_i)\chi(g_i^{-1})|Z|,\end{aligned}$$

where $K_i = \sum_{x \in \mathrm{Cl}(g_i)} x$ and $\omega = \omega_\chi$. It follows that

$$|G:Z|/\chi(1) = \sum_{i=1}^{r} \omega(K_i)\chi(g_i^{-1}),$$

an algebraic integer which is a rational number. The result now follows. ∎

As a combined application of Theorems 3.8 and 3.12, we prove the following.

(3.13) THEOREM Let G have a faithful irreducible character of degree p^a, where p is a prime and suppose that a Sylow p-subgroup of G is abelian. Then p^a is the exact power of p dividing $|G:\mathbf{Z}(G)|$.

Proof Let χ be the given faithful character of G. By Theorem 3.12, $p^a = \chi(1)$ divides $|G:\mathbf{Z}(G)|$.

Let $P \in \mathrm{Syl}_p(G)$ and let $x \in P$. Thus $P \subseteq \mathbf{C}(x)$ and hence $(\chi(1), |\mathrm{Cl}(x)|) = 1$. By Theorem 3.8, $\chi(x) = 0$ if $x \notin \mathbf{Z}(\chi) = \mathbf{Z}(G)$. Let $Z = P \cap \mathbf{Z}(G)$ so that χ vanishes on $P - Z$. Now by Problem 2.16, we conclude that $|P:Z| \mid p^a$. Since $P/Z \cong P\mathbf{Z}(G)/\mathbf{Z}(G)$, which is a Sylow subgroup of $G/\mathbf{Z}(G)$, the result follows. ∎

The above result is typical of a number of theorems about "complex linear groups," that is, groups of nonsingular matrices over $\mathbb{C}$. In these theorems, one is given the degree n of a finite linear group G and the object is to control the structure of G in terms of n, often under certain additional assumptions such as the irreducibility of G. In Theorem 3.13 we are given a group having a faithful representation of known degree, and this is obviously equivalent to being given a linear group of that degree.

If G is a complex linear group, let S be the subgroup consisting of the elements of G which are scalar matrices (that is, of the form εI). Observe that if G is irreducible, then $S = \mathbf{Z}(G)$. The group G/S is called the *collineation group* associated with G. Frequently, the object of a theorem about linear groups of given degree is to obtain information about the associated collineation group. Theorem 3.13 is of this nature.

The reason for this situation may be seen from the following. Let G be a linear group of degree n and let C be a group of $n \times n$ scalar matrices. Let $G^* = GC$ and let S^* be the scalar subgroup of G^*. The linear group G^* may be much larger than G but $G/S \cong G^*/S^*$.

We have already seen several situations in which a character value is forced to be zero. We shall now prove that every nonlinear irreducible character vanishes somewhere. We begin with a preliminary result.

(3.14) LEMMA Let G be a cyclic group and let χ be a (possibly reducible) character of G. Let $S = \{g \in G \mid G = \langle g \rangle\}$ and assume that $\chi(s) \neq 0$ for all $s \in S$. Then

$$\sum_{s \in S} |\chi(s)|^2 \geq |S|.$$

Proof Let $n = |G|$ and let E be the splitting field for the polynomial $x^n - 1$ over $\mathbb{Q}$ in $\mathbb{C}$. Let $\mathscr{G}$ be the Galois group of E over $\mathbb{Q}$. If $\sigma \in \mathscr{G}$ and ε is an nth root of 1, then $\varepsilon^\sigma = \varepsilon^m$ for some $m \in \mathbb{Z}$, $(m, n) = 1$. Now $\chi(s) = \varepsilon_1 + \cdots + \varepsilon_f$ where $\varepsilon_i{}^n = 1$ and hence $\chi(s)^\sigma = \varepsilon_1{}^m + \cdots + \varepsilon_f{}^m = \chi(s^m)$.

The group $\mathscr{G}$ is abelian and the restriction of complex conjugation to E is an element of $\mathscr{G}$. It follows that $\overline{\alpha^\sigma} = (\bar{\alpha})^\sigma$ for all $\alpha \in E$ and $\sigma \in \mathscr{G}$, and thus $|\alpha^\sigma|^2 = \alpha^\sigma \overline{\alpha^\sigma} = \alpha^\sigma \bar{\alpha}^\sigma = (|\alpha|^2)^\sigma$. Therefore $(|\chi(s)|^2)^\sigma = |\chi(s^m)|^2$, where $(m, n) = 1$ and m depends only on σ.

Observe that if $s \in S$ and $(m, n) = 1$, then $s^m \in S$. Also, the map $x \mapsto x^m$ is one-to-one on G and therefore effects a permutation of S. It follows that $\prod_{s \in S} |\chi(s)|^2$ is invariant under $\mathscr{G}$ and hence is rational. Since it is an algebraic integer, it must lie in $\mathbb{Z}$, and since χ does not vanish on S, we have

$$\prod_{s \in S} |\chi(s)|^2 \geq 1.$$

Now we use the fact that for any positive real numbers $r_1, r_2, \ldots, r_k$, we have

$$\frac{1}{k} \sum r_i \geq \left(\prod r_i\right)^{1/k}$$

and we conclude that

$$\frac{1}{|S|} \sum_{s \in S} |\chi(s)|^2 \geq 1$$

and the proof is complete. ∎

(3.15) THEOREM (*Burnside*) Let $\chi \in \mathrm{Irr}(G)$ with $\chi(1) > 1$. Then $\chi(g) = 0$ for some $g \in G$.

Proof Partition G into equivalence classes by calling two elements of G equivalent if they generate the same cyclic subgroup of G. Assume $\chi(g) \neq 0$ for all $g \in G$. Then by Lemma 3.14, we have

$$\sum_{s \in S} |\chi(s)|^2 \geq |S|$$

for every equivalence class S. Sum this inequality over all equivalence classes of nonidentity elements to obtain

$$\sum_{g \neq 1} |\chi(g)|^2 \geq |G| - 1$$

and thus

$$|G| = \sum_{g \in G} |\chi(g)|^2 \geq |G| - 1 + \chi(1)^2.$$

This forces $\chi(1) \leq 1$ which contradicts the hypothesis. ∎

The next topic we shall discuss does not, strictly speaking, depend on algebraic integers. Nevertheless it seems appropriate to include it here.

Let G and H be finite groups and suppose $\mathbb{C}[G] \cong \mathbb{C}[H]$, where this is a $\mathbb{C}$-algebra isomorphism. What can we infer about the relationship between G and H? Clearly, $|G| = |H|$ and there exists a degree-preserving one-to-one correspondence between $\text{Irr}(G)$ and $\text{Irr}(H)$. We cannot conclude, however, that G and H are isomorphic or even that they have identical character tables. Indeed, if G is abelian, it follows from the results of Chapter 1 that $\mathbb{C}[G]$ is the direct sum of $|G|$ copies of $\mathbb{C}$. Thus if G and H are abelian and $|G| = |H|$, then $\mathbb{C}[G] \cong \mathbb{C}[H]$.

The situation becomes more interesting if we make the weaker assumption that $\mathbb{Z}[G] \cong \mathbb{Z}[H]$, where $\mathbb{Z}[G]$ represents the group ring of G over $\mathbb{Z}$ and may be identified with the ring of $\mathbb{Z}$-linear combinations of elements of G in $\mathbb{C}[G]$. It is conjectured that if $\mathbb{Z}[G] \cong \mathbb{Z}[H]$ for finite G and H, then $G \cong H$. The best result in this direction that has been proved as of this writing is due to A. Whitcomb. It asserts that if G is metabelian and $\mathbb{Z}[G] \cong \mathbb{Z}[H]$, then $G \cong H$.

If $\mathbb{Z}[G] \cong \mathbb{Z}[H]$, then it is clear that we may view H as a multiplicative subgroup of $\mathbb{Z}[G] \subseteq \mathbb{C}[G]$ and that H spans $\mathbb{C}[G]$ over $\mathbb{C}$. In particular, $|H| \geq |G|$ and hence by symmetry $|H| = |G|$ and H is a basis for $\mathbb{C}[G]$.

(3.16) DEFINITION Let $H \subseteq \mathbb{C}[G]$ be such that H is a multiplicative group which is a basis for $\mathbb{C}[G]$. Suppose that every element of H is a $\mathbb{Z}$-linear combination of elements of G. Then H is an *integral group basis* in $\mathbb{C}[G]$.

If H is an integral group basis in $\mathbb{C}[G]$, we identify $\mathbb{C}[G]$ with $\mathbb{C}[H]$ in the natural manner. An important result used in studying the isomorphism problem is due to G. Glauberman. Most of the rest of this chapter is devoted to its proof and some consequences. We use the notation $\Sigma\mathscr{K}$ to denote the sum of the elements of the conjugacy class $\mathscr{K}$, computed in the group algebra.

(3.17) THEOREM (*Glauberman*) Let H be an integral group basis in $\mathbb{C}[G]$. Then there exists a one-to-one correspondence between the sets of conjugacy classes of H and G such that if $\mathscr{L}$ corresponds to $\mathscr{K}$, then $|\mathscr{L}| = |\mathscr{K}|$ and $\Sigma\mathscr{L} = \pm\Sigma\mathscr{K}$.

We need a lemma.

(3.18) LEMMA Let α, $\beta \in \mathbf{Z}(\mathbb{C}[G])$ and view the characters of G as being defined on all of $\mathbb{C}[G]$. Define

$$\langle \alpha, \beta \rangle = \sum_{\chi \in \mathrm{Irr}(G)} \chi(\alpha)\overline{\chi(\beta)}.$$

Then

(a) $\langle c_1\alpha_1 + c_2\alpha_2, \beta \rangle = c_1\langle \alpha_1, \beta \rangle + c_2\langle \alpha_2, \beta \rangle$;
(b) $\langle \beta, \alpha \rangle = \overline{\langle \alpha, \beta \rangle}$;
(c) $\langle K_i, K_j \rangle = 0$ if $i \neq j$;
(d) $\langle K_i, K_i \rangle = |G||\mathscr{K}_i|$;

where the $\mathscr{K}_i$ are the classes of G and $K_i = \Sigma\mathscr{K}_i$.

Proof Statements (a) and (b) are immediate since characters are linear functions. To prove (c) and (d), let $x \in \mathscr{K}_i$ and $y \in \mathscr{K}_j$. Then $\chi(K_i) = |\mathscr{K}_i|\chi(x)$ and $\chi(K_j) = |\mathscr{K}_j|\chi(y)$. The result is now immediate from the second orthogonality relation. ∎

Proof of Theorem 3.17 Identify $\mathbb{C}[H]$ with $\mathbb{C}[G]$ so that $k = \dim(\mathbf{Z}(\mathbb{C}[G]))$ is the common number of classes of H and G. Let $\mathscr{L}_1, \mathscr{L}_2, \ldots, \mathscr{L}_k$ be the classes of H and $\mathscr{K}_1, \mathscr{K}_2, \ldots, \mathscr{K}_k$ the classes of G. Write $K_i = \Sigma\mathscr{K}_i$ and $L_i = \Sigma\mathscr{L}_i$. Since the K_i are a basis for $\mathbf{Z}(\mathbb{C}[G])$, we may write

$$L_i = \sum_j c_{ij}K_j.$$

Since L_i is a $\mathbb{Z}$-linear combination of elements of G, we conclude that all $c_{ij} \in \mathbb{Z}$.

The form $\langle\ ,\ \rangle$ defined in Lemma 3.18 depends only on the algebra $\mathbb{C}[G] = \mathbb{C}[H]$ and not on the particular groups G and H. This is because the irreducible characters, defined on $\mathbb{C}[G]$, are simply the traces of the irreducible representations of the algebra. We therefore have

$$(1) \qquad |H||\mathscr{L}_i| = \langle L_i, L_i \rangle = \sum_j (c_{ij})^2|G||\mathscr{K}_j|.$$

Sum (1) over i to obtain

$$(2) \qquad |H|^2 = |G|\sum_j \left(\sum_i (c_{ij})^2\right)|\mathscr{K}_j|.$$

Since the L_i span $\mathbf{Z}(\mathbb{C}[G])$ but no proper subset of $\{K_j\}$ spans this space, it follows that for each j, there exists i with $c_{ij} \neq 0$. Thus $\sum_i (c_{ij})^2 \geq 1$. Since $|H| = |G|$, Equation (2) yields

$$|G| = \sum_j \left(\sum_i (c_{ij})^2\right)|\mathscr{K}_j| \geq \sum_j |\mathscr{K}_j| = |G|$$

and thus we have equality and $\sum_i (c_{ij})^2 = 1$ for all j. Therefore, for each j there is a unique i with $c_{ij} \neq 0$. This defines a map$\{\mathscr{K}_j\} \to \{\mathscr{L}_i\}$. Since for each i there exists j with $c_{ij} \neq 0$, this map is onto and hence is one-to-one. Because all nonzero $c_{ij} = \pm 1$, we have $L_i = \pm K_j$ when $\mathscr{L}_i$ and $\mathscr{K}_j$ correspond. Finally, Equation (1) yields $|H||\mathscr{L}_i| = |G||\mathscr{K}_j|$ and thus $|\mathscr{L}_i| = |\mathscr{K}_j|$ and the proof is complete. ▮

(3.19) COROLLARY Let H be an integral group basis in $\mathbb{C}[G]$. Then there exists an integral group basis H^* in $\mathbb{C}[G]$ such that $H \cong H^*$ and the class sums of H^* equal the class sums of G.

Proof Let δ be the linear extension of the principal character of G to $\mathbb{C}[G]$. (This is usually called the *augmentation* map.) Note that $\delta(\sum a_g g) = \sum a_g$ and δ is an algebra homomorphism $\mathbb{C}[G] \to \mathbb{C}$.

If $h \in H$, then $\delta(h) \in \mathbb{Z}$ and since $\delta(h)\delta(h^{-1}) = \delta(1) = 1$, we have $\delta(h) = \pm 1$.

Let $h^* = \delta(h)h \in \mathbb{C}[G]$ for $h \in H$ and note that $\delta(h^*) = \delta(h)^2 = 1$. Put $H^* = \{h^* | h \in H\}$ and note that H^* is a group and the map $h \mapsto h^*$ is an isomorphism. Clearly, H^* is an integral group basis in $\mathbb{C}[G]$.

Let $\mathscr{L}$ be a class of H^* so that $\Sigma\mathscr{L} = \pm\Sigma\mathscr{K}$ for some class $\mathscr{K}$ of G by Theorem 3.17. Since $\delta(h^*) = 1$ for $h^* \in H^*$, we have $\delta(\Sigma\mathscr{L}) = |\mathscr{L}|$. Since $\delta(\Sigma\mathscr{K}) = |\mathscr{K}|$, the ambiguous sign above must be positive and the proof is complete. ▮

(3.20) COROLLARY Let H be an integral group basis in $\mathbb{C}[G]$. Then G and H have identical character tables.

Proof By Corollary 3.19, we may assume that there exists a one-to-one correspondence between the classes of H and the classes of G such that if $\mathscr{L}$ and $\mathscr{K}$ correspond, then $\Sigma\mathscr{L} = \Sigma\mathscr{K}$ and $|\mathscr{L}| = |\mathscr{K}|$. Let χ be the trace of an irreducible representation of $\mathbb{C}[H] = \mathbb{C}[G]$. Let $\mathscr{L}$ and $\mathscr{K}$ correspond and let $h \in \mathscr{L}$ and $g \in \mathscr{K}$. It suffices to show that $\chi(h) = \chi(g)$. However

$$\chi(h)|\mathscr{L}| = \chi(\Sigma\mathscr{L}) = \chi(\Sigma\mathscr{K}) = \chi(g)|\mathscr{K}|$$

and since $|\mathscr{L}| = |\mathscr{K}|$, the result follows. ▮

We have seen that if $\mathbb{Z}[G] \cong \mathbb{Z}[H]$, then $\mathbb{C}[G]$ has an integral group basis isomorphic to H. The converse is true but is less obvious.

(3.21) THEOREM Let H be an integral group basis in $\mathbb{C}[G]$. Then $\mathbb{Z}[G] = \mathbb{Z}[H]$ in $\mathbb{C}[G]$.

Proof Here, $\mathbb{Z}[H]$ denotes the ring of $\mathbb{Z}$-linear combinations of elements of H. (It is isomorphic to the abstract integral group ring.) Since $H \subseteq \mathbb{Z}[G]$, we have $\mathbb{Z}[H] \subseteq \mathbb{Z}[G]$. We show that $G \subseteq \mathbb{Z}[H]$.

Write $G = \{g_i | 1 \leq i \leq n\}$ and $H = \{h_i | 1 \leq i \leq n\}$, where $n = |G| = |H|$.

We have $h_i = \sum_i a_{ij} g_j$ for $a_{ij} \in \mathbb{Z}$. We shall show that the matrix $A = (a_{ij})$ has an inverse with entries in $\mathbb{Z}$ and this will complete the proof. Write $h_j^{-1} = \sum_i b_{ij} g_i^{-1}$ and $B = (b_{ij})$, an integer matrix.

Let ρ be the character of the regular representation of $\mathbb{C}[G] = \mathbb{C}[H]$ so that $\rho(h_i h_j^{-1}) = n\delta_{ij}$. Now express $h_i h_j^{-1}$ as a linear combination of elements of G and observe that the coefficient of 1 is $\sum_\nu a_{i\nu} b_{\nu j}$. It follows that

$$n\delta_{ij} = \rho(h_i h_j^{-1}) = n \sum_\nu a_{i\nu} b_{\nu j}$$

and thus $AB = I$, the identity matrix. The result now follows. ∎

Problems

(3.1) Let α be an algebraic integer and suppose that $f(\alpha) = 0$, where $f(x) \in \mathbb{Q}[x]$ is irreducible and monic. Show that $f(x) \in \mathbb{Z}[x]$.

(3.2) Let G be a group, $g \in G$, and let χ be a character of G. Suppose $|\chi(g)| = 1$. Show that $\chi(g)$ is a root of unity.

Hint Let $E \subseteq \mathbb{C}$ be the splitting field for $x^n - 1$ over $\mathbb{Q}$. For integral $\alpha \in E$, with $|\alpha| = 1$, let $f_\alpha \in \mathbb{Z}[x]$ be the polynomial of Problem 3.1. Show that only finitely many polynomials can arise this way. Do this by bounding the degree and the coefficients of f_α.

(3.3) Show that no simple group can have an irreducible character of degree 2.

Hint Problem 2.3 is relevant.

(3.4) Let G be a simple group and suppose $\chi \in \mathrm{Irr}(G)$ with $\chi(1) = p$, a prime. Show that a Sylow p-subgroup of G has order p.

Hint If the Sylow p-subgroup P is nonabelian, then $\mathbf{Z}(P) \subseteq \mathbf{Z}(\chi)$.

(3.5) Suppose $A \subseteq G$ is abelian and $|G:A|$ is a prime power. Show that $G' < G$.

(3.6) Let G be a p-group and suppose $\chi \in \mathrm{Irr}(G)$. Show that $\chi(1)^2 \mid |G:\mathbf{Z}(\chi)|$.

(3.7) Let $\chi \in \mathrm{Irr}(G)$ be faithful and suppose $\chi(1) = p^a$ for some prime p. Let $P \in \mathrm{Syl}_p(G)$ and suppose that $\mathbf{C}_G(P) \nsubseteq P$. Show that $G' < G$.

Hint Let $Q \subseteq \mathbf{C}_G(P)$ be a p'-subgroup, $Q \neq 1$. Show that $Q \cap \mathbf{Z}(\chi) \neq 1$ and consider $\det \chi$.

(3.8) Let χ be a (possibly reducible) character of G which is constant on $G - \{1\}$. Show that $\chi = a1_G + b\rho_G$, where $a, b \in \mathbb{Z}$ and ρ is the regular character of G. Show that if $G \neq \ker \chi$, then $\chi(1) \geq |G| - 1$.

Hint First show that $\chi = a1_G + b\rho_G$ for some $a, b \in \mathbb{C}$.

(3.9) Let $\mathscr{K}_1, \ldots, \mathscr{K}_k$ be the conjugacy classes of a group G and let $K_1, \ldots, K_k$ be the corresponding class sums. Choose representatives $g_i \in \mathscr{K}_i$ and let $a_{ij\nu}$ be the integers defined by

$$K_i K_j = \sum_\nu a_{ij\nu} K_\nu.$$

Show that

$$a_{ij\nu} = \frac{|\mathscr{K}_i||\mathscr{K}_j|}{|G|} \sum_{\chi \in \mathrm{Irr}(G)} \frac{\chi(g_i)\chi(g_j)\overline{\chi(g_\nu)}}{\chi(1)}.$$

Hint Use the ω_χ and the second orthogonality relation.

Notes This formula shows that the $a_{ij\nu}$ can be calculated from the character table. Therefore, the character table of G can be used to answer questions such as: Is an element $g \in \mathscr{K}_\nu$ the product of an element of $\mathscr{K}_i$ with one of $\mathscr{K}_j$?

The fact that the $a_{ij\nu}$ are nonnegative rational integers imposes another necessary condition on an array of complex numbers that the array be a character table for some group.

Since the K_i are a basis for $\mathbf{Z}(\mathbb{C}[G])$, it follows that the character table of G determines $\mathbf{Z}(\mathbb{C}[G])$ up to algebra isomorphism.

(3.10) We write $[x, y]$ for the *commutator* $x^{-1}y^{-1}xy$ of x and y in a group G.

(a) Let $g \in G$ and fix $x \in G$. Show that g is conjugate to $[x, y]$ for some $y \in G$ iff

$$\sum_{\chi \in \mathrm{Irr}(G)} \frac{|\chi(x)|^2\overline{\chi(g)}}{\chi(1)} \neq 0.$$

(b) Show that $g = [x, y]$ for some $x, y \in G$ iff

$$\sum_{\chi \in \mathrm{Irr}(G)} \frac{\chi(g)}{\chi(1)} \neq 0.$$

Note We already knew that the character table of a group determines the commutator subgroup. Problem 3.10 says more than this, since the commutator subgroup usually does not consist entirely of commutators.

(3.11) Let $g \in G$ be a commutator. Suppose $m \in \mathbb{Z}$, $(m, o(g)) = 1$. Show that g^m is a commutator.

Hint See the hint to Problem 2.12.

(3.12) Let $\chi \in \mathrm{Irr}(G)$ and $g, h \in G$. Show

$$\chi(g)\chi(h) = \frac{\chi(1)}{|G|} \sum_{z \in G} \chi(gh^z).$$

Hint Let $\mathfrak{X}$ afford χ and let K_i and K_j be the class sums in $\mathbb{C}[G]$ which contain g and h. Use the fact that $\mathfrak{X}(K_i)\mathfrak{X}(K_j) = \mathfrak{X}(K_iK_j)$.

(3.13) (*Brauer*) Let $K_1, \ldots, K_k$ be the class sums in $\mathbb{C}[G]$. Suppose there exists $c \in \mathbb{C}$ such that $\sum K_i = c \prod K_i$. Show $G = G'$.

Hint Let $1_G \neq \chi \in \text{Irr}(G)$. Show that $\omega_\chi(K_i) = 0$ for some i.

(3.14) (*Brauer*) In the notation of the previous problem, show that if $G = G'$, then there exists $c \in \mathbb{Q}$ such that $\sum K_i = c \prod K_i$.

Hint To show that two elements $a, b \in \mathbf{Z}(\mathbb{C}[G])$ are equal, it suffices to prove that $\omega_\chi(a) = \omega_\chi(b)$ for all $\chi \in \text{Irr}(G)$.

(3.15) (*Thompson*) Let E be a Galois extension of $\mathbb{Q}$ with Galois group $\mathscr{G}$. Let $\alpha \in E$ be an algebraic integer with the property that α^σ is real and positive for all $\sigma \in \mathscr{G}$. A theorem of Siegel (*Ann. of Math.* **46** (1945) p. 303, Theorem III) asserts that if $\alpha \neq 1$, then

$$\frac{1}{|\mathscr{G}|} \sum_{\sigma \in \mathscr{G}} \alpha^\sigma \geq \tfrac{3}{2}.$$

Use this to show that if $\chi \in \text{Irr}(G)$ then $\chi(x)$ is either zero or a root of unity for more than a third of the elements $x \in G$.

Hints Mimic the proof of Theorem 3.15. Use Problem 3.2.

(3.16) (*Burnside*) Let $|G|$ be odd and suppose $\chi \in \text{Irr}(G)$ is not principal. Show that $\chi \neq \bar{\chi}$.

Hint Using orthogonality, show that if $\chi \neq 1_G$ and $\chi = \bar{\chi}$, then $\chi(1) = 2\alpha$ for some algebraic integer α.

(3.17) (*Burnside*) Let $|G|$ be odd and suppose that G has exactly k conjugacy classes. Show that

$$|G| \equiv k \bmod 16.$$

Hint If n is an odd integer, then $n^2 \equiv 1 \bmod 8$.

4 Products of characters

Let χ and ψ be characters of G. The fact that $\chi + \psi$ is a character is a triviality. We may define a new class function $\chi\psi$ on G by setting $(\chi\psi)(g) = \chi(g)\psi(g)$. It is true but somewhat less trivial that $\chi\psi$ is a character. [If either χ or ψ is linear, this is Problem 2.6(a).]

Let V and W be $\mathbb{C}[G]$-modules. We shall construct a new $\mathbb{C}[G]$-module $V \otimes W$ called the *tensor product* of V and W. Choose bases $\{v_1, \ldots, v_n\}$ for V and $\{w_1, \ldots, w_m\}$ for W. Let $V \otimes W$ be the $\mathbb{C}$-space spanned by the mn symbols $v_i \otimes w_j$. [More precisely, $V \otimes W$ is the set of formal sums of the form $\sum a_{ij}(v_i \otimes w_j)$, with $a_{ij} \in \mathbb{C}$.] If $v \in V$ and $w \in W$, suppose $v = \sum a_i v_i$ and $w = \sum b_j w_j$. We define

$$v \otimes w = \sum a_i b_j (v_i \otimes w_j) \in V \otimes W.$$

Note that not every element of $V \otimes W$ has the form $v \otimes w$ for $v \in V$ and $w \in W$ (except in the special case that n or $m = 1$).

We define an action of G on $V \otimes W$ by setting

$$(v_i \otimes w_j)g = v_i g \otimes w_j g$$

and extending this by linearity to all of $V \otimes W$. The reader should check that if $v \in V$, $w \in W$, and $g \in G$, then $(v \otimes w)g = vg \otimes wg$. It follows that $(xg_1)g_2 = x(g_1 g_2)$ for $x \in V \otimes W$ and $g_i \in G$.

Next we give $V \otimes W$ the structure of a $\mathbb{C}[G]$-module by extending the action of G by linearity in $\mathbb{C}[G]$. In other words, for $x \in V \otimes W$ we define

$$x\left(\sum a_g g\right) = \sum a_g(xg).$$

It is routine to check that this really makes $V \otimes W$ into a $\mathbb{C}[G]$-module.

A few words of caution are in order here. If $\alpha \in \mathbb{C}[G]$, it is not necessarily true that $(v_i \otimes w_j)\alpha = v_i\alpha \otimes w_j\alpha$. It is for this reason that we first defined the action of G on $V \otimes W$ and then extended it to $\mathbb{C}[G]$. If A is an arbitrary algebra with modules V and W, it is not generally possible to define the structure of an A-module on $V \otimes W$.

We have not yet shown that the $\mathbb{C}[G]$-module $V \otimes W$ is determined (up to isomorphism) by V and W, independently of the choice of bases. This is, in fact, not hard to prove for group algebras over any field. We shall leave the general situation to the problems.

(4.1) THEOREM Let $\mathbb{C}[G]$-modules V and W afford characters χ and ψ, respectively. Choose bases in V and W and construct $V \otimes W$. Then $V \otimes W$ affords the character $\chi\psi$ and is independent of the choice of bases.

Proof Let $\{v_i \mid 1 \le i \le n\}$ and $\{w_r \mid 1 \le r \le m\}$ be bases for V and W, respectively, and let $g \in G$. Write

$$v_i g = \sum_{j=1}^{n} a_{ij} v_j \qquad \text{and} \qquad w_r g = \sum_{s=1}^{m} b_{rs} w_s,$$

with $a_{ij}, b_{rs} \in \mathbb{C}$.

Then $\chi(g) = \sum_{i=1}^{n} a_{ii}$ and $\psi(g) = \sum_{r=1}^{m} b_{rr}$. Let ϑ be the character afforded by $V \otimes W$. Since

$$(v_i \otimes w_r)g = \sum_{j,s} a_{ij} b_{rs} (v_j \otimes w_s),$$

we have

$$\vartheta(g) = \sum_{i,r} a_{ii} b_{rr} = \sum_i a_{ii} \sum_r b_{rr} = \chi(g)\psi(g).$$

We now know that the character afforded by $V \otimes W$ is independent of the choice of bases and the result follows by Corollary 2.9. ∎

(4.2) COROLLARY Products of characters are characters.

Corollary 4.2 provides another necessary condition for an array of complex numbers to be a character table. It asserts that the inner product of any row with the product of any two rows is a nonnegative integer. In this connection, the following formula is relevant:

$$[\chi_1\psi, \chi_2] = \frac{1}{|G|} \sum \chi_1(g)\psi(g)\overline{\chi_2(g)} = [\chi_1, \bar{\psi}\chi_2].$$

In general, products of irreducible characters are not irreducible. For instance, if $\chi \in \operatorname{Irr}(G)$, then $\bar{\chi} \in \operatorname{Irr}(G)$; nevertheless, 1_G is a constituent of $\chi\bar{\chi}$ since

$$[\chi\bar{\chi}, 1_G] = [\chi, \chi 1_G] = 1.$$

(4.3) THEOREM (*Burnside–Brauer*) Let χ be a faithful character of G and suppose $\chi(g)$ takes on exactly m different values for $g \in G$. Then every $\psi \in \operatorname{Irr}(G)$ is a constituent of one of the characters $(\chi)^j$ for $0 \le j < m$.

Note In the special case $\chi = \rho$, the regular character of G, we have $m = 2$. The theorem is already known to be true in this case.

Proof Let $\alpha_1, \alpha_2, \ldots, \alpha_m$ be the distinct values taken on by χ. Let

$$G_i = \{g \in G \mid \chi(g) = \alpha_i\}.$$

Assume $\alpha_1 = \chi(1)$ so that $G_1 = \ker \chi = \{1\}$. Fix $\psi \in \operatorname{Irr}(G)$ and let $\beta_i = \sum_{g \in G_i} \overline{\psi(g)}$. Now for $j \ge 0$, we have

$$[\chi^j, \psi] = \frac{1}{|G|} \sum_{i=1}^{m} (\alpha_i)^j \beta_i.$$

If ψ is not a constituent of χ^j for any j, $0 \le j < m$, we have

$$\sum_i (\alpha_i)^j \beta_i = 0, \qquad 0 \le j < m.$$

The determinant of this system of m equations in the m "unknowns" β_i is the so-called "Vandermonde determinant" and is equal to $\pm \prod_{i<j} (\alpha_i - \alpha_j) \ne 0$. It follows that all $\beta_i = 0$.

On the other hand, $\beta_1 = \psi(1) \ne 0$ and this contradiction proves the theorem. ∎

Let $g \in G$ and let $n > 0$ be an integer. We ask how many nth roots g has in G. Let

$$\vartheta_n(g) = |\{h \in G \mid h^n = g\}|.$$

Observe that if $(|G|, n) = 1$, we may choose an integer m such that $nm \equiv 1 \mod |G|$. Thus if $h^n = k^n$, then $h = h^{nm} = k^{nm} = k$ and we have $\vartheta_n(g) \le 1$ for all $g \in G$. Since the map $h \mapsto h^n$ is one-to-one on G, it must be onto and it follows that $\vartheta_n(g) = 1$ for all $g \in G$.

The situation becomes considerably more interesting if we drop the assumption that n is prime to $|G|$. Since ϑ_n is clearly a class function on G, we may write

$$\vartheta_n = \sum_{\chi \in \operatorname{Irr}(G)} v_n(\chi)\chi,$$

where $v_n(\chi)$ is a uniquely determined complex number.

(4.4) LEMMA $v_n(\chi) = (1/|G|) \sum_{g \in G} \chi(g^n)$.

Proof By the orthogonality relations we have

$$\nu_n(\chi) = [\vartheta_n, \chi] = \frac{1}{|G|}\sum_{g \in G} \vartheta_n(g)\overline{\chi(g)}.$$

Since $\vartheta_n(g)\overline{\chi(g)} = \sum_{h \in G;\, h^n = g} \overline{\chi(h^n)}$, we have

$$\nu_n(\chi) = \frac{1}{|G|}\sum_{h \in G} \overline{\chi(h^n)}.$$

Finally, replace h by h^{-1} to obtain the desired result. ∎

Let φ be any class function of G and let n be a positive integer. We define a new function $\varphi^{(n)}$ by $\varphi^{(n)}(g) = \varphi(g^n)$. Note that $\varphi^{(n)}$ is a class function. Lemma 4.4 asserts that $\nu_n(\chi) = [\chi^{(n)}, 1_G]$ for $\chi \in \text{Irr}(G)$. It is not, in general, true that $\chi^{(n)}$ is a character for $\chi \in \text{Irr}(G)$, however, it is always a difference of two characters (see problems) and thus $\nu_n(\chi) \in \mathbb{Z}$.

(4.5) THEOREM (*Frobenius–Schur*) Let $\chi \in \text{Irr}(G)$. Then

(a) $\chi^{(2)}$ is a difference of characters;
(b) $\nu_2(\chi) = 1, -1$, or 0;
(c) $\nu_2(\chi) \neq 0$ iff χ is real valued.

Proof Let V be a $\mathbb{C}[G]$-module which affords χ and let $v_1, v_2, \ldots, v_n$ be a basis for V. Let $W = V \otimes V$ and define the linear map $*$: $W \to W$ by $(v_i \otimes v_j)^* = v_j \otimes v_i$. Let

$$W_S = \{w \in W \mid w^* = w\} \qquad \text{and} \qquad W_A = \{w \in W \mid w^* = -w\}.$$

These subspaces are called the *symmetric* and *antisymmetric* parts of W, respectively.

If $w \in W$, we have $w + w^* \in W_S$ and $w - w^* \in W_A$. Since

$$w = \frac{w + w^*}{2} + \frac{w - w^*}{2},$$

it follows that $W = W_S \dot{+} W_A$.

We claim that if $w \in W$ and $g \in G$, then $(wg)^* = w^*g$. It suffices to check this as w runs over the basis $v_i \otimes v_j$, and thus we need to show that

$$(v_i g \otimes v_j g)^* = v_j g \otimes v_i g.$$

In fact, it is true that

$$(x \otimes y)^* = y \otimes x$$

for all $x, y \in V$. This is seen by expanding x and y in terms of the v_i.

It now follows that W_S and W_A are $\mathbb{C}[G]$-submodules of W and we have

$$\chi^2 = \chi_S + \chi_A,$$

where χ_S and χ_A are the characters afforded by W_S and W_A. We compute χ_A using the basis

$$w_{ij} = v_i \otimes v_j - v_j \otimes v_i, \qquad i < j,$$

for W_A.

Suppose $v_i g = \sum a_{ir} v_r$. We have

$$w_{ij} g = \sum_{r,s} (a_{ir} a_{js} - a_{jr} a_{is}) v_r \otimes v_s = \sum_{r<s} (a_{ir} a_{js} - a_{jr} a_{is}) w_{rs}.$$

Therefore

$$\chi_A(g) = \sum_{i<j} a_{ii} a_{jj} - a_{ji} a_{ij}.$$

This yields

$$2\chi_A(g) = \sum_{i \neq j} a_{ii} a_{jj} - \sum_{i \neq j} a_{ij} a_{ji} = \left(\sum_i a_{ii}\right)\left(\sum_j a_{jj}\right) - \sum_{i,j} a_{ij} a_{ji}.$$

Let $\mathfrak{X}$ be the representation corresponding to the basis $\{v_i\}$ of V, so that $\mathfrak{X}(g) = (a_{ij}) = M$. We have

$$2\chi_A(g) = (\operatorname{tr} M)^2 - \operatorname{tr}(M^2) = \chi(g)^2 - \chi(g^2)$$

since $M^2 = \mathfrak{X}(g^2)$.

This yields $\chi^{(2)}(g) = \chi(g)^2 - 2\chi_A(g)$ and thus $\chi^{(2)} = \chi^2 - 2\chi_A$, a difference of characters, as claimed.

Now $\nu_2(\chi) = [\chi^{(2)}, 1_G] = [\chi^2, 1_G] - 2[\chi_A, 1_G]$. If χ is not real valued, then $0 = [\chi, \bar{\chi}] = [\chi^2, 1_G]$ and thus $[\chi_A, 1_G] = 0$ since χ_A is a constituent of χ^2. It follows that $\nu_2(\chi) = 0$ in this case. If χ is real valued, then $1 = [\chi^2, 1_G]$ thus $[\chi_A, 1_G] = 0$ or 1 and $\nu_2(\chi) = 1$ or -1. The proof is now complete. ∎

(4.6) COROLLARY Let G have exactly t involutions. Then

$$1 + t = \sum_{\chi \in \operatorname{Irr}(G)} \nu_2(\chi)\chi(1),$$

where $\nu_2(\chi) = 0$ if $\chi \neq \bar{\chi}$ and $\nu_2(\chi) = \pm 1$ if $\chi = \bar{\chi}$.

Proof This is immediate since $\vartheta_2(1) = 1 + t$. ∎

If $N \lhd G$, we have identified $\operatorname{Irr}(G/N)$ with a subset of $\operatorname{Irr}(G)$. The following lemma shows that $\nu_n(\chi)$ is well defined under this identification.

(4.7) LEMMA Let $N \lhd G$ and let $\chi \in \operatorname{Irr}(G)$ with $N \subseteq \ker \chi$. Let $\nu_n(\chi)$ be as above and let $\hat{\nu}_n(\chi)$ be the corresponding number computed in G/N. Then $\nu_n(\chi) = \hat{\nu}_n(\chi)$.

Proof We have

$$\hat{\nu}_n(\chi) = \frac{1}{|G:N|}\sum_{Ng\in G/N}\chi((Ng)^n) = \frac{1}{|G:N|}\frac{1}{|N|}\sum_{g\in G}\chi(Ng^n)$$

$$= \frac{1}{|G|}\sum_{g\in G}\chi(g^n) = \nu_n(\chi). \quad ∎$$

(4.8) LEMMA Let $\chi \in \mathrm{Irr}(G)$ and let λ be a linear character of G with $\lambda^n = 1_G$. Then $\nu_n(\chi) = \nu_n(\lambda\chi)$.

Proof Recall that $\lambda\chi \in \mathrm{Irr}(G)$ by Problem 2.6(b). Now

$$\nu_n(\lambda\chi) = \frac{1}{|G|}\sum_{g\in G}(\lambda\chi)(g^n) = \frac{1}{|G|}\sum_{g}\lambda(g^n)\chi(g^n) = \frac{1}{|G|}\sum\chi(g^n) = \nu_n(\chi)$$

since $\lambda(g^n) = \lambda(g)^n = \lambda^n(g) = 1$. ∎

As an application of the Frobenius–Schur theorem, we prove the following.

(4.9) THEOREM (*Alperin–Feit–Thompson*) Let G be a 2-group containing exactly t involutions. If $t \equiv 1 \bmod 4$, then either G is cyclic or $|G:G'| = 4$.

Proof If G is abelian, then it clearly must be cyclic. We suppose G is not abelian and show that $|G:G'| = 4$.

If $\mathbf{Z}(G)$ is not cyclic, choose $K \subseteq \mathbf{Z}(G)$ elementary abelian of order 4. The set $\{x \in G \mid x^2 = 1\}$ is a union of cosets of K and hence $4 \mid (t+1)$, a contradiction. Therefore, $\mathbf{Z}(G)$ is cyclic and G contains the unique minimal subgroup Z of order 2.

Since $G' > 1$, we have $Z \subseteq G'$. Also G/Z is not cyclic since G is not abelian. If G/Z satisfies the hypotheses of the theorem, then $|G:G'| = |(G/Z):(G/Z)'| = 4$ by induction and we are done. We may therefore assume that the number of involutions in G/Z is not $\equiv 1 \bmod 4$.

We have

$$\sum_{\chi\in\mathrm{Irr}(G)}\nu_2(\chi)\chi(1) = t + 1 \equiv 2 \bmod 4$$

and

$$\sum_{\chi\in\mathrm{Irr}(G);\, Z\subseteq\ker\chi}\nu_2(\chi)\chi(1) \not\equiv 2 \bmod 4,$$

where the second statement follows via Lemma 4.7. We conclude that

$$(*) \qquad \sum_{\chi\in\mathrm{Irr}(G);\, Z\nsubseteq\ker\chi}\nu_2(\chi)\chi(1) \not\equiv 0 \bmod 4.$$

Now let C be the group of linear characters λ of G which satisfy $\lambda^2 = 1_G$. For $\chi \in \mathrm{Irr}(G)$, we have $\lambda\chi \in \mathrm{Irr}(G)$ and since $Z \subseteq \ker\lambda$, we conclude that

$Z \not\subseteq \ker \chi$ iff $Z \not\subseteq \ker(\lambda\chi)$. Therefore C permutes $\{\chi \in \mathrm{Irr}(G) \mid Z \not\subseteq \ker \chi\}$ and partitions this set into orbits $\mathcal{O}_i$. By Lemma 4.8, $\nu_2(\chi)$ is constant on each orbit as is $\chi(1)$.

Since $|\mathcal{O}_i|$ is a power of 2, we conclude from Equation $(*)$ that there exists $\chi \in \mathrm{Irr}(G)$ such that

(i) $Z \not\subseteq \ker \chi$,
(ii) $\nu_2(\chi) \neq 0$,
(iii) $\chi(1)|\mathcal{O}| \leq 2$,

where $\mathcal{O}$ is the orbit containing χ.

Since $Z \not\subseteq \ker \chi$, we have $\ker \chi = 1$. Since G is not abelian, $\chi(1) > 1$ and hence $\chi(1) = 2$ and $|\mathcal{O}| = 1$. Therefore, $\lambda\chi = \chi$ for all $\lambda \in C$.

Now let $F = \Phi(G)$, the Frattini subgroup. If $g \in G - F$, then there exists $\lambda \in C$ for which $\lambda(g) \neq 1$ and it follows that $\chi(g) = 0$. By Lemma 2.29, we now have

$$4 = \chi(1)^2 \geq [\chi_F, \chi_F] = |G : F|.$$

Since G is not cyclic, $|G : F| \geq 4$ and we have equality above. This forces $\chi_F = 2\mu$, where μ is a faithful linear character of F.

Since $\nu_2(\chi) \neq 0$, we know that χ is real valued. Therefore, μ is also, and so $|F| \leq 2$. Thus $|G| \leq 8$ and hence $G \cong D_8$ or Q_8. In either case, $|G : G'| = 4$ as desired. ∎

A theorem of O. Taussky [Theorem III 11.9(a) in Huppert] asserts that the only nonabelian 2-groups G for which $|G : G'| = 4$ are the dihedral, semidihedral, and generalized quaternion groups. In each of these groups, the number of involutions is in fact $\equiv 1 \bmod 4$.

As another application of Corollary 4.6, we prove some results of Brauer and Fowler that were originally obtained in a different way.

(4.10) LEMMA Let $a_1, a_2, \ldots, a_n$ be arbitrary real numbers. Then

$$\sum a_i^2 \geq \frac{1}{n}\left(\sum a_i\right)^2.$$

Proof The well-known Schwarz inequality asserts that

$$\left(\sum a_i b_i\right)^2 \leq \left(\sum a_i^2\right)\left(\sum b_i^2\right)$$

for real a_i and b_i. The lemma follows by setting all $b_i = 1$. ∎

(4.11) THEOREM Assume $|G| = g$ is even and that G contains exactly t involutions. Let $\alpha = (g - 1)/t$. Then

(a) There exists $x \in G$, $x \neq 1$, with $|G : \mathbf{C}(x)| \leq \alpha^2$.
(b) There exists real $\chi \in \operatorname{Irr}(G)$, $\chi \neq 1_G$, with $\chi(1) \leq \alpha$.

Proof By Corollary 4.6,

$$0 < (g - 1)/\alpha = t \leq \sum_{\chi \in \mathscr{S}} \chi(1),$$

where

$$\mathscr{S} = \{\chi \in \operatorname{Irr}(G) \mid \chi = \bar{\chi} \quad \text{and} \quad \chi \neq 1_G\}.$$

In particular, $s = |\mathscr{S}| \neq 0$. Now

$$t^2 \leq \left(\sum_{\mathscr{S}} \chi(1)\right)^2 \leq s \sum_{\mathscr{S}} \chi(1)^2 \leq s(g - 1)$$

by Lemma 4.10. It follows that

$$g - 1 \leq s\alpha^2$$

and thus

$$s\alpha^2 \geq \sum_{\mathscr{S}} \chi(1)^2.$$

Therefore, $\chi(1) \leq \alpha$ for some $\chi \in \mathscr{S}$, proving (b).

Since $s \leq k - 1$, where $k = |\operatorname{Irr}(G)|$ is the total number of conjugacy classes of G, we have

$$(k - 1)\alpha^2 \geq g - 1$$

and hence some nonidentity class of G has size $\leq \alpha^2$. This proves (a). ∎

Recall that an element $x \in G$ is said to be *real* if x is conjugate in G to x^{-1}. It is a fact (which will be proved later) that the number of classes of real elements of G is equal to the number of real irreducible characters.

Assuming this, it is immediate from the inequality $s\alpha^2 \geq g - 1$ in the above proof that statement (a) can be strengthened to guarantee that a *real* $x \in G$ exists with $x \neq 1$ and $|G : \mathbf{C}(x)| \leq \alpha^2$.

(4.12) COROLLARY Let n be a positive integer. There exist at most finitely many simple groups containing an involution with centralizer of order n.

Proof Let G be such a group with $|G| = g$. Then G contains at least g/n involutions and hence $\alpha < n$ in the notation of Theorem 4.11.

By (a) of that theorem, there exists $x \in G$ with $1 < |G : \mathbf{C}(x)| < n^2$. Therefore, G is isomorphic to a subgroup of the alternating group $A_{n^2 - 1}$. The result follows. ∎

(4.13) COROLLARY Let G have even order, $g > 2$. Then G contains a proper subgroup of order $> (g)^{1/3}$.

Proof We use induction on $|G|$. First assume $\mathbf{Z}(G) = Z > 1$. If $|G:Z|$ is even, then there exists $H/Z < G/Z$ with

$$|G:H| = |(G/Z):(H/Z)| < |G/Z|^{2/3} < g^{2/3}$$

and the result follows. If $|G:Z|$ is odd, then G has a central Sylow 2-subgroup and thus has a normal 2-complement by a transfer theorem of Burnside. (Or see Theorem 5.6.) It follows that G has a subgroup of index 2 and the result follows.

Now assume $\mathbf{Z}(G) = 1$ and let α be as in Theorem 4.11. Then

$$1 < |G:\mathbf{C}(x)| \leq \alpha^2$$

for some $x \in G$ and we are done if $\alpha^2 < g^{2/3}$. Assume, then, that $\alpha \geq g^{1/3}$ and let τ be an involution in G. Then

$$1 < |G:\mathbf{C}(\tau)| \leq (g-1)/\alpha < g/\alpha \leq g^{2/3}$$

and the result follows. ∎

We now wish to discuss the question of which real $\chi \in \text{Irr}(G)$ satisfy $\nu_2(\chi) = +1$. The answer is interesting but does not seem to be very important in the applications of the Frobenius–Schur theorem. The solution to this problem involves some matrix theory.

Let U and V be $\mathbb{C}[G]$-modules with bases $\{u_1, \ldots, u_n\}$ and $\{v_1, \ldots, v_m\}$, respectively. An element $w \in U \otimes V$ is uniquely of the form

$$w = \sum_{i,j} a_{ij}(u_i \otimes v_j)$$

and this defines the $n \times m$ matrix (a_{ij}). We write $M(w) = (a_{ij})$. For $g \in G$, we compute $M(wg)$.

Let $\mathfrak{X}$ and $\mathfrak{Y}$ be representations of G corresponding to U and V respectively, with respect to the given bases. Write

$$\mathfrak{X}(g) = (b_{rs}), \qquad \mathfrak{Y}(g) = (c_{pq}).$$

Then

$$(u_i \otimes v_j)g = u_i g \otimes v_j g = \sum_{s,q} b_{is} c_{jq}(u_s \otimes v_q).$$

Thus

$$wg = \sum_{i,j,s,q} b_{is} a_{ij} c_{jq}(u_s \otimes v_q)$$

and, therefore,

$$M(wg) = \mathfrak{X}(g)^{\mathsf{T}} M(w)\mathfrak{Y}(g).$$

(4.14) THEOREM Let $\chi \in \mathrm{Irr}(G)$ be real valued and let $\mathfrak{X}$ be a representation which affords χ. Then there exists a nonzero matrix M such that

$$\mathfrak{X}(g)^{\mathsf{T}} M\mathfrak{X}(g) = M$$

for all $g \in G$. Furthermore, for any such matrix, $M^{\mathsf{T}} = \nu_2(\chi)M$.

Proof Let V be a $\mathbb{C}[G]$-module corresponding to $\mathfrak{X}$ and write $W = V \otimes V$. Let $M : W \to M_n(\mathbb{C})$ be as previously stated so that

$$M(wg) = \mathfrak{X}(g)^{\mathsf{T}} M(w)\mathfrak{X}(g)$$

for $w \in W$ and $g \in G$.

Now W affords the character χ^2 and $[\chi^2, 1_G] = [\chi, \chi] = 1$ and hence there is a one-dimensional space of G-fixed points, $W_0 \subseteq W$. The matrices M satisfying $\mathfrak{X}(g)^{\mathsf{T}} M\mathfrak{X}(g) = M$ are precisely the $M(x)$ for $x \in W_0$.

In the proof of Theorem 4.5, we had a map $*: W \to W$ and we decomposed $W = W_S \dotplus W_A$, where $W_S = \{w \in W \mid w^* = w\}$ and $W_A = \{w \in W \mid w^* = -w\}$. Also, the proof of Theorem 4.5 showed that $\nu_2(\chi) = 1$ iff $[\chi_A, 1_G] = 0$, where χ_A is the character afforded by W_A.

We must have one of the following two situations:

(i) $W_0 \subseteq W_A$, $[\chi_A, 1_G] = 1$, $\nu_2(\chi) = -1$;
(ii) $W_0 \subseteq W_S$, $[\chi_A, 1_G] = 0$, $\nu_2(\chi) = 1$.

Thus if $x \in W_0$, we have $x^* = \nu_2(\chi)x$.

Now, for $w \in W$ we clearly have $M(w^*) = M(w)^{\mathsf{T}}$. Thus for $x \in W_0$ we have

$$M(x)^{\mathsf{T}} = \nu_2(\chi)M(x)$$

and the result follows. ∎

(4.15) COROLLARY Suppose $\chi \in \mathrm{Irr}(G)$ is afforded by a real representation. Then $\nu_2(\chi) = +1$.

Proof Let $\mathfrak{X}$ be an $\mathbb{R}$-representation which affords χ and let

$$M = \sum_{h \in G} \mathfrak{X}(h)^{\mathsf{T}} \mathfrak{X}(h).$$

It is clear that $M^{\mathsf{T}} = M$ and $\mathfrak{X}(g)^{\mathsf{T}} M\mathfrak{X}(g) = M$ for all $g \in G$. All that remains is to show that $M \neq 0$. For any $n \times n$ real matrix X, is it clear that the diagonal entries of $X^{\mathsf{T}} X$ are ≥ 0, and at least one of them is >0 unless $X = 0$. The result now follows. ∎

By Problem 2.5(b), we know that it is possible that a real-valued irreducible character χ is not afforded by any real representation. Such characters are exactly those for which $\nu_2(\chi) = -1$. The proof of this seems to require some nontrivial matrix theory.

If M is a square, complex matrix, we write $M^* = \overline{M}^{\top}$. and say M is *unitary* if $MM^* = I$ and M is *normal* if $MM^* = M^*M$. A standard theorem of linear algebra asserts that if M is normal, then there exists a unitary matrix, U, such that $U^{-1}MU$ is diagonal.

(4.16) LEMMA Let D be a diagonal matrix. Then $D = E^2$ for some diagonal matrix E such that every matrix which commutes with D also commutes with E.

Proof Suppose $\alpha_1, \ldots, \alpha_n$ are distinct complex numbers and $\beta_1, \ldots, \beta_n \in \mathbb{C}$ are arbitrary. Then there exists a polynomial f such that $f(\alpha_i) = \beta_i$. We may therefore choose f such that $f(\alpha)^2 = \alpha$ for every diagonal entry α of D. Then $E = f(D)$ has the desired properties. ▮

(4.17) THEOREM Let $\mathfrak{X}$ be a $\mathbb{C}$-representation of a group, G. Then $\mathfrak{X}$ is similar to a representation $\mathfrak{Y}$ such that $\mathfrak{Y}(g)$ is unitary for all $g \in G$.

Proof Let $M = \sum_{g \in G} \mathfrak{X}(g)^*\mathfrak{X}(g)$. Then $M^* = M$ and M is normal. We may choose a unitary matrix U such that $U^{-1}MU$ is diagonal. Since $U^{-1} = U^*$, we have

$$(U^{-1}\mathfrak{X}(g)U)^* = U^{-1}\mathfrak{X}(g)^*U \qquad \text{and} \qquad U^{-1}MU = \sum_{g \in G} \mathfrak{X}_1(g)^*\mathfrak{X}_1(g),$$

where $\mathfrak{X}_1 = U^{-1}\mathfrak{X}U$. It is therefore no loss to assume that M is diagonal.

The diagonal entries of $\mathfrak{X}(g)^*\mathfrak{X}(g)$ are all real and nonnegative and since $\mathfrak{X}(1) = I$, it follows that the diagonal entries of M are positive. We may therefore write $M = P^2$ where P is a nonsingular real diagonal matrix.

Since $\mathfrak{X}(g)^*M\mathfrak{X}(g) = M$ for all $g \in G$, we have

$$(P^{-1}\mathfrak{X}(g)^*P)(P\mathfrak{X}(g)P^{-1}) = I.$$

Since $(P\mathfrak{X}(g)P^{-1})^* = P^{-1}\mathfrak{X}(g)^*P$, it follows that $\mathfrak{Y} = P\mathfrak{X}P^{-1}$ is the desired representation. ▮

(4.18) LEMMA Let $P^*P = I$ and $P^{\top} = P$ for a square matrix P. Then $P = \overline{Q}Q^{-1}$ for some matrix Q.

Proof Since P is unitary, it is normal and we may write $U^{-1}PU = D = E^2$, where U is unitary, D and E are diagonal, and every matrix which commutes with D also commutes with E. Now $D^{\top} = D$, $P^{\top} = P$, and $U^{\top} = \overline{U}^{-1}$. This yields

$$D = D^{\top} = (U^{-1}PU)^{\top} = \overline{U}^{-1}P\overline{U} = \overline{U}^{-1}UDU^{-1}\overline{U}.$$

Therefore, $U^{-1}\bar{U}$ commutes with D and hence also with E. It follows that $E = \bar{U}^{-1}UEU^{-1}\bar{U}$ and

$$UEU^{-1} = \bar{U}E\bar{U}^{-1}.$$

From $P^*P = I$, we obtain $I = D^*D = \bar{D}D$ and it follows that the diagonal entries of D and hence of E have absolute value 1 and $\bar{E} = E^{-1}$. Now let $Q = U\bar{E}U^{-1}$. Then

$$\bar{Q}Q^{-1} = (\bar{U}E\bar{U}^{-1})(UEU^{-1}) = (UEU^{-1})^2 = UDU^{-1} = P$$

and the proof is complete. ∎

(4.19) THEOREM Let $\chi \in \operatorname{Irr}(G)$ and suppose $\nu_2(\chi) = 1$. Then χ is afforded by some real representation of G.

Proof By Theorem 4.17, χ is afforded by a representation $\mathfrak{X}$, such that $\mathfrak{X}(g)$ is unitary for all $g \in G$. Since $\chi = \bar{\chi}$, $\mathfrak{X}$ and $\bar{\mathfrak{X}}$ are similar and we may choose P such that $P^{-1}\bar{\mathfrak{X}}P = \mathfrak{X}$. Now

$$\mathfrak{X}(g)^{\mathsf{T}} P\mathfrak{X}(g) = \overline{\mathfrak{X}(g)}^{-1}\overline{\mathfrak{X}(g)}P = P.$$

Since $\nu_2(\chi) = 1$, it follows that $P^{\mathsf{T}} = P$ by Theorem 4.14.

From the equation $\overline{\mathfrak{X}(g)}P = P\mathfrak{X}(g)$, we obtain $P^*\overline{\mathfrak{X}(g)}^* = \mathfrak{X}(g)^*P^*$ and since $\mathfrak{X}(g)^* = \mathfrak{X}(g)^{-1}$, this yields $P^*\overline{\mathfrak{X}(g)}^{-1} = \mathfrak{X}(g)^{-1}P^*$. Thus $\mathfrak{X}P^* = P^*\bar{\mathfrak{X}}$ and

$$\mathfrak{X}P^*P = P^*\bar{\mathfrak{X}}P = P^*P\mathfrak{X}.$$

By Schur's lemma, it follows that $P^*P = \alpha I$ for some α. Clearly, α is real and positive and we may therefore replace P by $P/\alpha^{1/2}$ and assume $P^*P = I$.

Now Lemma 4.18 applies to yield $P = \bar{Q}Q^{-1}$ for some Q and we have

$$\bar{Q}Q^{-1}\mathfrak{X} = \bar{\mathfrak{X}}\bar{Q}Q^{-1}$$

or $Q^{-1}\mathfrak{X}Q = \overline{Q^{-1}\mathfrak{X}Q}$. In other words, $\mathfrak{Y} = Q^{-1}\mathfrak{X}Q$ is a real representation which affords χ and the proof is complete. ∎

We may summarize the situation as follows for $\chi \in \operatorname{Irr}(G)$:

(a) $\nu_2(\chi) = 0$ iff χ is not real;
(b) $\nu_2(\chi) = 1$ iff χ is afforded by a real representation;
(c) $\nu_2(\chi) = -1$ iff χ is real but is not afforded by any real representation.

One further remark should be made. If $\chi \in \operatorname{Irr}(G)$ and $\nu_2(\chi) = -1$, then $\chi(1)$ must be even. This may be seen as follows. The nonzero matrix M in Theorem 4.14 is nonsingular by Schur's lemma. If $\nu_2(\chi) = -1$ then $M^{\mathsf{T}} = -M$ and

$$\det(M) = \det(M^{\mathsf{T}}) = \det(-M) = (-1)^{\chi(1)} \det M.$$

It follows that $(-1)^{\chi(1)} = 1$ and $\chi(1)$ is even.

We close this chapter with a discussion of the characters of direct products.

(4.20) DEFINITION Let $G = H \times K$ and let φ and ϑ be class functions on H and K, respectively. Define $\chi = \varphi \times \vartheta$ by $\chi(hk) = \varphi(h)\vartheta(k)$ for $h \in H$ and $k \in K$.

Observe that $\varphi \times \vartheta$ is a class function of G. If φ is a character of H, then under the isomorphism $H \cong G/K$, there is a corresponding character $\hat{\varphi}$ of G with $K \subseteq \ker \hat{\varphi}$ and $\hat{\varphi}(hk) = \varphi(h)$. Similarly, if ϑ is a character of K, there is a corresponding character $\hat{\vartheta}$ of G, with $\hat{\vartheta}(hk) = \vartheta(k)$. It follows that $\varphi \times \vartheta = \hat{\varphi}\hat{\vartheta}$ is a character of G.

(4.21) THEOREM Let $G = H \times K$. Then the characters $\varphi \times \vartheta$ for $\varphi \in \mathrm{Irr}(H)$ and $\vartheta \in \mathrm{Irr}(K)$ are exactly the irreducible characters of G.

Proof Let $\varphi, \varphi_1 \in \mathrm{Irr}(H)$ and $\vartheta, \vartheta_1 \in \mathrm{Irr}(K)$. Let $\chi = \varphi \times \vartheta$ and $\chi_1 = \varphi_1 \times \vartheta_1$. Then

$$[\chi, \chi_1] = \frac{1}{|G|} \sum_{g \in G} \chi(g)\overline{\chi_1(g)} = \frac{1}{|H||K|} \sum_{h \in H;\, k \in K} \varphi(h)\vartheta(k)\overline{\varphi_1(h)}\,\overline{\vartheta_1(k)}$$

$$= \left(\frac{1}{|H|} \sum_{h \in H} \varphi(h)\overline{\varphi_1(h)}\right)\left(\frac{1}{|K|} \sum_{k \in K} \vartheta(k)\overline{\vartheta_1(k)}\right) = [\varphi, \varphi_1][\vartheta, \vartheta_1].$$

It follows that the $\varphi \times \vartheta$ are all distinct and irreducible.

Now

$$\sum_{\varphi \in \mathrm{Irr}(H);\, \vartheta \in \mathrm{Irr}(K)} (\varphi \times \vartheta)(1)^2 = \sum_{\varphi, \vartheta} \varphi(1)^2\vartheta(1)^2 = \left(\sum_{\varphi} \varphi(1)^2\right)\left(\sum_{\vartheta} \vartheta(1)^2\right)$$
$$= |H||K| = |G|,$$

and thus the $\varphi \times \vartheta$ are all of $\mathrm{Irr}(G)$. ∎

Problems

(4.1) Let χ and ψ be characters of G. Express $\det(\chi\psi)$ in terms of $\det \chi$ and $\det \psi$.

Hint It suffices to assume that G is cyclic.

(4.2) (*Garrison*) Let $\mathscr{K}$ be a class of G which is not contained in any proper normal subgroup. Let K be the corresponding class sum in $\mathbb{C}[G]$ and let m be the number of distinct values of $\omega_\chi(K)$ for $\chi \in \mathrm{Irr}(G)$. Show that every element of G can be written as a product of fewer than m elements of $\mathscr{K}$.

Hint Mimic the proof of Theorem 4.3 using the second orthogonality relation.

(4.3) Let $G = H \times K$. Let $\varphi \in \mathrm{Irr}(H)$ and $\vartheta \in \mathrm{Irr}(K)$ be faithful. Show that $\varphi \times \vartheta$ is faithful iff $(|\mathbf{Z}(H)|, |\mathbf{Z}(K)|) = 1$.

(4.4) Suppose $G = HK$ with $H \subseteq \mathbf{C}(K)$.

(a) Let $\chi \in \mathrm{Irr}(G)$. Show that $\chi_H = \vartheta(1)\varphi$ and $\chi_K = \varphi(1)\vartheta$ for some $\varphi \in \mathrm{Irr}(H)$ and $\vartheta \in \mathrm{Irr}(K)$.

(b) Let $\varphi \in \mathrm{Irr}(H)$ and $\vartheta \in \mathrm{Irr}(K)$ and suppose that $\varphi_{H \cap K}$ and $\vartheta_{H \cap K}$ have a common constituent. Show that there exists a unique $\chi \in \mathrm{Irr}(G)$ with $\chi_H = \vartheta(1)\varphi$ and $\chi_K = \varphi(1)\vartheta$.

Hint G is a homomorphic image of $H \times K$.

(4.5) Let $G = H \times A$ where A is abelian and let $n > 0$ with $(|H|, n) = 1$. Show that $\chi^{(n)} \in \mathrm{Irr}(G)$ for every $\chi \in \mathrm{Irr}(G)$.

(4.6) Let $n > 0$ and assume that $\chi^{(n)} \in \mathrm{Irr}(G)$ for every $\chi \in \mathrm{Irr}(G)$. Show that $G = H \times A$, where A is abelian and $(|H|, n) = 1$.

Hints (a) Let $d = (|G|, n)$. Show that it is no loss to assume that $(|G|/d, n) = 1$.

(b) Let $A = \bigcap_{\chi \in \mathrm{Irr}(G)} \ker \chi^{(n)}$. Show that $A = \{g \in G \mid g^n = 1\}$ and $|A| = d$.

(c) Let $H = \bigcap\{\ker \chi \mid \chi \in \mathrm{Irr}(G),\ \chi^{(n)} = 1_G\}$. Show $|G : H| = d$.

(4.7) Let V be a $\mathbb{C}[G]$-module with basis $\{v_1, \ldots, v_n\}$ and let p be a prime. Let $W = V \otimes \cdots \otimes V$ (p times) and define α on W so that $(x_1 \otimes \cdots \otimes x_p)^\alpha = x_2 \otimes x_3 \otimes \cdots \otimes x_p \otimes x_1$ for all $x_i \in V$. Let ε be a primitive pth root of 1 in $\mathbb{C}$. Let $W_1 = \{w \in W \mid w^\alpha = \varepsilon w\}$. Show that W_1 is a $\mathbb{C}[G]$-submodule of W which affords the character $\chi_1 = (\chi^p - \chi^{(p)})/p$, where χ is the character afforded by V. Conclude that $\chi^{(m)}$ is a difference of characters for all $m \geq 1$.

Hints Let $\mathscr{W} = \{v_{i_1} \otimes \cdots \otimes v_{i_p} \mid \text{not all } i_j \text{ are equal}\}$. For $w \in \mathscr{W}$, let

$$\bar{w} = w + \varepsilon^{-1}w^\alpha + \varepsilon^{-2}w^{\alpha^2} + \cdots + \varepsilon^{-(p-1)}w^{\alpha^{p-1}}.$$

Now $\langle\alpha\rangle$ permutes $\mathscr{W}$ into orbits of size p. Let $\mathscr{W}_0$ be a set of representatives of these orbits. Then $\{\bar{w} \mid w \in \mathscr{W}_0\}$ is a basis for W_1.

(4.8) Let G be a p-group with Frattini factor group of order $\geq p^{2a-1}$. Show that the number of elements of order p in G is $\equiv -1 \bmod p^a$.

Hint Mimic the proof of Theorem 4.9. Use the fact that $v_p(\chi)$ is an integer (Problem 4.7).

(4.9) If $\chi \in \text{Irr}(G)$, then $|\nu_2(\chi)| \leq 1$. Show that no absolute bound on $\nu_n(\chi)$ exists for any $n > 2$.

Hint There exist groups of prime exponent $p \neq 2$ and of exponent 4 with irreducible characters of arbitrarily large degree.

(4.10) Let V, W, V_1, and W_1 be $F[G]$-modules for an arbitrary field F. Fix a particular F-basis in each of these modules and construct $V \otimes W$ and $V_1 \otimes W_1$ as was done for $\mathbb{C}[G]$. Now suppose $\alpha: V \to V_1$ and $\beta: W \to W_1$ are module homomorphisms. Show that there exists a module homomorphism $\gamma: V \otimes W \to V_1 \otimes W_1$ such that $\gamma(v \otimes w) = \alpha(v) \otimes \beta(w)$ for $v \in V$ and $w \in W$. Conclude that $V \otimes W$ is determined up to module isomorphism independently of the choice of bases.

(4.11) Let G be simple and let $S \in \text{Syl}_2(G)$ be elementary abelian of order q. Suppose $S = \mathbf{C}_G(x)$ for all x, $1 \neq x \in S$. Show that $\nu_2(\chi) = 1$ for all $\chi \in \text{Irr}(G)$ and that $\chi(x) = \chi(1) - q$ for $\chi \neq 1_G$ and $x \in S$.

Hint First show that G contains exactly $|G|/q$ involutions and that the remaining elements of G have odd order. This is done without characters using the fact that if x, y are nonconjugate involutions, then $|\mathbf{Z}(\langle x, y \rangle)|$ is even.

Note This problem is continued as Problem 5.20 where it is proved that $|G| = q(q - 1)(q + 1)$.

(4.12) Let $\chi, \psi, \vartheta \in \text{Irr}(G)$. Show that $[\chi\psi, \vartheta] \leq \vartheta(1)$.

5 Induced characters

Let $H \subseteq G$ be a subgroup. Given a character χ of G, we obtained the character χ_H of H by restriction. In this chapter we study an approximately dual process, where characters of G are induced from characters of H.

(5.1) DEFINITION Let $H \subseteq G$ and let φ be a class function of H. Then φ^G, the *induced class function* on G, is given by

$$\varphi^G(g) = \frac{1}{|H|} \sum_{x \in G} \varphi^\circ(xgx^{-1}),$$

where φ° is defined by $\varphi^\circ(h) = \varphi(h)$ if $h \in H$ and $\varphi^\circ(y) = 0$ if $y \notin H$.

Observe that φ^G is really a class function on G and that $\varphi^G(1) = |G : H| \varphi(1)$. Another useful formula for $\varphi^G(g)$ is obtained by choosing a transversal T for the right cosets of H in G (that is, a set of representatives for these cosets). It is then easy to see that

$$\varphi^G(g) = \sum_{t \in T} \varphi^\circ(tgt^{-1}).$$

(5.2) LEMMA (*Frobenius Reciprocity*) Let $H \subseteq G$ and suppose that φ is a class function on H and that ϑ is a class function on G. Then

$$[\varphi, \vartheta_H] = [\varphi^G, \vartheta].$$

Proof We have

$$[\varphi^G, \vartheta] = \frac{1}{|G|} \sum_{g \in G} \varphi^G(g)\overline{\vartheta(g)} = \frac{1}{|G|}\frac{1}{|H|} \sum_{g \in G} \sum_{x \in G} \varphi^\circ(xgx^{-1})\overline{\vartheta(g)}.$$

Setting $y = xgx^{-1}$ and observing that $\vartheta(g) = \vartheta(y)$, we obtain

$$[\varphi^G, \vartheta] = \frac{1}{|G|}\frac{1}{|H|}\sum_{y\in G}\sum_{x\in G}\varphi^\circ(y)\overline{\vartheta(y)} = \frac{1}{|H|}\sum_{y\in H}\varphi(y)\overline{\vartheta(y)} = [\varphi, \vartheta_H]. \quad \blacksquare$$

(5.3) COROLLARY Let $H \subseteq G$ and suppose φ is a character of H. Then φ^G is a character of G.

Proof Let $\chi \in \mathrm{Irr}(G)$. Then χ_H is a character of H and thus $[\varphi^G, \chi] = [\varphi, \chi_H]$ is a nonnegative integer. This proves the result since $\varphi^G \neq 0$. ∎

(5.4) COROLLARY Let $H \subseteq G$ and suppose $\varphi \in \mathrm{Irr}(H)$. Then there exists $\chi \in \mathrm{Irr}(G)$ such that φ is a constituent of χ_H.

Proof Take χ to be an irreducible constituent of φ^G. Then

$$0 \neq [\varphi^G, \chi] = [\varphi, \chi_H]$$

and the result follows. ∎

(5.5) COROLLARY Let G be abelian and suppose $H \subseteq G$. Then every $\lambda \in \mathrm{Irr}(H)$ is the restriction of some $\mu \in \mathrm{Irr}(G)$.

Induction of characters is closely related to the transfer map from a group into an abelian factor group of a subgroup. Many of the results provable by transfer can also be proved using induced characters. We give one example of this now.

(5.6) THEOREM Let G have an abelian Sylow p-subgroup. Then

$$|G' \cap \mathbf{Z}(G)|$$

is not divisible by p.

Proof Assume the contrary and choose $U \subseteq G' \cap \mathbf{Z}(G)$ of order p, with $U \subseteq P \in \mathrm{Syl}_p(G)$. Let $\lambda \in \mathrm{Irr}(U)$, $\lambda \neq 1_U$, and choose $\mu \in \mathrm{Irr}(P)$ with $\mu_U = \lambda$. Now write

$$\mu^G = \sum_{\chi\in\mathrm{Irr}(G)} a_\chi \chi$$

and observe that $\mu^G(1) = \mu(1)|G:P| = |G:P|$ is prime to p. Therefore, there exists $\chi \in \mathrm{Irr}(G)$ such that $a_\chi \neq 0$ and $p \nmid \chi(1)$. We have

$$0 \neq [\mu^G, \chi] = [\mu, \chi_P]$$

and thus λ is a constituent of χ_U. Since $U \subseteq \mathbf{Z}(G)$, we have $\chi_U = \chi(1)\lambda$ and $(\det \chi)_U = \lambda^{\chi(1)}$. (See Problem 2.3.) Since $U \subseteq G'$, we have $(\det \chi)_U = 1_U$ and thus $\lambda^{\chi(1)} = 1_U$. Since $p \nmid \chi(1)$, $\lambda \neq 1_U$, and $|U| = p$, this is a contradiction and proves the theorem. ∎

Explicit computation of induced characters is extremely useful for the construction of character tables, despite the fact that φ^G is usually reducible even if φ is irreducible. [Note that if φ is reducible, then φ^G is necessarily reducible since $(\varphi_1 + \varphi_2)^G = \varphi_1{}^G + \varphi_2{}^G$.]

Given $H \subseteq G$, φ a character of H and $g \in G$, an efficient way to compute $\varphi^G(g)$ explicitly is to choose representatives, $x_1, \ldots, x_m$ for the classes of H contained in $\mathrm{Cl}(g)$ in G and to use the formula

$$\varphi^G(g) = |\mathbf{C}_G(g)| \sum_{i=1}^{m} \frac{\varphi(x_i)}{|\mathbf{C}_H(x_i)|},$$

where it is understood that $\varphi^G(g) = 0$ if $H \cap \mathrm{Cl}(g) = \varnothing$. This formula is immediate from the definition of φ^G since as x runs over G, $xgx^{-1} = x_i$ for exactly $|\mathbf{C}_G(g)|$ values of x.

We shall now construct the character table of $G = A_5$. There are five conjugacy classes; one each of elements of order 1, 2, and 3 and two of elements of order 5. It is convenient to label these classes 1, 2, 3, 5_1, and 5_2 in the table. Let $\chi_1 = 1_G$. We have:

Class $\mathscr{K}$:	1	2	3	5_1	5_2
$\lvert\mathbf{C}(x)\rvert$ for $x \in \mathscr{K}$:	60	4	3	5	5
$\lvert\mathscr{K}\rvert$:	1	15	20	12	12
χ_1:	1	1	1	1	1

Now let $H = A_4 \subseteq G$ and compute $(1_H)^G$. We have

$$(1_H)^G\colon\quad 5 \quad 1 \quad 2 \quad 0 \quad 0$$

using the above formula. [For instance, if $o(g) = 3$, there are two classes in H of elements conjugate in G to g. If x_1 and x_2 are representatives of these classes, we have $1_H(x_i) = 1$ and $|\mathbf{C}_G(g)| = 3 = |\mathbf{C}_H(x_i)|$ and thus $(1_H)^G(g) = 2$ as listed.]

Now $[(1_H)^G, 1_G] = [1_H, 1_H] = 1$ by Frobenius reciprocity and thus $(1_H)^G - 1_G$ is a character. We call this character χ_4. Thus

$$\chi_4 = (1_H)^G - 1_G\colon\quad 4 \quad 0 \quad 1 \quad -1 \quad -1.$$

Direct computation now yields $[\chi_4, \chi_4] = 1$ and we have found an irreducible character.

Next, let λ be a nonprincipal linear character of $H = A_4$ so that $\lambda(x) = 1$ if $o(x) = 1$ or 2 and $\lambda(x) = \omega = e^{2\pi i/3}$ for one class of elements of order 3 in H and $\lambda(x) = \omega^2$ for the other class. We get

$$\chi_5 = \lambda^G\colon\quad 5 \quad 1 \quad -1 \quad 0 \quad 0$$

since $\omega + \omega^2 = -1$. We compute $[\chi_5, \chi_5] = 1$ and so $\chi_5 \in \mathrm{Irr}(G)$.

Now let K be a cyclic subgroup of G of order 5 and let $\mu \in \mathrm{Irr}(K)$, $\mu \neq 1_K$. The class 5_1 of G contains a pair of inverse elements of K, say x and x^{-1}. Choose μ so that $\mu(x) = \varepsilon = e^{2\pi i/5}$. Since x^2 and x^{-2} lie in the class 5_2 we have

$$\mu^G\colon \quad 12 \quad 0 \quad 0 \quad \varepsilon + \varepsilon^4 \quad \varepsilon^2 + \varepsilon^3.$$

Computation yields $[\mu^G, \chi_5] = 1$ and thus

$$\mu^G - \chi_5\colon \quad 7 \quad -1 \quad 1 \quad \varepsilon + \varepsilon^4 \quad \varepsilon^2 + \varepsilon^3$$

is a character.

Further computation yields that $[\mu^G - \chi_5, \chi_4] = 1$ and hence

$$\chi_3 = \mu^G - \chi_5 - \chi_4\colon \quad 3 \quad -1 \quad 0 \quad \varepsilon + \varepsilon^4 + 1 \quad \varepsilon^2 + \varepsilon^3 + 1$$

is a character and $[\chi_3, \chi_3] = 1$ so that χ_3 is irreducible.

Finally, replacing μ above, by $v \in \mathrm{Irr}(K)$ with $v(x) = \varepsilon^2$ yields

$$\psi_3\colon \quad 3 \quad -1 \quad 0 \quad \varepsilon^2 + \varepsilon^3 + 1 \quad \varepsilon + \varepsilon^4 + 1$$

is irreducible and we have found all five irreducible characters of G.

We now consider the module theoretic interpretation of induced characters.

(5.7) DEFINITION Let V be a $F[G]$-module. Suppose $V = W_i \dotplus \cdots \dotplus W_k$ where the W_i are subspaces of V which are transitively permuted by the action of G. Then $V = W_1 \dotplus \cdots \dotplus W_k$ is an *imprimitivity decomposition* of V. If V is irreducible and has no proper imprimitivity decomposition (that is, with $k > 1$), then V is a *primitive* $F[G]$-module.

We emphasize that if $V = W_1 \dotplus \cdots \dotplus W_k$ is an imprimitivity decomposition of V with $k > 1$, then the W_i are definitely not $F[G]$-submodules of V. In this situation, let $H = \{g \in G \mid W_1 h = W_1\}$, then W_1 is a $F[H]$-module.

(5.8) THEOREM Let $V = W_1 \dotplus \cdots \dotplus W_k$ be an imprimitivity decomposition of the $F[G]$-module V. Assume $F \subseteq \mathbb{C}$. Let H be the stabilizer of W_1. Suppose V affords the character χ of G and W_1 affords the character ϑ of H. Then $\chi = \vartheta^G$.

Proof Let T be a right transversal for H in G and write $W = W_1$. Then Wt runs over the W_i as t runs over T and

$$V = \sum_{t \in T} \cdot Wt.$$

Choose a basis, $w_1, \ldots, w_m$ for W so that $\{w_i t \mid 1 \leq i \leq m, t \in T\}$ is a basis for V. Let $g \in G$. We compute $\chi(g)$ using this basis.

For $t \in T$, the contribution of the basis vectors $w_i t$ to $\chi(g)$ will be zero unless $Wtg = Wt$, that is, $tgt^{-1} \in H$. Assuming $tgt^{-1} = h \in H$, write $w_i h = \sum a_{ij} w_j$ so that $\sum_{i=1}^{m} a_{ii} = \vartheta(tgt^{-1})$. Now, $(w_i t)g = w_i ht = \sum a_{ij} w_j t$ and hence the contribution of $w_i t$ to $\chi(g)$ is a_{ii}. Therefore, the total contribution of the $w_i t$, $1 \le i \le m$, is $\vartheta(tgt^{-1})$. It follows that $\chi(g) = \sum_{t \in T} \vartheta^{\circ}(tgt^{-1}) = \vartheta^G(g)$ as desired. ∎

(5.9) THEOREM Let $H \subseteq G$ and let W be an $F[H]$-module. Then there exists an $F[G]$-module V having an imprimitivity decomposition $V = \sum \cdot W_i$, where H is the stabilizer of W_1 and $W_1 \cong W$ as $F[H]$-modules.

Proof Let V be the external direct sum of $|G:H|$ copies of W as an F-vectorspace. Let T be a right transversal for H in G and assume $1 \in T$. Let $W \otimes t$, for $t \in T$, denote subspaces of V, each isomorphic to W, such that $V = \sum \cdot_{t \in T}(W \otimes t)$. Fix isomorphisms α_t: $W \to W \otimes t$ and let $w \otimes t$ denote $\alpha_t(w)$ so that $W \otimes t = \{w \otimes t \mid w \in W\}$.

Now for $w \in W$ and $g \in G$, define $w \otimes g \in V$ as follows. Write $g = ht$, with $h \in H$ and $t \in T$, and set $w \otimes g = wh \otimes t$. Note that for $w \in W$, $x \in H$, and $g \in G$, we have $wx \otimes g = w \otimes xg$. (The usual notation for what we have constructed is $V = W \otimes_{F[H]} F[G]$.)

We now let G act on V by defining $(w \otimes t)g = w \otimes tg$ for $g \in G$. Observe that $(w \otimes g_1)g_2 = w \otimes g_1 g_2$ for all $g_1, g_2 \in G$. Extending this definition by linearity gives V the structure of an $F[G]$-module and

$$V = \sum_{t \in T} \cdot W \otimes t$$

is an imprimitivity decomposition.

Clearly the stabilizer of $W \otimes 1$ is H and the map α_1: $W \to W \otimes 1$ is an isomorphism of $F[H]$-modules. The proof is complete. ∎

If $s \in Ht$, then $W \otimes s = W \otimes t$ in the above construction and it is reasonably clear that the module V is independent of the choice of coset representatives. We write $V = W^G$. Note, however, that we have not yet proved that W^G is the unique (up to $F[G]$-isomorphism) $F[G]$-module which satisfies the conclusion of Theorem 5.9.

In the special case that $F = \mathbb{C}$, let W afford the character ϑ. Then by Theorem 5.8, it follows that any $\mathbb{C}[G]$-module which satisfies the conclusion of 5.9 affords $\chi = \vartheta^G$. This proves uniqueness in this case. The general case is left to the problems.

Note that Theorems 5.8 and 5.9 provide an alternate proof of the fact that induced characters are characters. Another consequence of these results is that if $\chi \in \mathrm{Irr}(G)$, then χ is afforded by a primitive module iff $\chi \neq \vartheta^G$ for any character ϑ of a proper subgroup of G. Such χ are called *primitive* characters.

(5.10) DEFINITION Let χ be a character of G. Then χ is *monomial* if $\chi = \lambda^G$, where λ is a linear character of some (not necessarily proper) subgroup of G. The group G is an *M-group* if every $\chi \in \text{Irr}(G)$ is monomial.

Note that if χ is monomial, then by Theorems 5.8 and 5.9, χ is afforded by a module which has an imprimitivity decomposition into subspaces of dimension 1. It follows that χ is afforded by a representation $\mathfrak{X}$ with the property that each row and column of $\mathfrak{X}(g)$ has exactly one nonzero entry for each $g \in G$.

(5.11) LEMMA Let ϑ be a character of $H \subseteq G$. Then

$$\ker(\vartheta^G) = \bigcap_{x \in G} (\ker \vartheta)^x.$$

Proof Let $\chi = \vartheta^G$. Then $g \in \ker \chi$ iff

$$\sum_{x \in G} \vartheta^\circ(xgx^{-1}) = \sum_{x \in G} \vartheta(1).$$

Since $|\vartheta^\circ(xgx^{-1})| \leq \vartheta(1)$, we conclude that $g \in \ker \chi$ iff $\vartheta^\circ(xgx^{-1}) = \vartheta(1)$ for all $x \in G$. This happens iff $g \in (\ker \vartheta)^x$ for all x and the proof is complete. ∎

(5.12) THEOREM Let G be an M-group and let $1 = f_1 < f_2 < \cdots < f_k$ be the distinct degrees of the irreducible characters of G. Let $\chi \in \text{Irr}(G)$ with $\chi(1) = f_i$. Then $G^{(i)} \subseteq \ker \chi$, where $G^{(i)}$ denotes the ith term of the derived series of G.

Proof If $i = 1$, then χ is linear and $G' \subseteq \ker \chi$. Assume $i > 1$ and work by induction on i. If $\psi \in \text{Irr}(G)$ and $\psi(1) < \chi(1)$, then $\psi(1) = f_j$ for $j < i$ and $G^{(i-1)} \subseteq G^{(j)} \subseteq \ker \psi$.

Since G is an M-group, we may choose $H < G$ and linear $\lambda \in \text{Irr}(H)$ with $\chi = \lambda^G$. Now $[(1_H)^G, 1_G] = [1_H, 1_H] = 1$ and thus $(1_H)^G$ is not irreducible. If ψ is any irreducible constituent of $(1_H)^G$, then $\psi(1) < (1_H)^G(1) = |G:H| = \lambda^G(1) = \chi(1)$ and thus $G^{(i-1)} \subseteq \ker \psi$. It follows that $G^{(i-1)} \subseteq \ker((1_H)^G) \subseteq H$ by Lemma 5.11.

Now $G^{(i)} \subseteq H' \subseteq \ker \lambda$ and since $G^{(i)} \lhd G$, we have

$$G^{(i)} \subseteq \bigcap_{x \in G} (\ker \lambda)^x = \ker \chi$$

and the proof is complete. ∎

(5.13) COROLLARY (*Taketa*) Let G be an M-group. Then G is solvable.

It is not the case that all solvable groups are M-groups. [The smallest counterexample is SL(2, 3) of order 24.] In Chapter 6 we shall discuss some sufficient conditions. There is no known characterization of M-groups other than in terms of characters.

We shall now discuss some of the connections between character theory and the theory of permutation groups. Suppose the group G acts on the finite set Ω. In other words, for each $\alpha \in \Omega$ and $g \in G$, an element $\alpha^g \in \Omega$ is defined such that $\alpha^1 = \alpha$ and $(\alpha^g)^h = \alpha^{gh}$ for $\alpha \in \Omega$ and $g, h \in G$. For $\alpha \in \Omega$, we shall use the notation $G_\alpha = \{g \in G \mid \alpha^g = \alpha\}$ so that $|G : G_\alpha| = |\mathcal{O}|$, where $\mathcal{O}$ is the orbit containing α.

Let G act on Ω. For $g \in G$, let $\chi(g) = |\{\alpha \in \Omega \mid \alpha^g = \alpha\}|$. The nonnegative integer valued function χ is called the *permutation character* associated with the action. To see that χ really is a character, let V be a $\mathbb{C}$-space with a basis identified with Ω. Let G act on V by permuting the basis. This makes V into a $\mathbb{C}[G]$-module called a *permutation module*. It is clear that V affords χ.

(5.14) LEMMA Let G act transitively on Ω, let $\alpha \in \Omega$ and let $H = G_\alpha$. Then $(1_H)^G$ is the permutation character of the action.

Proof Let V be the permutation module with basis Ω. Then $V = \sum^{\cdot}_{\beta \in \Omega} \mathbb{C}\beta$ is an imprimitivity decomposition for V. The result is now immediate from Theorem 5.8. ∎

(5.15) COROLLARY Let G act on Ω with permutation character χ. Suppose Ω decomposes into exactly k orbits under the action of G. Then $[\chi, 1_G] = k$.

Proof Write $\Omega = \bigcup_{i=1}^{k} \mathcal{O}_i$ where the $\mathcal{O}_i$ are orbits. Let χ_i be the permutation character of G on $\mathcal{O}_i$ so that $\chi = \sum \chi_i$. For $\alpha \in \mathcal{O}_i$, we have $\chi_i = (1_{G\alpha})^G$ by Lemma 5.14. Thus $[\chi_i, 1_G] = [(1_{G_\alpha})^G, 1_G] = [1_{G_\alpha}, 1_{G_\alpha}] = 1$. Therefore, $[\chi, 1_G] = \sum [\chi_i, 1_G] = k$ and the proof is complete. ∎

Corollary 5.15 can also be proved as follows. Let $\mathcal{O}_i$ be as above with $\alpha \in \mathcal{O}_i$. Then $|\mathcal{O}_i| = |G|/|G_\alpha|$. Now let $\mathcal{S} = \{(\alpha, g) \mid \alpha \in \Omega, g \in G, \alpha^g = \alpha\}$ and compute $|\mathcal{S}|$ two ways. We have

$$\sum_{g \in G} \chi(g) = |\mathcal{S}| = \sum_{\alpha \in \Omega} |G_\alpha| = \sum_i \sum_{\alpha \in \mathcal{O}_i} |G_\alpha| = \sum_i \sum_{\alpha \in \mathcal{O}_i} \frac{|G|}{|\mathcal{O}_i|} = \sum_i |G| = k|G|$$

and the result follows.

(5.16) COROLLARY Let G act transitively on Ω with permutation character χ. Suppose that $\alpha \in \Omega$ and that G_α has exactly r orbits on Ω. (One of these is $\{\alpha\}$.) Then $[\chi, \chi] = r$.

Proof We have

$$r = [\chi_{G_\alpha}, 1_{G_\alpha}] = [\chi, (1_{G_\alpha})^G] = [\chi, \chi]$$

by 5.15, 5.14 and Frobenius reciprocity. ∎

The integer r in Corollary 5.16 is called the *rank* of the transitive action. Recall that G is doubly transitive (or 2-transitve) on Ω if $|\Omega| \geq 2$ and G_α is transitive on $\Omega - \{\alpha\}$, that is, if $r = 2$.

(5.17) COROLLARY Let G act on Ω with permutation character χ. Then the action is 2-transitive iff $\chi = 1_G + \psi$ where $\psi \in \text{Irr}(G)$ and $\psi \neq 1_G$.

Corollary 5.17 is often useful for finding irreducible characters. For instance, it shows that the symmetric group Σ_n has an irreducible character of degree $n - 1$ for $n \geq 2$, and that this character restricts irreducibly to A_n for $n \geq 4$.

We now consider the question of finding (not necessarily normal) subgroups of a group G from information about its characters.

Suppose we wish to prove that G has a subgroup H of fairly small index. This could be done if the character $(1_H)^G$ could be recognized as such. Because $(1_H)^G$ is a transitive permutation character (of the action on the set of right cosets of H), there are a number of necessary conditions it must satisfy.

(5.18) THEOREM Let $H \subseteq G$ and let $\chi = (1_H)^G$. Then

(a) $\chi(1) \mid |G|$;
(b) $[\chi, \psi] \leq \psi(1)$ for $\psi \in \text{Irr}(G)$;
(c) $[\chi, 1_G] = 1$;
(d) $\chi(g)$ is a nonnegative rational integer for all $g \in G$;
(e) $\chi(g) \leq \chi(g^m)$ for $g \in G$ and integers m;
(f) $\chi(g) = 0$ if $o(g) \nmid |G|/\chi(1)$;
(g) $\chi(g)|\text{Cl}(g)|/\chi(1)$ is integral for all $g \in G$.

Proof Most of these assertions are immediate from the fact that χ is a transitive permutation character. Statement (b) follows since

$$[\chi, \psi] = [(1_H)^G, \psi] = [1_H, \psi_H] \leq \psi(1).$$

Statement (f) follows since $|H| = |G|/\chi(1)$ and if $o(g) \nmid |H|$, then no conjugate of g lies in H and so $(1_H)^G(g) = 0$.

To prove (g), let Ω be the set of right cosets of H so that χ is the character of the action of G on Ω. Let $\mathscr{K} = \text{Cl}(g)$ and let $\mathscr{S} = \{(\alpha, x) | \alpha \in \Omega, x \in \mathscr{K},$ and $\alpha^x = \alpha\}$. Since χ is constant on $\mathscr{K}$, we have

$$|\mathscr{K}|\chi(g) = |\mathscr{S}| = \sum_{\alpha \in \Omega} |\mathscr{K} \cap G_\alpha|.$$

Since all G_α are conjugate in G, $|\mathscr{K} \cap G_\alpha| = m$ is independent of α and we have $|\mathscr{K}|\chi(g) = m|\Omega| = m\chi(1)$, proving the result. ∎

The necessary conditions given in Theorem 5.18 are definitely not sufficient to guarantee that χ is a permutation character. For example, the simple group M_{22}, of order 443,520, has an irreducible character of degree 55. The character obtained by adding the principal character to it satisfies the conditions (a)–(g) and yet M_{22} has no subgroup of index 56.

The following result can often be used to find proper subgroups of a group.

(5.19) THEOREM (*Brauer*) Let χ be a character of G with $[\chi, 1_G] = 0$. Let $A, B \subseteq G$ and suppose

$$[\chi_A, 1_A] + [\chi_B, 1_B] > [\chi_{A \cap B}, 1_{A \cap B}].$$

Then A and B generate a proper subgroup of G.

Proof Let U be a $\mathbb{C}[G]$-module which affords χ and let V, W and Y be the subspaces of fixed points of U under A, B and $A \cap B$ respectively. Then $V \subseteq Y, W \subseteq Y$, and

$$\dim V + \dim W = [\chi_A, 1_A] + [\chi_B, 1_B] > [\chi_{A \cap B}, 1_{A \cap B}] = \dim Y.$$

It follows that $V \cap W \neq 0$ and thus $\langle A, B \rangle$ has nontrivial fixed points on U. Since $[\chi, 1_G] = 0$, we conclude that $\langle A, B \rangle < G$. ∎

An example showing how Theorem 5.19 can be used is the following proof that A_6 is the unique simple group of its order. [Since $|A_6| = |\mathrm{PSL}(2, 9)|$, it follows that $A_6 \cong \mathrm{PSL}(2, 9)$.] As is often the case when trying to identify a particular group up to isomorphism, there is a great deal of tedious and detailed work involved. Since the result hardly seems to be worth that much effort, the proof which follows is somewhat more sketchy than most in this book.

(5.20) THEOREM Let G be simple and suppose $|G| = 360 = 2^3 \cdot 3^2 \cdot 5$. Then $G \cong A_6$.

Proof Suppose $G_0 \subseteq G$ with $|G : G_0| = k > 1$. Then G is isomorphic to a subgroup of the alternating group A_k. Therefore $k \geq 6$ and the result follows if we can find G_0 with $k = 6$.

The only divisors of 360 which are $\equiv 1 \bmod 3$ are 1, 4, 10, and 40. Let $P \in \mathrm{Syl}_3(G)$ and $N = \mathbf{N}_G(P)$. We conclude that $|G : N| = 10$ or 40. By Burnside's transfer theorem, $N > P$ and hence $|G : N| = 10$.

Now suppose $P \subseteq M < G$. We claim $M \subseteq N$. If not, then $1 < |M : M \cap N| \leq 10$ and hence $|M : M \cap N| = 4$ or 10 by Sylow's theorem. Since the Sylow 3-subgroups of G generate all of G, we have $|\mathrm{Syl}_3(M)| < 10$ and thus $|M : M \cap N| = 4$ and $|G : M|$ divides 10. Thus $|G : M| = 10$ and $M \cap N = P$. Now application of Burnside's theorem to M yields $K \lhd M$ with $|K| = 4$. Thus $8 \,|\, |\mathbf{N}_G(K)|$ and $P \subseteq \mathbf{N}_G(K)$ so that $|G : \mathbf{N}_G(K)| \leq 5$. This contradiction shows $M \subseteq N$ as claimed.

Now let $P \neq P_1 \in \mathrm{Syl}_3(G)$ and set $D = P \cap P_1$. Then $P_1 \subseteq \mathbf{N}(D)$ so that $\mathbf{N}(D) \nsubseteq N$. However, $P \subseteq \mathbf{N}(D)$ and we conclude that $\mathbf{N}(D) = G$ so that $D = 1$.

Next we claim that G contains no element of order 6. If x is an even permutation on 10 points and $o(x) = 6$, then x must contain a 2-cycle and x^2 fixes at least two points. Since no element of G of order 3 lies in two distinct Sylow 3-subgroups, the claim follows. In particular, N/P acts faithfully on P. Since the automorphism group of a cyclic group of order 9 has order 6, it follows that P is elementary abelian. Furthermore, each involution in N inverts all elements of P.

Now $P = \mathbf{C}(y)$ for each $1 \neq y \in P$ and it follows that N has exactly six conjugacy classes and thus exactly two nonlinear irreducible characters each of degree 4. If $1_G \neq \chi \in \mathrm{Irr}(G)$, then χ_N must have one of these as a constituent since $N' \nsubseteq \ker \chi$. Therefore $\chi(1) \geq 4$.

Let $\chi \in \mathrm{Irr}(G)$ with $\chi(1)$ odd. Since $4 \nmid \chi(1)$, it follows that χ_N has a linear constituent λ and thus χ is a constituent of λ^G. Therefore, $\chi(1) \leq \lambda^G(1) = 10$ and if $\chi \neq 1_G$ we have $\chi(1) = 5$ or 9.

We will show that G has an irreducible character of degree 5. From $\sum_{\chi \in \mathrm{Irr}(G)} \chi(1)^2 = 360 \equiv 0 \mod 4$, it follows that there exist at least three nonprincipal irreducible characters of odd degree. It suffices to show that there is at most one of degree 9. In fact we show that if $\chi \in \mathrm{Irr}(G)$ and $\chi(1) = 9$, then χ is a constituent of $(1_N)^G$. Indeed, χ is a constituent of λ^G for some linear character λ of N. Thus $\lambda^G = \chi + \mu$, where $\mu(1) = 1$ and thus $\mu = 1_G$. Since $[\mu_N, \lambda] \neq 0$, it follows that $\lambda = 1_N$ as claimed.

Now fix $\chi \in \mathrm{Irr}(G)$ with $\chi(1) = 5$. Then χ_N has a linear constituent and so $[\chi_P, 1_P] \neq 0$. Let v be a nonprincipal linear constituent of χ_P. Since every element of P is real, χ_P is real valued and $[\chi_P, v] = [\chi_P, \bar{v}]$. Let $U = \ker v$, so that $|U| = 3$. It follows that $[\chi_U, 1_U] \geq 3$ and hence by Theorem 5.19, any two conjugates of U generate a proper subgroup of G. Note that we can find ten conjugates of U such that no two of them centralize each other.

Let V be a conjugate of U which does not centralize U, and let $H = \langle U, V \rangle < G$. If $|G:H| = 6$, we are done, so assume that $|H| < 60$.

Now $U \in \mathrm{Syl}_3(H)$ or else $P \subseteq H$ and thus $H \subseteq N$. This would imply $V \subseteq P$ which is not the case. Since $U, V \in \mathrm{Syl}_3(H)$, it follows that 3×4 or 3×10 divides $|H|$ and thus $|H| = 12, 24$, or 30.

If $|H| = 30$, then there exists $S \lhd H$ with $|S| = 5$. Therefore, $|G:\mathbf{N}(S)| \equiv 1 \mod 5$ and divides $|G:H| = 12$. Thus $|G:\mathbf{N}(S)| = 6$. We may assume that this does not occur and hence $|H| = 12$ or 24. The only group of order 24 which is generated by elements of order 3 has a normal subgroup of order 8 and thus contains elements of order 6. We conclude $|H| = 12$ and there exists $K \lhd H$, elementary of order 4.

Next, we claim $[\chi_H, 1_H] = 2$. To see this, we use the fact that the unique nonlinear irreducible character ϑ of H has degree 3 and is real. It follows that $[\vartheta_U, 1_U] = 1$ and thus $\chi_H = \vartheta + \lambda_1 + \lambda_2$ where $(\lambda_i)_U = 1_U$. It follows that $\lambda_i = 1_H$ and the claim is established. Also $[\chi_U, 1_U] = 3$.

Now choose W conjugate to U with $W \not\subseteq H$ and W not commuting with U. Let $L = \langle U, W \rangle$. We may assume $|L| < 60$ and thus $|L| = 12$ and $H \cap L = U$. Since $[\chi_H, 1_H] = 2 = [\chi_L, 1_L]$, Theorem 5.19 applies again to show $R = \langle H, L \rangle < G$. Since $R \not\subseteq N$, it follows that $9 \nmid |R|$. Reasoning as before, $|R| \neq 24$. Since $12 \mid |R|$, this forces $|R| = 60$ and the proof is complete. ∎

It is often convenient to be able to write a character of a group in terms of characters induced from linear characters of subgroups of a specified type. For example, an important theorem of Brauer (which we will prove in Chapter 8) asserts that every character is a $\mathbb{Z}$-linear combination of characters induced from linear characters of nilpotent subgroups. For now we consider only rational valued characters and cyclic subgroups. We prove the following.

(5.21) THEOREM (*Artin*) Let χ be a rational valued character of G. Then

$$\chi = \sum \frac{a_H}{|\mathbf{N}(H) : H|} 1_H{}^G,$$

where H runs over the cyclic subgroups of G and $a_H \in \mathbb{Z}$.

The proof of Artin's theorem depends on a well-known result of algebraic number theory, namely that the cyclotomic polynomials are irreducible over $\mathbb{Q}$. In other words, for each positive integer n, all of the primitive nth roots of unity are conjugate over $\mathbb{Q}$. (This fact is also needed for Problem 3.11.)

(5.22) LEMMA Let χ be a rational valued character of G and let $x, y \in G$ with $\langle x \rangle = \langle y \rangle$. Then $\chi(x) = \chi(y)$.

Proof Let $n = o(x) = o(y)$ and let ε be a primitive nth root of unity. We have $y = x^m$, with $(m, n) = 1$ and thus ε^m is a primitive nth root of unity and $\varepsilon^m = \varepsilon^\sigma$ for some automorphism σ of $\mathbb{Q}[\varepsilon]$.

Now $\chi(x) = \sum_{i=1}^{\chi(1)} \varepsilon_i$, where each ε_i is an nth root of unity and hence is a power of ε. We have

$$\chi(y) = \chi(x^m) = \sum \varepsilon_i{}^m = \sum \varepsilon_i{}^\sigma = \chi(x)^\sigma.$$

Since $\chi(x)$ is rational, we have $\chi(x)^\sigma = \chi(x)$ and the proof is complete. ∎

Proof of 5.21 Define an equivalence relation $\equiv$ on G by $x \equiv y$ if $\langle x \rangle$ and $\langle y \rangle$ are conjugate in G. It follows from Lemma 5.22 that χ is constant on the equivalence classes under $\equiv$ and thus χ is a $\mathbb{Z}$-linear combination of the characteristic functions of these classes.

Let $\mathscr{C}_1, \mathscr{C}_2, \ldots, \mathscr{C}_m$ be the distinct ($\equiv$)-classes of G and let Φ_i be the characteristic function of $\mathscr{C}_i$ so that $\Phi_i(x) = 1$ if $x \in \mathscr{C}_i$ and $\Phi_i(x) = 0$, otherwise. Choose representatives $x_i \in \mathscr{C}_i$ and let $H_i = \langle x_i \rangle$ and $n_i = |H_i|$. We have

$$(1) \qquad |\mathscr{C}_i| = |G : \mathbf{N}(H_i)|\,\varphi(n_i).$$

We prove by induction on n_i that

$$(2) \qquad |\mathbf{N}(H_i)|\,\Phi_i = \sum_j a_j n_j (1_{H_j})^G$$

for suitable $a_j \in \mathbb{Z}$ where $a_j = 0$ unless H_j is conjugate to a subgroup of H_i. The result will then follow since $|\mathbf{N}(H_i)|$ divides $|\mathbf{N}(H_j)|$ when $a_j \neq 0$.

If $n_i = 1$, then $\mathscr{C}_i = \{1\}$ and $|G|\Phi_i = (1_{H_i})^G$ and (2) holds in this case. Suppose $n_i > 1$. Write

$$(1_{H_i})^G = \sum b_j \Phi_j$$

and compute the coefficients b_j as follows using Frobenius reciprocity and the fact that $[\Phi_j, \Phi_k] = \delta_{jk}|\mathscr{C}_j|/|G|$:

$$b_j|\mathscr{C}_j|/|G| = [(1_{H_i})^G, \Phi_j] = [1_{H_i}, (\Phi_j)_{H_i}].$$

Now $(\Phi_j)_{H_i} = 0$ unless H_j is conjugate to a subgroup of H_i. In that case, $(\Phi_j)_{H_i}$ takes on the value 1 on $\varphi(n_j)$ elements of H_i and vanishes elsewhere. Thus $[1_{H_i}, (\Phi_j)_{H_i}] = \varphi(n_j)/n_i$ and we conclude that

$$b_j = |G|\varphi(n_j)/n_i|\mathscr{C}_j| = |\mathbf{N}(H_j)|/n_i$$

using (1).

Therefore,

$$n_i(1_{H_i})^G = \sum_j |\mathbf{N}(H_j)|\,\Phi_j,$$

where j runs over the set of subscripts for which H_j is conjugate to a subgroup of H_i. Solving for $|\mathbf{N}(H_i)|\Phi_i$ and applying the inductive hypothesis yields (2) and proves the theorem. ∎

(5.23) COROLLARY Let χ be any rational valued character of G. Then $|G|\chi = \sum_H a_H 1_H{}^G$, where H runs over the cyclic subgroups of G and $a_H \in \mathbb{Z}$.

Problems

(5.1) Let $H \subseteq K \subseteq G$ and suppose that φ is a class function of H. Show that $(\varphi^K)^G = \varphi^G$.

(5.2) Let $H, K \subseteq G$ with $HK = G$ and suppose φ is a class function of H. Show that $(\varphi^G)_K = (\varphi_{H \cap K})^K$.

(5.3) Let $H \subseteq G$ and suppose φ is a class function of H and ψ is a class function of G. Show that $(\varphi\psi_H)^G = \varphi^G\psi$.

Note Problems (5.1)–(5.3) will be used frequently.

(5.4) Let $b(G) = \max\{\chi(1) \mid \chi \in \operatorname{Irr}(G)\}$. If $H \subseteq G$, show that $b(H) \le b(G) \le |G:H|\,b(H)$.

Note Since H is abelian iff $b(H) = 1$, the inequality $b(G) \le |G:H|\,b(H)$ generalizes Problem 2.9(b).

(5.5) Let F be an arbitrary field and let $H \subseteq G$. Let W be an $F[H]$-module. Show that there is a unique (up to $F[G]$-isomorphism) $F[G]$-module V such that V has an imprimitivity decomposition $V = \sum \cdot W_i$, where H is the stabilizer of W_1 and $W \cong W_1$ as $F[H]$-modules.

Notes By Theorem 5.9, $V \cong W^G$. In the following problem, if V is an $F[G]$-module and $H \subseteq G$, we use the notation V_H to denote the module V viewed as an $F[H]$-module.

(5.6) (*Mackey*) Let $H, K \subseteq G$ and let T be a set of double coset representatives so that

$$G = \bigcup_{t \in T} HtK$$

is a disjoint union. Let W be an $F[H]$-module for an arbitrary field F. Show that

$$(W^G)_K = \sum_{t \in T} \cdot (W_{H^t \cap K})^K.$$

Note Mackey's theorem generalizes Problem 5.2.

(5.7) Let $H \subseteq G$ and suppose ψ is a character of H. Let $K \subseteq G$ and assume $(\psi^G)_K \in \operatorname{Irr}(K)$. Show that $HK = G$.

Note Problem 5.7 is an immediate consequence of Problem 5.6. It can, however, be done without using modules. The key step is to show that $[(\psi^G)_K, (\psi_{H \cap K})^K] \ne 0$ and then conclude that $|G:H| \le |K:K \cap H|$.

(5.8) Let χ be a monomial character of G and suppose $K \subseteq G$ with $\chi_K \in \operatorname{Irr}(K)$. Show that χ_K is a monomial character of K.

(5.9) Suppose that F is a subfield of $\mathbb{C}$ and let $H \subseteq G$. Suppose that ϑ is a character of H which is afforded by an F-representation. Show that ϑ^G is afforded by an F-representation.

Note It follows that every monomial character of G is afforded by a $\mathbb{Q}[\varepsilon]$-representation, where ε is a primitive nth root of unity, n being the exponent of G (that is, the least common multiple of the orders of the elements of G). Clearly, an arbitrary character of G has values in $\mathbb{Q}[\varepsilon]$. In Chapter 10 we shall prove the theorem of Brauer which asserts that every character of G is afforded by a $\mathbb{Q}[\varepsilon]$-representation.

(5.10) Let $H \subseteq G$ and view $\mathbb{C}[H] \subseteq \mathbb{C}[G]$. Let I be a right ideal of $\mathbb{C}[H]$ and let $J = I\mathbb{C}[G]$. Suppose that I affords ϑ (viewing I as a $\mathbb{C}[H]$-module). Show that J affords ϑ^G.

(5.11) Let $N \lhd G$ and $\chi \in \mathrm{Irr}(G)$. Suppose that $[\chi_N, 1_N] \neq 0$. Show that $N \subseteq \ker \chi$.

Hint Use Lemma 5.11.

(5.12) Let $\varphi^G = \chi \in \mathrm{Irr}(G)$ with $\varphi \in \mathrm{Irr}(H)$. Show that $\mathbf{Z}(\chi) \subseteq H$.

(5.13) (*L. Solomon*) Show that each row sum in a character table is a nonnegative rational integer.

Hint Let G act on G by conjugation. Consider the permutation character.

Note The column sums are also integers. They can be negative.

(5.14) Let G be nonabelian and let $f = \min\{\chi(1) \mid \chi \in \mathrm{Irr}(G), \chi(1) > 1\}$. Show

(a) If $|G'| \leq f$, then $G' \subseteq \mathbf{Z}(G)$,
(b) if $|[G, G']| \leq f$, then G' is abelian,
(c) if $H \subseteq G$ and $|G : H| \leq f$, then $G' \subseteq H$.

Hint For (a). If $x \in G$, then $|\mathrm{Cl}(x)| \leq |G'|$.

(5.15) Let $H \subseteq G$ and suppose φ is a character of H with $\det \varphi = 1_H$. Let $\chi = \varphi^G$ and show $(\det \chi)^2 = 1_G$.

(5.16) Let H be a maximal subgroup of G and let $\chi = (1_H)^G$. Let ψ be a nonprincipal irreducible constituent of χ. Show $\ker \psi = \ker \chi$.

(5.17) Let $H \subseteq G$ and let $\chi = (1_H)^G$. Fix a positive integer n. For $g \in G$, let $m(g) = [\chi_{\langle g \rangle}, 1_{\langle g \rangle}]$ and define $\vartheta(g) = n^{m(g)}$. Show that ϑ is a character of G.

Hint Let Ω be the set of right cosets of H in G. Fix a set S with $|S| = n$ and let Λ be the set of all functions $\Omega \to S$. Then G acts on Λ.

Note This result provides another necessary condition to add to the list in Theorem 5.18.

(5.18) Let G be a group. Write $a(n) = |\{g \in G \mid o(g) = n\}|$. Then the polynomial

$$f(x) = (1/|G|) \sum_m a(m) x^{|G|/m}$$

takes on integer values whenever $x \in \mathbb{Z}$.

Hint Use Problem 5.17.

(5.19) Let G be doubly transitive on Ω and let $H \subseteq G$ with $|G:H| < |\Omega|$. Show that H is transitive on Ω.

(5.20) Assume the notation and hypotheses of Problem 4.11 and show that $|G| = q(q-1)(q+1)$.

Hints Let $N = \mathbf{N}(S)$ and show $|N| = q(q-1)$. Let $\lambda \in \mathrm{Irr}(N)$ with $\lambda \neq 1_N$ and $S \subseteq \ker \lambda$. Let $x \in S$, $x \neq 1$. Show $\lambda^G(x) = \lambda(1)$. Write $\lambda^G = \sum_{\chi \in \mathrm{Irr}(G)} a_\chi \chi$ and conclude

$$\lambda(1) = \sum a_\chi(\chi(1) - q) = |G:N|\lambda(1) - q \sum a_\chi.$$

Finish the problem by showing that if $a_\chi \neq 0$, then $\chi(1) \geq q + 1$ so that $\sum a_\chi \leq \lambda(1)$. To show $\chi(1) \geq q + 1$, compute $[\chi_S, 1_S]$ and use the fact that since $[\chi_N, \lambda] > 0$ and χ is real, then $[\chi_S, 1_S] \geq 2$.

Note Using the result of this problem, it is not very hard to prove that $G \cong \mathrm{SL}(2, q)$, a theorem originally due to Brauer, Suzuki, and Wall.

(5.21) Let $P(G)$ denote the set of all $\mathbb{Z}$-linear combinations of characters of of G of the form $1_H{}^G$ for $H \subseteq G$. Show that $P(G)$ is a ring.

(5.22) Let χ be a rational valued character of G and let $n_G(\chi)$ be the least positive integer such that $n_G(\chi)\chi \in P(G)$. Let $N = \ker \chi$ so that χ may be viewed as a character of G/N. Show that

$$n_G(\chi) = n_{G/N}(\chi).$$

Note The existence of $n_G(\chi)$ is guaranteed by Corollary 5.23.

Hint If $n\chi = \sum a_H(1_H)^G$, then $n\chi = \sum a_H(1_{NH})^G$. Show this by writing $(1_H)^G = (1_{NH})^G + \xi^{(H)}$, where $\xi^{(H)}$ is a character of G with the property that N is not contained in the kernel of any of its irreducible constituents.

(5.23) Let χ be any character of G. Show that

$$|G|\chi = \sum a_\lambda \lambda^G,$$

where λ runs over the set of linear characters of cyclic subgroups and $a_\lambda \in \mathbb{Z}$.

Hint Use Problem 5.3.

(5.24) Let G be a doubly transitive permutation group on Ω and let $\alpha, \beta \in \Omega$, with $\alpha \neq \beta$. Let $\varphi \in \mathrm{Irr}(G_\alpha)$ and assume that $\varphi_{G_{\alpha\beta}} \in \mathrm{Irr}(G_{\alpha\beta})$. Show that $[\varphi^G, \varphi^G] \leq 2$.

Hint Use Problem 5.6.

(5.25) Assume that every Sylow subgroup of G has a faithful irreducible character. Show that G has one also.

Hint Let S be the product of all of the minimal normal subgroups of G. (This is called the *socle* of G.) Then S is a direct product of simple groups. Find $\vartheta \in \mathrm{Irr}(S)$ such that $\ker \vartheta$ contains no nontrivial normal subgroup of G.

(5.26) (*Passman*) Let $X \subseteq G - \{1\}$ be a subset and let $n = |X|$. Assume for all primes $p < n$, that a Sylow p-subgroup of G is cyclic. Show that there exists $\chi \in \mathrm{Irr}(G)$ such that $X \cap \ker(\chi) = \varnothing$.

Hints Reduce to the case that G is abelian by observing that it suffices to assume that $G = \langle X \rangle$ and that if $[x, y] \notin \ker(\chi)$, then both x and $y \notin \ker \chi$. In the abelian case, let $N \subseteq G$ be maximal such that $N \cap X = \varnothing$. Show that G/N is cyclic.

Note If $n = 2$, the hypothesis on Sylow subgroups is vacuously satisfied.

6 Normal subgroups

Let $\chi \in \mathrm{Irr}(G)$ and $H \subseteq G$. In general, very little can be said about the restriction χ_H. The situation is quite different if H is normal in G.

Let $H \lhd G$. If ϑ is a class function of H and $g \in G$, we define $\vartheta^g\colon H \to \mathbb{C}$ by $\vartheta^g(h) = \vartheta(ghg^{-1})$. We say that ϑ^g is *conjugate* to ϑ in G.

(6.1) LEMMA Let $H \lhd G$ and let φ, ϑ be class functions of H and $x, y \in G$. Then

(a) φ^x is a class function;
(b) $(\varphi^x)^y = \varphi^{xy}$;
(c) $[\varphi^x, \vartheta^x] = [\varphi, \vartheta]$;
(d) $[\chi_H, \varphi^x] = [\chi_H, \varphi]$ for class functions χ of G;
(e) φ^x is a character if φ is.

Proof (a) If h and k are conjugate in H, then so are h^g and k^g for all $g \in G$. Take $g = x^{-1}$.

(b) Compute. [Note that had we defined $\vartheta^g(h)$ to be $\vartheta(h^g)$, then (b) would fail.]

(c) The two sums are identical since xhx^{-1} runs over H as h does.

(d) We have $(\chi_H)^x = \chi_H$. Apply (c).

(e) Let $\mathfrak{X}$ afford φ and define $\mathfrak{X}^x$ by $\mathfrak{X}^x(h) = \mathfrak{X}(xhx^{-1})$. Check that $\mathfrak{X}^x$ is a representation which affords φ^x. ∎

It follows from Lemma 6.1 that G permutes $\mathrm{Irr}(H)$ by $g\colon \vartheta \mapsto \vartheta^g$. Note that H acts trivially and thus G/H permutes $\mathrm{Irr}(H)$.

(6.2) THEOREM (*Clifford*) Let $H \triangleleft G$ and let $\chi \in \operatorname{Irr}(G)$. Let ϑ be an irreducible constituent of χ_H and suppose $\vartheta = \vartheta_1, \vartheta_2, \ldots, \vartheta_t$ are the distinct conjugates of ϑ in G. Then

$$\chi_H = e \sum_{i=1}^{t} \vartheta_i,$$

where $e = [\chi_H, \vartheta]$.

Proof We compute $(\vartheta^G)_H$. For $h \in H$, we have

$$\vartheta^G(h) = \frac{1}{|H|} \sum_{x \in G} \vartheta^\circ(xhx^{-1}) = \frac{1}{|H|} \sum_{x \in G} \vartheta^x(h)$$

since $xhx^{-1} \in H$ for all $x \in G$. Thus $|H|(\vartheta^G)_H = \sum_{x \in G} \vartheta^x$ and hence if $\varphi \in \operatorname{Irr}(H)$ and $\varphi \notin \{\vartheta_i\}$, we have $0 = [\sum \vartheta^x, \varphi]$ and therefore $[(\vartheta^G)_H, \varphi] = 0$. Since χ is a constituent of ϑ^G by Frobenius reciprocity, it follows that $[\chi_H, \varphi] = 0$. Thus all irreducible constituents of χ_H are among the ϑ_i and $\chi_H = \sum_{i=1}^{t} [\chi_H, \vartheta_i]\vartheta_i$. However, $[\chi_H, \vartheta_i] = [\chi_H, \vartheta]$ by Lemma 6.1(d) and the proof is complete. ∎

Theorem 6.2 is so important that we digress to consider another proof in a more general, module theoretic setting. If $\mathfrak{X}$ is an F-representation of $H \triangleleft G$ for an arbitrary field F, we define the *conjugate* representation $\mathfrak{X}^g$ for $g \in G$ by $\mathfrak{X}^g(h) = \mathfrak{X}(ghg^{-1})$. Note that $\mathfrak{X}^g$ is an F-representation which is irreducible iff $\mathfrak{X}$ is. Also, if $\mathfrak{X}_1$ and $\mathfrak{X}_2$ are F-representations of H, then $\mathfrak{X}_1{}^g$ and $\mathfrak{X}_2{}^g$ are similar iff $\mathfrak{X}_1$ and $\mathfrak{X}_2$ are.

(6.3) DEFINITION Let $H \triangleleft G$ and let W_1 and W_2 be $F[H]$-modules. Then W_1 is *conjugate* to W_2 if there exist bases for the W_i such that the corresponding F-representations of H are conjugate.

(6.4) LEMMA Let V be an $F[G]$-module and let $H \triangleleft G$. Suppose $W \subseteq V$ is an $F[H]$-submodule.

(a) If $g \in G$, then Wg is an $F[H]$-submodule of V and is conjugate to W.

(b) If M is an $F[H]$-module conjugate to W, then $M \cong Wg$ for some $g \in G$.

(c) If $U \subseteq V$ is an $F[H]$-submodule isomorphic to W, then $Ug \cong Wg$ as $F[H]$-modules.

Proof We have $(Wg)h = W(ghg^{-1})g = Wg$ since $ghg^{-1} \in H$. Thus Wg is an $F[H]$-submodule. Let $w_1, w_2, \ldots, w_n$ be a basis for W. Then $w_1 g, \ldots, w_n g$ is a basis for Wg. Let $\mathfrak{X}$ and $\mathfrak{Y}$ be the F-representations of H corresponding to W and Wg respectively, with respect to these bases. We claim that $\mathfrak{Y} = \mathfrak{X}^g$.

If $\mathfrak{X}(ghg^{-1}) = (a_{ij})$, we have

$$w_i ghg^{-1} = \sum_j a_{ij} w_j$$

and thus

$$(w_i g)h = \sum_j a_{ij}(w_j g).$$

Therefore, $\mathfrak{Y}(h) = (a_{ij}) = \mathfrak{X}^g(h)$, establishing the claim. The proof of (a) is now complete.

If M is conjugate to W, then M corresponds to the representation $\mathfrak{X}^g$ for some $g \in G$. Since Wg corresponds to the same representation, we have $Wg \cong M$, proving (b). Finally, if $U \cong W$ then with respect to a suitable basis, U corresponds to $\mathfrak{X}$. Thus Ug and Wg both correspond to $\mathfrak{X}^g$ and (c) follows. ▮

(6.5) THEOREM (*Clifford*) Let $H \lhd G$ and let V be an irreducible $F[G]$-module. Let W be any irreducible $F[H]$-submodule of V. Then

(a) $V = \sum \cdot W_i$ where the W_i are irreducible $F[H]$-submodules of V.
(b) Each W_i is of the form Wg_i for some $g_i \in G$ and thus is conjugate to W.
(c) In the notation of Lemma 1.13, $n_W(V) = n_M(V)$ for every $F[H]$-module M conjugate to W.

Note Another way of saying (c) is that each isomorphism class of $F[H]$-modules conjugate to W is represented equally often among the W_i.

Proof of Theorem 6.5 Clearly, $\sum_{g \in G} Wg$ is G-invariant and hence

$$V = \sum_{g \in G} Wg$$

by the irreducibility of V. The $F[H]$-modules Wg are conjugate to W and hence are irreducible. By Lemma 1.11, it follows that V is the direct sum of some of the Wg. This proves (a) and (b).

Now let M be an $F[H]$-module conjugate to W. Since $\dim(M) = \dim(W)$, it follows from Lemma 1.13(b) that it suffices to show that $\dim(M(V)) = \dim(W(V))$ in order to prove that $n_M(V) = n_W(V)$.

Since M is conjugate to W, it follows by Lemma 6.4(b) that $M \cong Wg$ for some $g \in G$. Now 6.4(c) yields $W(V)g \subseteq M(V)$ and $M(V)g^{-1} \subseteq W(V)$. Thus $W(V)g = M(V)$ and the result follows. ▮

(6.6) COROLLARY Let $H \lhd G$ and let V be an irreducible $F[G]$-module for some field F. Then when viewed as an $F[H]$-module, V is completely reducible.

Proof Use Theorems 1.10 and 6.5. ▮

Note that Corollary 6.6 is independent of Maschke's theorem and is valid even if char(F) divides $|H|$.

Also note that Theorem 6.2 is a consequence of Theorem 6.5.

We now return to the study of $\mathbb{C}$-characters and obtain some consequences of Theorem 6.2.

(6.7) COROLLARY Let $H \lhd G$ and suppose that $\chi \in \mathrm{Irr}(G)$ and $[\chi_H, 1_H] \neq 0$. Then $H \subseteq \ker \chi$.

Proof We have $\chi_H = e \sum \vartheta_i$, where $\vartheta_1 = 1_H$ and the ϑ_i are conjugate. Since $(1_H)^g = 1_H$ for all $g \in G$, we have $\chi_H = \chi(1)1_H$ and $H \subseteq \ker \chi$. ∎

Corollary 6.7 can also be obtained as a consequence of Lemma 5.11 (see Problem 5.11) since χ is a constituent of $(1_H)^G$ and $\ker(1_H)^G = \bigcap H^g = H$.

An interesting class of problems in character theory arises from considering how the structure of a group G and the set $\{\chi(1) \mid \chi \in \mathrm{Irr}(G)\}$ are related. We give one result of this type as an application of Theorem 6.2. Many more results like this will be found in Chapter 12.

(6.8) LEMMA Let $H \lhd G$ and suppose $\chi \in \mathrm{Irr}(G)$ and $\vartheta \in \mathrm{Irr}(H)$ with $[\chi_H, \vartheta] \neq 0$. Then $\vartheta(1) \mid \chi(1)$.

Proof Since $\vartheta^g(1) = \vartheta(1)$, we conclude that $\chi(1) = et\vartheta(1)$ where $\chi_H = e\sum_{i=1}^t \vartheta_i$ as in Theorem 6.2 with $\vartheta = \vartheta_1$. ∎

(6.9) THEOREM Suppose $\chi(1)$ is a power of the prime p for every $\chi \in \mathrm{Irr}(G)$. Then G has a normal abelian p-complement.

Proof If $N \lhd G$, then by Lemma 6.8, $\vartheta(1)$ is a power of p for all $\vartheta \in \mathrm{Irr}(N)$. Working by induction on $|G|$, we see that it suffices to find a normal subgroup N of index p, since the normal abelian p-complement for N will be one for G.

We have

$$|G| = |G : G'| + \sum_{\chi \in \mathrm{Irr}(G);\ \chi(1) > 1} \chi(1)^2.$$

If G is abelian, the result is trivial. Otherwise, some $\chi \in \mathrm{Irr}(G)$ is nonlinear and thus $p \mid |G|$. Since the last sum is divisible by p, we see that $p \mid |G : G'|$. It follows that G has a normal subgroup of index p and the proof is complete. ∎

The converse of Theorem 6.9 is true and will be proved later in this chapter.

(6.10) DEFINITION Let $H \lhd G$ and let $\vartheta \in \text{Irr}(H)$. Then

$$I_G(\vartheta) = \{g \in G \mid \vartheta^g = \vartheta\}$$

is the *inertia group* of ϑ in G.

Since $I_G(\vartheta)$ is the stabilizer of ϑ in the action of G on $\text{Irr}(H)$, it follows that it is a subgroup and that $I_G(\vartheta) \supseteq H$. Also $|G : I_G(\vartheta)|$ is the size of the orbit of ϑ and so in the formula

$$\chi_H = e \sum_{i=1}^{t} \vartheta_i$$

of Theorem 6.2, we have $t = |G : I_G(\vartheta)|$. In particular, t divides $|G : H|$. It turns out that e divides $|G : H|$ also, but that is much more difficult to show. We shall prove some special cases in this chapter and the general result will be proved in Chapter 11, using projective representations.

The following result is of fundamental importance in the character theory of normal subgroups.

(6.11) THEOREM Let $H \lhd G$, $\vartheta \in \text{Irr}(H)$, and $T = I_G(\vartheta)$. Let

$$\mathscr{A} = \{\psi \in \text{Irr}(T) \mid [\psi_H, \vartheta] \neq 0\}, \qquad \mathscr{B} = \{\chi \in \text{Irr}(G) \mid [\chi_H, \vartheta] \neq 0\}.$$

Then

(a) If $\psi \in \mathscr{A}$, then ψ^G is irreducible;
(b) The map $\psi \mapsto \psi^G$ is a bijection of $\mathscr{A}$ onto $\mathscr{B}$;
(c) If $\psi^G = \chi$, with $\psi \in \mathscr{A}$, then ψ is the unique irreducible constituent of χ_T which lies in $\mathscr{A}$;
(d) If $\psi^G = \chi$, with $\psi \in \mathscr{A}$, then $[\psi_H, \vartheta] = [\chi_H, \vartheta]$.

Proof Let $\psi \in \mathscr{A}$ and choose any irreducible constituent χ of ψ^G. Then ψ is a constituent of χ_T and since ϑ is a constituent of ψ_H, we conclude $\chi \in \mathscr{B}$.

Let $\vartheta = \vartheta_1, \vartheta_2, \ldots, \vartheta_t$ be the distinct G-conjugates of ϑ. Then $t = |G : T|$ and $\chi_H = e \sum_{i=1}^{t} \vartheta_i$ for some integer e. Since ϑ is invariant in T, we conclude from Theorem 6.2 that $\psi_H = f\vartheta$ for some f. Since ψ is a constituent of χ_T, we have $f \leq e$.

Therefore

$$et\vartheta(1) = \chi(1) \leq \psi^G(1) = t\psi(1) = ft\vartheta(1) \leq et\vartheta(1)$$

and hence equality holds throughout. In particular, $\chi(1) = \psi^G(1)$ and we conclude that $\chi = \psi^G$ so that (a) follows. Also

$$[\chi_H, \vartheta] = e = f = [\psi_H, \vartheta]$$

and (d) is proved. Statement (c) follows from the last equality since if $\psi_1 \in \mathscr{A}$, $\psi_1 \neq \psi$, and ψ_1 is a constituent of χ_T, then

$$[\chi_H, \vartheta] \geq [(\psi + \psi_1)_H, \vartheta] = [\psi_H, \vartheta] + [(\psi_1)_H, \vartheta] > [\psi_H, \vartheta],$$

a contradiction.

The map in (b) is well defined by (a) and its image lies in $\mathscr{B}$ by (d). It is one-to-one by (c). To prove that it maps onto $\mathscr{B}$, let $\chi \in \mathscr{B}$. Since ϑ is a constituent of χ_H, there must be some irreducible constituent ψ of χ_T with $[\psi_H, \vartheta] \neq 0$. Thus $\psi \in \mathscr{A}$ and χ is a constituent of ψ^G. Therefore $\chi = \psi^G$ and the proof is complete. ∎

(6.12) COROLLARY Let $\chi \in \text{Irr}(G)$ be primitive. Then for every $N \lhd G$, χ_N is a multiple of an irreducible character of N.

Proof We have $\chi_N = e \sum_{i=1}^{t} \vartheta_i$, where the ϑ_i are distinct and $t = |G : I_G(\vartheta_1)|$. By Theorem 6.11, $\chi = \psi^G$ for some $\psi \in \text{Irr}(I_G(\vartheta_1))$. Since χ is primitive, it follows that $I_G(\vartheta_1) = G$ and $t = 1$. Thus $\chi_N = e\vartheta_1$ and the proof is complete. ∎

The converse of Corollary 6.12 is false. For instance the irreducible character of A_5 of degree 5 is imprimitive and yet the conclusion of the corollary is (trivially) valid for this character. Characters which are multiples of an irreducible are called *homogeneous* characters. Irreducible characters whose restriction to every normal subgroup is homogeneous are often called *quasi-primitive* characters.

(6.13) COROLLARY Suppose G has a faithful primitive character and let $A \lhd G$ be abelian. Then $A \subseteq \mathbf{Z}(G)$.

Proof Let $\chi \in \text{Irr}(G)$ be primitive and faithful. Then $\chi_A = e\lambda$ for some linear character λ. Thus $A \subseteq \mathbf{Z}(\chi) = \mathbf{Z}(G)$. ∎

(6.14) COROLLARY Every nilpotent group is an M-group.

Proof Let G be nilpotent and let $\chi \in \text{Irr}(G)$. Choose $H \subseteq G$ minimal, such that there exists $\psi \in \text{Irr}(H)$, with $\chi = \psi^G$. By transitivity of induction (Problem 5.1), ψ is a primitive character of H. It follows that ψ is a faithful primitive character of $H/\ker(\psi) = \bar{H}$ and thus every normal abelian subgroup of $\bar{H}$ is central. Since every nilpotent group contains a normal self-centralizing subgroup, we conclude that $\bar{H}$ is abelian. Then ψ is linear and the proof is complete. ∎

We shall soon give a much more general sufficient condition for a group to be an M-group. The preceding result is included here simply to illustrate the usefulness of Theorem 6.11.

(6.15) THEOREM (*Ito*) Let $A \lhd G$ be abelian. Then $\chi(1)$ dividies $|G:A|$ for all $\chi \in \text{Irr}(G)$.

Proof Let $\lambda \in \text{Irr}(A)$, with $[\chi_A, \lambda] \neq 0$, and let $T = I_G(\lambda)$. Then for some $\psi \in \text{Irr}(T)$, we have $\chi = \psi^G$ and $\psi_A = e\lambda$. Thus $A \subseteq \mathbf{Z}(\psi)$ and hence $\psi(1)$ divides $|T:A|$ by Theorem 3.12. Since $\chi(1) = |G:T|\psi(1)$, we conclude that $\chi(1)$ divides $|G:A|$ and the proof is complete. ∎

Note that the converse of Theorem 6.9 is immediate from Theorem 6.15. Therefore, the purely group theoretic condition that G has an abelian normal p-complement is precisely equivalent to the condition that $\chi(1)$ is a power of p for all $\chi \in \text{Irr}(G)$.

Suppose $N \lhd G$ and $\vartheta \in \text{Irr}(N)$. Write $\vartheta^G = \sum e_i\chi_i$ for $\chi_i \in \text{Irr}(G)$ and $e_i > 0$. Then $(\chi_i)_N = e_i \sum_{j=1}^{t} \vartheta_j$, where $\vartheta = \vartheta_1, \ldots, \vartheta_t$ are the distinct conjugates of ϑ in G. Let $T = I_G(\vartheta)$. By Theorem 6.11 it follows that $\vartheta^T = \sum e_i\psi_i$ with $\psi_i \in \text{Irr}(T)$ and $\psi_i{}^G = \chi_i$. It follows that for the purpose of investigating the nature of the integers e_i, it suffices to assume $T = G$, that is, that ϑ is invariant in G.

For the remainder of this discussion we assume that ϑ is invariant. We have

$$\vartheta^G = \sum e_i\chi_i \qquad \text{and} \qquad (\chi_i)_N = e_i\vartheta.$$

Therefore, $\chi_i(1) = e_i\vartheta(1)$ and

$$|G:N|\vartheta(1) = \vartheta^G(1) = \sum e_i\chi_i(1) = \sum e_i{}^2\vartheta(1),$$

so that

$$\sum e_i^2 = |G:N|.$$

It was remarked earlier (without proof) that e_i divides $|G:N|$. (Note that if N is abelian then $e_i = \chi_i(1)$ and in this case we know that $e_i \mid |G:N|$ by Theorem 6.15.)

We see that in some respects, the integers e_i behave like character degrees for G/N. If they were really character degrees, then one of them, say e_1, would equal 1. That would mean that $(\chi_1)_N = \vartheta$. In other words, that ϑ is *extendible* to G. (Note that if ϑ is extendible then it is automatically invariant.)

A consequence of the next result is that if some $e_i = 1$, then the e_i are exactly the degrees of the irreducible characters of G/N.

(6.16) THEOREM Let $N \lhd G$ and let φ, $\vartheta \in \text{Irr}(N)$ be invariant in G. Assume $\varphi\vartheta$ is irreducible and that ϑ extends to $\chi \in \text{Irr}(G)$. Let $\mathscr{S} = \{\beta \in \text{Irr}(G) \mid [\varphi^G, \beta] \neq 0\}$ and $\mathscr{T} = \{\psi \in \text{Irr}(G) \mid [(\varphi\vartheta)^G, \psi] \neq 0\}$. Then $\beta \mapsto \beta\chi$ defines a bijection of $\mathscr{S}$ onto $\mathscr{T}$.

Proof We have $(\varphi^G)_N$ is a multiple of φ and hence $(\varphi^G)_N = |G:N|\varphi$ by comparing degrees. Thus $[\varphi^G, \varphi^G] = [\varphi, (\varphi^G)_N] = |G:N|[\varphi, \varphi] = |G:N|$. Similarly, $[(\varphi\vartheta)^G, (\varphi\vartheta)^G] = |G:N|$. Also, by Problem 5.3, $(\varphi\vartheta)^G = \varphi^G\chi$.

Now write $\varphi^G = \sum_{\beta \in \mathscr{S}} e_\beta \beta$. We have $[\varphi^G, \varphi^G] = |G:N| = [\varphi^G\chi, \varphi^G\chi]$ and hence

$$\sum_\beta e_\beta{}^2 = \sum_{\beta, \gamma \in \mathscr{S}} e_\beta e_\gamma [\beta\chi, \gamma\chi].$$

Since $[\beta\chi, \gamma\chi] \geq 0$ and $[\beta\chi, \beta\chi] \geq 1$ and all $e_\beta > 0$, this forces $[\beta\chi, \gamma\chi] = 0$ if $\beta \neq \gamma$ and $[\beta\chi, \beta\chi] = 1$. Thus the $\beta\chi$ are distinct irreducible characters for distinct β. These are all of the irreducible constituents of $\varphi^G\chi = (\varphi\vartheta)^G$. The proof is complete. ∎

What is probably the most important special case of Theorem 6.16 is when $\varphi = 1_N$.

(6.17) COROLLARY (*Gallagher*) Let $N \lhd G$ and let $\chi \in \mathrm{Irr}(G)$ be such that $\chi_N = \vartheta \in \mathrm{Irr}(N)$. Then the characters $\beta\chi$ for $\beta \in \mathrm{Irr}(G/N)$ are irreducible, distinct for distinct β and are all of the irreducible constituents of ϑ^G.

Proof This is exactly Theorem 6.16 in the case $\varphi = 1_N$. ∎

Note that in Corollary 6.17 we have identified $\mathrm{Irr}(G/N)$ with a subset of $\mathrm{Irr}(G)$. In this situation, we see that the e_i are exactly the $\beta(1)$ for $\beta \in \mathrm{Irr}(G/N)$.

There are other situations where we have control of the e_i's. The following "going down" theorem is useful in studying the characters of solvable groups.

(6.18) THEOREM Let K/L be an abelian chief factor of G. (That is, $K, L \lhd G$ and no $M \lhd G$ exists with $L < M < K$.) Suppose $\vartheta \in \mathrm{Irr}(K)$ is invariant in G. Then one of the following holds:

(a) $\vartheta_L \in \mathrm{Irr}(L)$;
(b) $\vartheta_L = e\varphi$ for some $\varphi \in \mathrm{Irr}(L)$ and $e^2 = |K:L|$;
(c) $\vartheta_L = \sum_{i=1}^t \varphi_i$ where the $\varphi_i \in \mathrm{Irr}(L)$ are distinct and $t = |K:L|$.

Proof Let φ be an irreducible constituent of ϑ_L and let $T = I_G(\varphi)$. Since ϑ is invariant in G, every G-conjugate of φ is a constituent of ϑ_L and hence is K-conjugate to φ. It follows that $|G:T| = |K:K \cap T|$ and hence $KT = G$. Since K/L is abelian, $K \cap T \lhd KT = G$ and thus either $K \cap T = K$ or $K \cap T = L$.

If $K \cap T = L$, then $\vartheta_L = e\sum_{i=1}^t \varphi_i$, where $t = |K:L|$, $\varphi_1 = \varphi$, and the φ_i are distinct. Thus $\vartheta(1) = e|K:L|\varphi(1)$. Since ϑ is a constituent of φ^K, we have $\vartheta(1) \leq |K:L|\varphi(1)$ and therefore $e = 1$. This is situation (c).

Now assume $K \cap T = K$ so that φ is invariant in K and $\vartheta_L = e\varphi$ for some e. Let $\lambda \in \mathrm{Irr}(K/L)$. Since λ is linear, $\lambda\vartheta \in \mathrm{Irr}(K)$. Also $(\lambda\vartheta)_L = \vartheta_L = e\varphi$. Suppose that all of the characters $\lambda\vartheta$ are distinct as λ runs over $\mathrm{Irr}(K/L)$. Each of these $|K:L|$ characters is an irreducible constituent of φ^K with multiplicity e and we have

$$e|K:L|\vartheta(1) \leq \varphi^K(1) = |K:L|\varphi(1).$$

Therefore,

$$e^2\varphi(1) = e\vartheta(1) \leq \varphi(1)$$

and $e = 1$. This is situation (a).

In the remaining case, $\lambda\vartheta = \mu\vartheta$ for some $\lambda, \mu \in \mathrm{Irr}(K/L)$ with $\lambda \neq \mu$. Let $U = \ker(\lambda\bar{\mu})$. Then $L \subseteq U < K$ and ϑ vanishes on $K - U$. Since ϑ is invariant in G, it follows that ϑ vanishes on $K - U^g$ for all $g \in G$. Since $\bigcap_{g \in G} U^g = L$, we conclude that ϑ vanishes on $K - L$. By Lemma 2.29 we have

$$|K:L| = |K:L|[\vartheta, \vartheta] = [\vartheta_L, \vartheta_L] = e^2$$

and the proof is complete. ∎

(6.19) COROLLARY Let $N \lhd G$ with $|G:N| = p$, a prime. Suppose $\chi \in \mathrm{Irr}(G)$. Then either

(a) χ_N is irreducible or
(b) $\chi_N = \sum_{i=1}^{p} \vartheta_i$, where the ϑ_i are distinct and irreducible.

Proof Take $K = G$ and $L = N$ and apply Theorem 6.18. Case (b) of that theorem cannot occur since p is not a square. ∎

(6.20) COROLLARY Let $N \lhd G$ and suppose $|G:N| = p$, a prime. Let $\vartheta \in \mathrm{Irr}(N)$ be invariant in G. Then ϑ is extendible to G.

Proof Let χ be an irreducible constituent of ϑ^G. Then $\chi_N = e\vartheta$ for some e. Comparison with Corollary 6.19 yields $e = 1$. ∎

Actually, Corollary 6.20 would still be true if the hypothesis that $|G:N|$ is prime were weakened to G/N cyclic. This stronger result will be proved in Chapter 11.

(6.21) DEFINITION Let $N \lhd G$ and let $\chi \in \mathrm{Irr}(G)$. Then χ is a *relative M-character* with respect to N if there exists H with $N \subseteq H \subseteq G$ and $\psi \in \mathrm{Irr}(H)$ such that $\psi^G = \chi$ and $\psi_N \in \mathrm{Irr}(N)$. If every $\chi \in \mathrm{Irr}(G)$ is a relative M-character with respect to N, then G is a *relative M-group* with respect to N.

Note that $\chi \in \mathrm{Irr}(G)$ is a relative M-character with respect to 1 iff it is a monomial character, and G is a relative M-group with respect to 1 iff it is an

M-group. Also, it is clear that if G is a relative M-group with respect to N, then G/N is an M-group.

The converse of the last statement is false. If $G = \mathrm{SL}(2, 3)$ and $N = \mathbf{Z}(G)$, then $G/N \cong A_4$, which is an M-group. However, G is not a relative M-group with respect to N since G has an irreducible character of degree 2 and has no subgroup of index 2 as would be required by the definition.

(6.22) THEOREM Suppose $N \lhd G$ and G/N is solvable. Suppose, furthermore, that every chief factor of every subgroup of G/N has nonsquare order. Then G is a relative M-group with respect to N.

Note The hypotheses on G/N, above, are automatically satisfied if G/N is nilpotent or supersolvable since then all chief factors of subgroups have prime order.

Proof of Theorem 6.22 Let $\chi \in \mathrm{Irr}(G)$. We must show that χ is a relative M-character with respect to N. If χ_N is irreducible, this is clearly the case and hence we assume χ_N reduces. Let $K \lhd G$ be minimal such that $K \supseteq N$ and χ_K is irreducible. Then $K > N$ and we may choose $L \lhd G$, $L \supseteq N$ such that K/L is a chief factor of G. In particular, K/L is abelian of nonsquare order.

We apply Theorem 6.18 to the chief factor K/L and the G-invariant character $\chi_K \in \mathrm{Irr}(K)$. Since χ_L is reducible and $|K:L|$ is nonsquare, we conclude that $\chi_L = \sum_{i=1}^{t} \varphi_i$, where the $\varphi_i \in \mathrm{Irr}(L)$ are distinct and $t = |K:L| > 1$.

Let $T = I_G(\varphi_1) \supseteq L \supseteq N$. By Theorem 6.11, we conclude that $\chi = \psi^G$ for some $\psi \in \mathrm{Irr}(T)$. Since $T/N < G/N$, we may apply induction on $|G:N|$ to conclude that T is a relative M-group with respect to N and that $\psi = \vartheta^T$ for some $\vartheta \in \mathrm{Irr}(H)$ where $N \subseteq H \subseteq T$ and $\vartheta_N \in \mathrm{Irr}(N)$. Now $\chi = \psi^G = (\vartheta^T)^G = \vartheta^G$ by Problem 5.1 and the proof is complete. ∎

Note that Corollary 6.14 follows immediately from this result as does the somewhat more general fact that all supersolvable groups are M-groups. We prove a still more general sufficient condition.

(6.23) THEOREM Let $N \lhd G$ and suppose that all Sylow subgroups of N are abelian. Assume that G is solvable and is a relative M-group with respect to N. Then G is an M-group.

Proof Let $\chi \in \mathrm{Irr}(G)$. Since χ is a relative M-character with respect to N, choose $H \subseteq G$ with $N \subseteq H$, $\psi \in \mathrm{Irr}(H)$, $\psi_N \in \mathrm{Irr}(N)$ and $\psi^G = \chi$. Now choose $U \subseteq H$, minimal such that there exists $\vartheta \in \mathrm{Irr}(U)$, with $\vartheta^H = \psi$. By Problem 5.1, $\vartheta^G = (\vartheta^G)^H = \psi^G = \chi$ and it suffices to prove that ϑ is linear.

Let $M = U \cap N$. Since $(\vartheta^{NU})^H = \psi$ and ψ_{NU} is irreducible, it follows that $\psi_{NU} = \vartheta^{NU}$ and thus $(\vartheta^{NU})_N = \psi_N \in \mathrm{Irr}(N)$. By Problem 5.2, $(\vartheta_M)^N = (\vartheta^{NU})_N \in \mathrm{Irr}(N)$ and hence $\vartheta_M \in \mathrm{Irr}(M)$.

By the minimality of U and Problem 5.1, ϑ is a primitive character of U. Let $K = \ker \vartheta$, $\bar{U} = U/K$ and $\bar{M} = MK/K$. Then ϑ is a faithful primitive character of $\bar{U}$ and $\bar{M}$ has all of its Sylow subgroups abelian. Now let $Z = \mathbf{Z}(\bar{M}) \lhd \bar{U}$. If $Z < \bar{M}$, let A/Z be a chief factor of $\bar{U}$ with $A \subseteq \bar{M}$. Then A/Z is a p-group for some prime p. Let $P \in \mathrm{Syl}_p(A)$. Then P is abelian and $A = PZ$, so that A is abelian. Since $A \lhd \bar{U}$, we conclude from Corollary 6.13 that $A \subseteq \mathbf{Z}(\bar{U})$. This contradicts $A > Z$.

We conclude that $Z = \bar{M}$, and hence $\bar{M}$ is abelian. Since ϑ_M is irreducible, so is ϑ_{MK}. Thus $\vartheta_{MK} \in \mathrm{Irr}(\bar{M})$ and hence $\vartheta(1) = 1$. The proof is complete. ∎

We introduce some notation. If χ is a character of G, let $\det \chi = \lambda$ be the uniquely defined linear character of Problem 2.3. Now write $o(\chi) = o(\lambda)$, the order of λ as an element of the group of linear characters of G. (We call $o(\chi)$ the *determinantal order* of χ.) Since $|G : \ker \lambda| = o(\lambda)$, it follows that $o(\chi)$ divides $|G|$.

Now let $N \lhd G$ and suppose $\vartheta \in \mathrm{Irr}(N)$ is invariant in G. We wish to find sufficient conditions that ϑ be extendible to G. If $(|G : N|, \vartheta(1)) = 1$, then ϑ is extendible iff $\det \vartheta$ is extendible. We shall prove this now under the additional hypothesis that G/N is solvable and defer the general proof to Chapter 8. As will be seen, there is a gain in replacing the problem of extending ϑ by that of extending the linear character, $\det \vartheta$.

(6.24) LEMMA Let $N \lhd G$ and suppose $\vartheta \in \mathrm{Irr}(N)$ is extendible to G and that $(|G : N|, \vartheta(1)) = 1$. Let $\lambda = \det \vartheta$ and let μ be an extension of λ to G. Then there exists a unique extension χ of ϑ to G such that $\det \chi = \mu$.

Proof Let η be an extension of ϑ to G and let $\nu = \det \eta$ so that $\nu_N = \lambda$. Let $\mu\bar{\nu} = \alpha$ so that $\alpha_N = 1_N$ and hence $\alpha^{|G:N|} = 1_G$. Write $f = \vartheta(1)$ so that $\det(\alpha^b \eta) = \alpha^{bf} \det \eta = \alpha^{bf} \nu$ for $b \in \mathbb{Z}$. Since $(f, |G : N|) = 1$, we can choose $b \in \mathbb{Z}$ such that $bf \equiv 1 \bmod |G:N|$ and thus $\alpha^{bf} = \alpha$. Let $\chi = \alpha^b \eta$. Then $\det \chi = \alpha^{bf} \nu = \alpha \nu = \mu$. Since $\alpha_N = 1_N$, we have $\chi_N = \eta_N = \vartheta$ as desired.

To prove uniqueness, suppose χ_0 is an extension of ϑ with $\det(\chi_0) = \mu$. By Corollary 6.17, we have $\chi_0 = \chi\beta$ for some $\beta \in \mathrm{Irr}(G/N)$. Since $\chi_0(1) = \vartheta(1) = \chi(1)$, we have $\beta(1) = 1$ and

$$\mu = \det(\chi_0) = \beta^f \det \chi = \beta^f \mu$$

and $\beta^f = 1_G$. Since $(f, |G : N|) = 1$, this forces $\beta = 1_G$ and $\chi_0 = \chi$. The proof is now complete. ∎

(6.25) THEOREM Let $N \lhd G$ and suppose G/N is solvable. Let $\vartheta \in \mathrm{Irr}(N)$ be invariant in G and suppose $(\vartheta(1), |G : N|) = 1$. Then ϑ is extendible to G iff $\det \vartheta$ is extendible to G.

Proof If ϑ is extendible, then obviously so is det ϑ. We prove the converse. Let μ be an extension of $\lambda = \det \vartheta$ to G. We work by induction on $|G:N|$. (We may assume $G > N$.) Let G_0 be a maximal normal subgroup of G with $G_0 \supseteq N$. Since λ extends to μ_{G_0}, the inductive hypothesis guarantees that ϑ is extendible to G_0. By Lemma 6.24, there is a unique extension, $\chi_0 \in \mathrm{Irr}(G_0)$ with $\det(\chi_0) = \mu_{G_0}$.

We claim χ_0 is invariant in G. For $g \in G$, we have $((\chi_0)^g)_N = \vartheta^g = \vartheta$ and $\det((\chi_0)^g) = \det(\chi_0)^g = (\mu_{G_0})^g = \mu_{G_0}$. The uniqueness of χ_0 now yields $(\chi_0)^g = \chi_0$ and establishes the claim.

Since G/N is solvable, $|G:G_0|$ is prime and by Corollary 6.20, χ_0 is extendible to G. The proof is now complete. ▮

We now discuss some sufficient conditions for extending linear characters. A version of the following theorem is true without the assumption of linearity. The general result will be derived from this special case in Chapter 11.

(6.26) THEOREM Let $N \lhd G$ and suppose λ is a linear character of N which is invariant in G. For each prime $p \mid o(\lambda)$, choose $H_p \subseteq G$ with $H_p/N \in \mathrm{Syl}_p(G/N)$, and assume that λ is extendible to H_p. Then λ is extendible to G.

Note If $p \nmid |G:N|$, then $H_p = N$ and λ is automatically extendible to H_p. It is only necessary, therefore, to check the hypothesis for primes dividing $|G:N|$.

Proof Let $m = o(\lambda)$. For each $p \mid m$, we may choose λ_p, a power of λ, such that $\lambda = \prod \lambda_p$ and $o(\lambda_p)$ is a power of p. We shall show that λ_p is extendible to $\mu_p \in \mathrm{Irr}(G)$. Then $\mu = \prod \mu_p$ is an extension of λ.

Since λ_p is a power of λ, which is extendible to H_p, it follows that λ_p is extendible to H_p. Also λ_p is invariant in G. We now see that it is no loss to assume that m is a power of p.

Let ν be an extension of λ to H_p. Since $p \nmid |G:H_p|$ and $\nu(1) = 1$, it follows that $p \nmid \nu^G(1)$ and hence there exists an irreducible constituent χ of ν^G with $p \nmid \chi(1)$. Let $\chi(1) = f$.

We have $[\chi_{H_p}, \nu] \neq 0$ and hence $[\chi_N, \lambda] \neq 0$. Since λ is invariant in G, we conclude that $\chi_N = f\lambda$ and thus $(\det \chi)_N = \lambda^f$. Let $\delta = \det \chi$. Since $p \nmid f$ and m is a power of p, we may choose $b \in \mathbb{Z}$ with $fb \equiv 1 \bmod m$. Then $(\delta^b)_N = \lambda^{fb} = \lambda$ and the proof is complete. ▮

(6.27) COROLLARY Let $N \lhd G$ and suppose λ is a linear character of N which is invariant in G. Assume $(|G:N|, o(\lambda)) = 1$. Then λ has a unique extension μ to G with the property that $(|G:N|, o(\mu)) = 1$. In fact, $o(\mu) = o(\lambda)$.

Proof The existence of an extension is immediate from Theorem 6.26 since the hypotheses are trivially satisfied. Let ν be an extension of λ and choose $b \in \mathbb{Z}$, with $b|G:N| \equiv 1 \bmod (o(\lambda))$. Let $\mu = \nu^{b|G:N|}$. Then $\mu_N = \lambda^{b|G:N|} = \lambda$. In particular, $o(\mu) \geq o(\lambda)$. On the other hand, $(\nu^{o(\lambda)})_N = 1_N$, so that $\nu^{o(\lambda)} \in \mathrm{Irr}(G/N)$ and

$$\mu^{o(\lambda)} = (\nu^{o(\lambda)b})^{|G:N|} = 1_G.$$

Therefore, $o(\mu) = o(\lambda)$.

For uniqueness, suppose τ is an extension of λ with $(o(\tau), |G:N|) = 1$. Then $(o(\mu\bar{\tau}), |G:N|) = 1$ and $(\mu\bar{\tau})_N = 1_N$ so that $\mu\bar{\tau} \in \mathrm{Irr}(G/N)$. It follows that $\mu\bar{\tau} = 1_G$ and $\mu = \tau$ as desired. ∎

(6.28) COROLLARY Let $N \lhd G$ with G/N solvable and suppose $\vartheta \in \mathrm{Irr}(N)$ is invariant in G. Assume $(|G:N|, o(\vartheta)\vartheta(1)) = 1$. Then ϑ has a unique extension, $\chi \in \mathrm{Irr}(G)$ with $(|G:N|, o(\chi)) = 1$. In fact $o(\chi) = o(\vartheta)$.

Proof By Corollary 6.27, let μ be the unique extension of $\lambda = \det \vartheta$ to G with $(|G:N|, o(\mu)) = 1$. By Theorem 6.25 and Lemma 6.24, let χ be the unique extension of ϑ to G with $\det \chi = \mu$. Then $o(\chi) = o(\mu) = o(\lambda) = o(\vartheta)$.

If χ_0 extends ϑ and $(|G:N|, o(\chi_0)) = 1$, let $\mu_0 = \det(\chi_0)$. Thus

$$(|G:N|, o(\mu_0)) = 1$$

and hence $\mu_0 = \mu$ by the uniqueness of μ. Then $\chi_0 = \chi$ by the uniqueness of χ. This completes the proof. ∎

Note that since $o(\vartheta)$ and $\vartheta(1)$ divide $|N|$, the condition that

$$(|G:N|, o(\vartheta)\vartheta(1)) = 1$$

in Corollary 6.28 will be automatic if $(|G:N|, |N|) = 1$. The hypothesis of solvability in Corollary 6.28 will be removed in Chapter 8.

We shall give one further extendibility criterion now. It can be used when G/N is a p-group for $p \neq 2$. Unlike the previous results, this condition is independent of character degree.

(6.29) DEFINITION Let χ be a character of G and let p be a prime. Then χ is *p-rational* if there exists an integer r with $p \nmid r$, such that $\chi(g) \in \mathbb{Q}[e^{2\pi i/r}]$ for every $g \in G$.

We use the notation $\mathbb{Q}_k = \mathbb{Q}[e^{2\pi i/k}]$ for $1 \leq k \in \mathbb{Z}$. As is well known, $\mathbb{Q}_k \cap \mathbb{Q}_l = \mathbb{Q}_d$, where $d = (k, l)$. By Lemma 2.15, it follows that $\chi(g) \in \mathbb{Q}_{|G|}$ for every character χ of G and every $g \in G$. Write $|G| = n = p^a m$, with $p \nmid m$. It follows that χ is p-rational iff all of its values lie in $\mathbb{Q}_m$.

Let $\mathcal{G}_p(G)$ denote the Galois group $\mathcal{G}(\mathbb{Q}_n/\mathbb{Q}_m)$. If χ is a character of G, let χ^σ be defined by $\chi^\sigma(g) = \chi(g)^\sigma$ for $\sigma \in \mathcal{G}(\mathbb{Q}_n/\mathbb{Q})$. Then χ is p-rational iff $\chi^\sigma = \chi$ for all $\sigma \in \mathcal{G}_p(g)$ [Note that by Problem 2.2, χ^σ is necessarily a character. In particular, $\mathcal{G}_p(G)$ permutes $\mathrm{Irr}(G)$.]

From Galois theory we have $\mathscr{G}(\mathbb{Q}_n/\mathbb{Q}_m) \cong \mathscr{G}(\mathbb{Q}_{p^a}/\mathbb{Q})$, where as above $n = mp^a$ and $p \nmid m$. This isomorphism is the restriction map. It follows if $p \neq 2$ and $a > 0$, that $e^{2\pi i/p}$ is not invariant under $\mathscr{G}(\mathbb{Q}_n/\mathbb{Q}_m)$ and hence if $p \neq 2$ and λ is a linear character of G with $p \mid o(\lambda)$, then λ is not p-rational. Also, when $p \neq 2$, we have $\mathscr{G}_p(G) \cong \mathscr{G}(\mathbb{Q}_{p^a}/\mathbb{Q})$ is cyclic.

We have one more general remark. If $H \subseteq G$, then $\mathbb{Q}_{|H|} \subseteq \mathbb{Q}_{|G|}$ and so χ^σ is defined for characters χ of H with $\sigma \in \mathscr{G}_p(G)$. It follows that χ is p-rational iff $\chi^\sigma = \chi$ for all $\sigma \in \mathscr{G}_p(G)$.

(6.30) THEOREM Let $N \lhd G$ with G/N a p-group, $p \neq 2$. Suppose $\vartheta \in \mathrm{Irr}(N)$ is invariant in G and p-rational. Then ϑ^G has a unique p-rational irreducible constituent χ. Furthermore, $\chi_N = \vartheta$.

Proof Use induction on $|G:N|$. We suppose $|G:N| > 1$. Let $K \lhd G$ with $N \subseteq K$ and $|K:N| = p$. By Corollary 6.20, ϑ can be extended to $\eta \in \mathrm{Irr}(K)$. Since $p \neq 2$, $\mathscr{G}_p(K)$ is cyclic and we choose a generator σ. Now $\eta^\sigma \in \mathrm{Irr}(K)$ and $(\eta^\sigma)_N = \vartheta^\sigma = \vartheta$ since ϑ is p-rational. By Corollary 6.17, we have $\eta^\sigma = \eta\lambda$ for some (linear) $\lambda \in \mathrm{Irr}(K/N)$.

If $\lambda = 1_K$, then $\eta^\sigma = \eta$ and η is p-rational. Suppose that this is not the case. Then $p = o(\lambda)$ and since $p \neq 2$ we have $\lambda^\sigma \neq \lambda$ and so $\lambda^\sigma = \lambda^m$ for some $m \in \mathbb{Z}$, with $m \not\equiv 1 \bmod p$. Choose $b \in \mathbb{Z}$ with $(1-m)b \equiv 1 \bmod p$ and let $\psi = \lambda^b\eta$.

Since $\lambda_N = 1_N$, we have $\psi_N = \eta_N = \vartheta$. Also, $mb + 1 \equiv b \bmod p$ and thus

$$\psi^\sigma = (\lambda^b\eta)^\sigma = \lambda^{mb+1}\eta = \lambda^b\eta = \psi$$

and ψ is p-rational. Thus in any case, ϑ has a p-rational extension, $\psi \in \mathrm{Irr}(K)$.

If $\varphi \in \mathrm{Irr}(K)$ is any p-rational extension of ϑ, then $\varphi = \psi\mu$ for some unique $\mu \in \mathrm{Irr}(K/N)$ by Corollary 6.17. We have $\varphi = \varphi^\sigma = (\psi\mu)^\sigma = \psi^\sigma\mu^\sigma = \psi\mu^\sigma$ and thus $\mu^\sigma = \mu$. Since $p \neq 2$, it follows that $p \nmid o(\mu)$ and thus $\mu = 1_K$. Therefore $\varphi = \psi$.

Now let $g \in G$. Then $(\psi^g)^\sigma = (\psi^\sigma)^g = \psi^g$ and $\vartheta^g = \vartheta$ so that ψ^g is a p-rational extension of ϑ to K and by the preceding paragraph, we have $\psi^g = \psi$ and ψ is invariant in G.

Since $|G:K| < |G:N|$, the inductive hypothesis guarantees that ψ^G has a unique p-rational irreducible constituent χ and that $\chi_K = \psi$ so that $\chi_N = \vartheta$.

Let χ_0 be any p-rational irreducible constituent of ϑ^G. We show $\chi_0 = \chi$. Let φ be an irreducible constituent of $(\chi_0)_K$. Since $(\chi_0)_N$ is a multiple of ϑ, it follows that $\varphi_N = \vartheta$ and $\varphi = \psi\nu$ for some $\nu \in \mathrm{Irr}(K/N)$. Since $K/N \subseteq \mathbf{Z}(G/N)$, we have $\nu^g = \nu$ for $g \in G$ and thus $\varphi^g = \psi^g\nu^g = \psi\nu = \varphi$ and φ is invariant in G. It follows that $(\chi_0)_K = e\varphi$ for some integer e and thus φ is p-rational since $\varphi(k) = (1/e)\chi_0(k)$ for $k \in K$. Therefore $\varphi = \psi$ and χ_0 is a constituent of ψ^G. It follows that $\chi_0 = \chi$ and the proof is complete. ∎

We can use some of our results on character extendibility to prove Tate's theorem. This result, which was originally proved in an entirely different way, serves as a "booster" for transfer theory, as will be explained.

We define $\mathbf{O}^p(G)$ to be the unique minimal normal subgroup of G such that $G/\mathbf{O}^p(G)$ is a p-group (for the fixed prime p). Similarly, let $\mathbf{A}^p(G)$ be the unique minimal normal subgroup of G such that $G/\mathbf{A}^p(G)$ is an abelian p-group. [Note that $\mathbf{A}^p(G) = G'\mathbf{O}^p(G)$.]

Let $P \in \mathrm{Syl}_p(G)$ and let $N \supseteq P$. Then N is said to *control p-transfer* if $N/\mathbf{A}^p(N) \cong G/\mathbf{A}^p(G)$, or equivalently, $N \cap \mathbf{A}^p(G) = \mathbf{A}^p(N)$. Several of the standard transfer theorems assert that under suitable hypotheses, certain subgroups N control p-transfer. [Usually, $N = \mathbf{N}(W)$ for some subgroup, $W \lhd P$.] Tate's theorem guarantees that whenever $N/\mathbf{A}^p(N) \cong G/\mathbf{A}^p(G)$, then also $N/\mathbf{O}^p(N) \cong G/\mathbf{O}^p(G)$.

(6.31) THEOREM (*Tate*) Let $P \in \mathrm{Syl}_p(G)$ and $N \supseteq P$. Suppose that $N \cap \mathbf{A}^p(G) = \mathbf{A}^p(N)$. Then $N \cap \mathbf{O}^p(G) = \mathbf{O}^p(N)$.

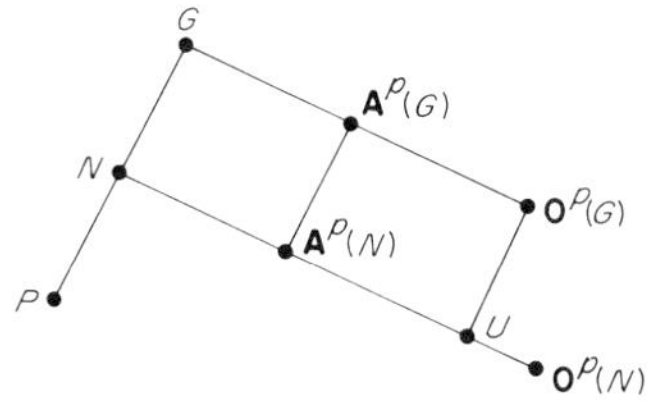

Proof (*Thompson*) Since $N \supseteq P$, we have $N\mathbf{O}^p(G) = G$. Let $U = N \cap \mathbf{O}^p(G)$ so that $U \lhd N$ and $N/U \cong G/\mathbf{O}^p(G)$ is a p-group. Thus $U \supseteq \mathbf{O}^p(N)$. Assume $U > \mathbf{O}^p(N)$ and choose $V \lhd N$, $V \supseteq \mathbf{O}^p(N)$ such that U/V is a chief factor of N. Let λ be a nonprincipal linear character of U/V. Since U/V is a chief factor of the p-group $N/\mathbf{O}^p(N)$, we conclude that $U/V \subseteq \mathbf{Z}(N/V)$ and hence λ is invariant in N.

Let $\vartheta = \lambda^{\mathbf{O}^p(G)}$. If χ is an irreducible constituent of ϑ and $x \in N$, then since $\lambda^x = \lambda$, we have

$$[\chi, \vartheta] = [\chi_U, \lambda] = [(\chi^x)_U, \lambda^x] = [(\chi^x)_U, \lambda] = [\chi^x, \vartheta].$$

Since $P \subseteq N$, it follows that we may write

$$\vartheta = \sum_{\Delta} a_\Delta \Delta$$

as Δ runs over the sums of the orbits of $\mathrm{Irr}(\mathbf{O}^p(G))$ under the action of P.

Now $\vartheta(1) = |\mathbf{O}^p(G) : U| = |G : N|$ is prime to p. It follows that for some Δ, we have $p \nmid a_\Delta$ and $p \nmid \Delta(1)$. Write $\Delta = \sum_{\chi \in \mathcal{O}} \chi$ where $\mathcal{O}$ is an orbit. Since $\chi(1)$

is constant for $\chi \in \mathcal{O}$, we conclude that $p \nmid |\mathcal{O}|$. Since $\mathcal{O}$ is an orbit under the p-group P, we have $|\mathcal{O}| = 1$ and $\Delta = \chi$, where $\chi \in \operatorname{Irr}(\mathbf{O}^p(G))$, χ is invariant under P, $p \nmid \chi(1)$ and $p \nmid [\vartheta, \chi]$.

Since $P\mathbf{O}^p(G) = G$ and χ is invariant under P, it follows that χ is invariant in G. We claim that χ is extendible to G. By Corollary 6.28, it suffices to show that $p \nmid \chi(1)o(\chi)$. Now $\mathbf{O}^p(G)$ has a normal subgroup, $\ker(\det \chi) = K$, such that $|\mathbf{O}^p(G) : K| = o(\chi)$ and $\mathbf{O}^p(G)/K$ is abelian. Since $\mathbf{O}^p(\mathbf{O}^p(G)) = \mathbf{O}^p(G)$, it follows that $\mathbf{O}^p(G)$ has no nontrivial p-factor group and thus $p \nmid o(\chi)$. Since $p \nmid \chi(1)$, Corollary 6.28 does apply.

Let ψ be an extension of χ to G and write

$$\psi_N = \sum_{\varphi \in \operatorname{Irr}(N)} b_\varphi \varphi.$$

We have

$$\sum_{\varphi} b_\varphi [\varphi_U, \lambda] = [\psi_U, \lambda] = [\chi_U, \lambda] = [\chi, \vartheta],$$

which is prime to p. We may therefore choose $\varphi \in \operatorname{Irr}(N)$ with $p \nmid [\varphi_U, \lambda]$. In particular, $[\varphi_U, \lambda] \neq 0$ and since $\lambda \in \operatorname{Irr}(U)$ is invariant in N, we conclude that $\varphi_U = e\lambda$ and thus $\varphi(1) = e = [\varphi_U, \lambda]$ is prime to p. Since $V \subseteq \ker \lambda$ and $\varphi_U = e\lambda$, we conclude that $V \subseteq \ker \varphi$ and $\varphi \in \operatorname{Irr}(N/V)$. Since N/V is a p-group, $\varphi(1)$ is a power of p and hence $\varphi(1) = 1$.

Now $N/\ker \varphi$ is abelian and thus $\mathbf{A}^p(N) \subseteq \ker \varphi$. However, $U = N \cap \mathbf{O}^p(G) \subseteq N \cap \mathbf{A}^p(G) = \mathbf{A}^p(N)$ and therefore $U \subseteq \ker \varphi$. Since $\varphi_U = e\lambda$, it follows that $\lambda = 1_U$. This contradicts the choice of λ and completes the proof of the theorem. ∎

We discuss one further topic in this chapter. Let $N \lhd G$ so that G permutes $\operatorname{Irr}(N)$. It is also true that G permutes the set of conjugacy classes of N and it is natural to consider the relationship between these two actions. It is not the case that they are necessarily permutation isomorphic although they are closely related.

(6.32) THEOREM (*Brauer*) Let A be a group which acts on $\operatorname{Irr}(G)$ and on the set of conjugacy classes of G. Assume that $\chi(g) = \chi^a(g^a)$ for all $\chi \in \operatorname{Irr}(G)$, $a \in A$ and $g \in G$; where g^a is an element of $\operatorname{Cl}(g)^a$. Then for each $a \in A$, the number of fixed irreducible characters of G is equal to the number of fixed classes.

Proof Let χ_i and $\mathcal{K}_j$ be the irreducible characters and conjugacy classes of G for $1 \leq i, j \leq k$. Choose $g_j \in \mathcal{K}_j$ and write $g_i^a = g_j$ if $\mathcal{K}_i^a = \mathcal{K}_j$. Let $X = (\chi_i(g_j))$, the character table of G, viewed as a matrix.

For $a \in A$, let $P(a) = (p_{ij})$, where $p_{ij} = 0$ unless $\chi_i{}^a = \chi_j$, in which case $p_{ij} = 1$. Similarly, define $Q(a) = (q_{ij})$, where $q_{ij} = 1$ if $\mathscr{K}_i{}^a = \mathscr{K}_j$ and is zero otherwise.

The (u, v) entry of the matrix $P(a)X$ is

$$\sum_i p_{ui}\chi_i(g_v) = \chi_u{}^a(g_v)$$

since only when $\chi_i = \chi_u{}^a$ is $p_{ui} \neq 0$. Similarly, the (u, v) entry of the matrix $XQ(a)$ is

$$\sum_j \chi_u(g_j)q_{jv} = \chi_u(g_v^{a^{-1}})$$

since only when $g_j = g_v^{a^{-1}}$ is $q_{jv} \neq 0$.

The hypothesis of the theorem now implies that $P(a)X = XQ(a)$. Thus $Q(a) = X^{-1}P(a)X$ since the orthogonality relations guarantee that X is nonsingular. We conclude that $\operatorname{tr} P(a) = \operatorname{tr} Q(a)$. Since $\operatorname{tr} P(a)$ is the number of $\chi \in \operatorname{Irr}(G)$ which a fixes and $\operatorname{tr} Q(a)$ is the number of a-fixed conjugacy classes, the proof is complete. ▮

(6.33) COROLLARY Under the hypotheses of Theorem 6.32 the numbers of orbits in the actions of A on the irreducible characters and conjugacy classes of G are equal.

Proof Apply Corollary 5.15 to the result of Theorem 6.32. ▮

We may apply Theorem 6.32 to obtain information about the characters of "Frobenius groups."

(6.34) THEOREM Let $N \lhd G$ and assume that $\mathbf{C}_G(x) \subseteq N$ for every $1 \neq x \in N$. Then

(a) For $\varphi \in \operatorname{Irr}(N)$, with $\varphi \neq 1_N$, we have $I_G(\varphi) = N$ and $\varphi^G \in \operatorname{Irr}(G)$.
(b) For $\chi \in \operatorname{Irr}(G)$ with $N \not\subseteq \ker \chi$, we have $\chi = \varphi^G$ for some $\varphi \in \operatorname{Irr}(N)$.

Proof Let $\varphi \in \operatorname{Irr}(N)$, $\varphi \neq 1_N$. To show that $\varphi^G \in \operatorname{Irr}(G)$, it suffices by Theorem 6.11 to show that $I_G(\varphi) = N$. In order to prove (a), therefore, it suffices to show that no element $g \in G - N$ can normalize any nontrivial conjugacy class of N and then apply Theorem 6.32.

Suppose then, $g \in G - N$ normalizes $\mathscr{K}$, a class of N. Let $x \in \mathscr{K}$. Then $x^g \in \mathscr{K}$ and thus $x^g = x^n$ for some $n \in N$. Therefore $gn^{-1} \in \mathbf{C}(x)$. Since $gn^{-1} \notin N$ and $x \in N$, the hypothesis yields $x = 1$ and thus $\mathscr{K} = \{1\}$. This proves (a).

Now let $\chi \in \operatorname{Irr}(G)$ with $N \not\subseteq \ker \chi$. Choose an irreducible constituent φ of χ_N with $\varphi \neq 1_N$. Then χ is a constituent of φ^N which is irreducible and so $\varphi^N = \chi$. The proof is now complete. ▮

It turns out that the groups G satisfying the hypotheses of Theorem 6.34 and for which $N < G$ are exactly the "Frobenius groups" which are discussed at some length in the next chapter.

Problems

(6.1) Let $N \lhd G$ and $\vartheta \in \mathrm{Irr}(N)$. Show that $\vartheta^G \in \mathrm{Irr}(G)$ iff $I_G(\vartheta) = N$.

(6.2) Let $N \lhd G$ and suppose G/N is abelian. Let C be the group of linear characters of G/N so that C acts on $\mathrm{Irr}(G)$ by multiplication. (See Problem 2.6.) Let $\vartheta \in \mathrm{Irr}(N)$. Show that

$$\varphi^G = f \sum_{i=1}^{s} \chi_i,$$

where f is an integer and the $\chi_i \in \mathrm{Irr}(G)$ constitute an orbit under C.

Hint Let $\chi \in \mathrm{Irr}(G)$. Then $(\chi_N)^G = \rho\chi$, where ρ is the regular character of G/N.

(6.3) Let $N \lhd G$ and let $\chi \in \mathrm{Irr}(G)$ and $\vartheta \in \mathrm{Irr}(N)$ with $[\chi_N, \vartheta] \neq 0$. Show that the following are equivalent:

(a) $\chi_N = e\vartheta$, with $e^2 = |G : N|$;
(b) χ vanishes on $G - N$ and ϑ is invariant in G;
(c) χ is the unique irreducible constituent of ϑ^G and ϑ is invariant in G.

Note In the situation of this problem, we say that χ and ϑ are *fully ramified* with respect to G/N.

(6.4) Define $\mathscr{H}(G) = \{H \subseteq G \mid \varphi^G \in \mathrm{Irr}(G)$ for some $\varphi \in \mathrm{Irr}(H)\}$. Let $\mathbf{H}(G) = \bigcap \mathscr{H}(G)$. Show that if G is an M-group, then $\mathbf{H}(G)$ is abelian.

Note By Problem 5.12, $\mathbf{Z}(G) \subseteq \mathbf{H}(G)$ for all groups, G. Of course, $\mathbf{H}(G)$ is characteristic in G.

(6.5) Let $\mathbf{H}(G)$ be as in the preceding problem. Show that $\mathbf{H}(G)$ centralizes N/N' for every $N \lhd G$ and conclude that if G is solvable, then $\mathbf{H}(G)$ is nilpotent.

Note In fact we may conclude that if G is solvable, then $\mathbf{H}(G)' \subseteq \mathbf{Z}(\mathbf{H}(G))$. This is so because a group which is nilpotent of class > 2 necessarily has a characteristic abelian noncentral subgroup; namely, the next to the last nontrivial term in its lower central series.

Hint If $\varphi \in \mathrm{Irr}(N)$ with $N \lhd G$, then $\mathbf{H}(G) \subseteq I_G(\varphi)$. Use this to show that $[N, \mathbf{H}(G)] \subseteq N'$.

(6.6) Let G be solvable and assume that every $\chi \in \mathrm{Irr}(G)$ is quasi-primitive. Show that G is abelian.

(6.7) Let $N \lhd G$ and assume that G/N is solvable. Let $\chi \in \mathrm{Irr}(G)$ and $\vartheta \in \mathrm{Irr}(N)$, with $[\chi_N, \vartheta] \neq 0$. Show that $\chi(1)/\vartheta(1)$ divides $|G:N|$.

Note The conclusion of this problem is valid even if G/N is not solvable.

(6.8) Suppose that G has exactly one nonlinear irreducible character. Show that G' is an elementary abelian p-group.

(6.9) (*Dornhoff*) Let G be an M-group and suppose $N \lhd G$ with $(|N|, |G:N|) = 1$. Show that N is an M-group.

Hint Let $\vartheta \in \mathrm{Irr}(N)$. Find $H \subseteq G$ and $\lambda \in \mathrm{Irr}(H)$, $\lambda(1) = 1$ such that $[(\lambda^{NH})_N, \vartheta] \neq 0$ and $\lambda^G \in \mathrm{Irr}(G)$. Use Problem 6.7 to show that $\lambda^{NH}(1) = \vartheta(1)$ and then use Problem 5.8.

(6.10) Let $N \lhd G$ with $(|G:N|, |N|) = 1$. Suppose that every subgroup of G/N is an M-group. Show that G is a relative M-group with respect to N.

(6.11) Let $A \lhd G$ with A abelian. Suppose $\chi \in \mathrm{Irr}(G)$ is a monomial character. Show that χ is a relative M-character with respect to A.

Hint Write $\chi = \lambda^G$, with λ a linear character of $H \subseteq G$. Then every irreducible constituent of $(\lambda_{A \cap H})^A$ has multiplicity 1.

Note Since Problem 6.11 is false if A is not abelian, it does not follow (and is not true) that an M-group is necessarily a relative M-group with respect to every normal subgroup.

(6.12) Let K/L be an abelian chief factor of G. Let $\varphi \in \mathrm{Irr}(L)$ and $T = I_G(\varphi)$. Assume that $KT = G$. Show that one of the following occurs:

(a) $\varphi^K \in \mathrm{Irr}(K)$;
(b) $\varphi^K = e\vartheta$ for some $\vartheta \in \mathrm{Irr}(K)$, where $e^2 = |K:L|$.
(c) $\varphi^K = \sum_{i=1}^{t} \vartheta_i$, where the $\vartheta_i \in \mathrm{Irr}(K)$ are distinct and $t = |K:L|$.

Hint Use Problem 6.2 and the ideas in the proof of Theorem 6.18.

(6.13) Show that the number of real classes of a group G is equal to the number of real valued $\chi \in \mathrm{Irr}(G)$.

Note It is easy to see that if $|G|$ is odd, then 1 is the only real element. It follows that Problem 6.13 generalizes Problem 3.16.

(6.14) Let F be a field with $\mathbb{Q} \subseteq F \subseteq \mathbb{C}$. Say that $g \in G$ is an F-element if $\chi(g) \in F$ for every $\chi \in \mathrm{Irr}(G)$. Let G be a p-group with $p \neq 2$ and show that the number of classes of F-elements of G is equal to the number of $\chi \in \mathrm{Irr}(G)$ with values in F.

Hint Let ε be a primitive $|G|$th root of unity. Let $\mathcal{G}$ be the Galois group of $F[\varepsilon]$ over F. Define actions of $\mathcal{G}$ on $\mathrm{Irr}(G)$ and on the set of classes of G.

Note Problem 6.14 is false without some assumption on G. Counterexamples with G a 2-group and G of odd order have been found by Thompson and Dade respectively.

(6.15) Let F be an arbitrary field and let $N \lhd G$. View $F[N] \subseteq F[G]$. Show that $J(F[N]) \subseteq J(F[G])$. (See Problem 1.4.)

(6.16) Let P be a p-group and Q a q-group, where p and q are distinct odd primes and $P \subseteq \mathrm{Aut}(Q)$. Show that $|P| < \frac{1}{2}|Q|$ by carrying out the following steps.

(a) It suffices to assume that Q is elementary abelian, that is, is an $F[P]$-module with $F = GF(q)$.
(b) It suffices to assume that Q is an irreducible $F[P]$-module.
(c) We may assume that P is not cyclic, so that by theorems on p-groups, it follows that there exists $A \lhd P$ with A elementary abelian of order p^2.
(d) Q is imprimitive as an $F[P]$-module and we may choose $H \subseteq P$, $|P:H| = p$, with $Q = \sum \cdot_{i=1}^{p} W_i$, each W_i an $F[H]$-module with all $|W_i|$ equal.
(e) Let N_i be the kernel of the action of H on W_i. Then

$$|P| \leq p \prod |H : N_i|.$$

(f) Complete the proof by induction.

Hint Clifford's theorem (6.5) is used in (d). Show that $F[A]$ cannot have a faithful irreducible module.

Notes The result of Problem 6.16 may be stated as a theorem in arithmetic. If Q is elementary abelian of order q^n, then $|\mathrm{Aut}(Q)| = \prod_{i=1}^{n-1} (q^n - q^i)$ and hence this number is not divisible by p^a if $p^a > \frac{1}{2}q^n$.

The induction in the previous step (f) would not go through to prove the weaker theorem that $|P| < |Q|$.

(6.17) Let $N \lhd G$ with G/N cyclic. Let $\vartheta \in \mathrm{Irr}(N)$ be invariant in G and assume that $(\vartheta(1), |H : N|) = 1$. Show that ϑ is extendible to G.

Hint Show that $G/\ker(\det \vartheta)$ is abelian.

Note The result of Problem 6.17 is true without the assumption that $(\vartheta(1), |G : N|) = 1$.

(6.18) Let $N \lhd G$ and suppose $G = NH$ with $N \cap H = 1$. Let $\vartheta \in \mathrm{Irr}(N)$ be invariant in G and assume $(\vartheta(1), |G:N|) = 1$. If H is solvable, show that ϑ is extendible to G.

Note The result of Problem 6.18 is true without the assumption that H is solvable. However, even if H is abelian, the condition $(\vartheta(1), |G:N|) = 1$ cannot be removed.

(6.19) Let $N \lhd G$ with G/N a p-group and let $\vartheta \in \mathrm{Irr}(N)$ be invariant and p-rational and assume $p \nmid o(\vartheta)$ and $p \nmid \vartheta(1)$. Let χ be the extension of ϑ to G with $p \nmid o(\chi)$ (which exists by Corollary 6.28.) Show that χ is p-rational.

(6.20) (*Thompson*) Let $\mathbf{E}^p(G)$ denote the minimal normal subgroup of G such that $G/\mathbf{E}^p(G)$ is an elementary abelian p-group. Suppose $N \subseteq G$ with $p \nmid |G:N|$ and $\mathbf{E}^p(N) = \mathbf{E}^p(G) \cap N$. Show that $\mathbf{O}^p(N) = \mathbf{O}^p(G) \cap N$.

Hints Sharpen the proof of Theorem 6.31 as follows. Choose λ so that $\lambda^p = 1_U$. Let $\mathscr{G} = \mathscr{G}(\mathbb{Q}_{|G|}/\mathbb{Q}_p)$ and let $A \in \mathrm{Syl}_p(\mathscr{G})$. Now A permutes $\mathrm{Irr}(H)$ for all $H \subseteq G$ and λ is invariant under A. Now proceed with the proof of Theorem 6.31 but arrange matters so that χ, ψ and φ are invariant under A.

(6.21) Let $1 = H_0 \lhd H_1 \lhd \cdots \lhd H_n = G$. Assume that H_i/H_{i-1} is nonabelian. Show that there exists $\chi \in \mathrm{Irr}(G)$ with $\chi(1) \geq 2^n$.

Hint Use Corollary 6.17.

7 T.I. sets and exceptional characters

Suppose we know that G has a subgroup H with certain specified properties and assume that we have some information about how H is embedded in G. How can we draw conclusions about G? We might try inducing the irreducible characters of H to G. Usually this gives little information since if $|H|$ is much smaller than $|G|$, the characters ϑ^G tend to have large numbers of irreducible constituents with large multiplicities, even if $\vartheta \in \mathrm{Irr}(H)$.

In certain situations, however, one can find a difference of two characters $\vartheta_1 - \vartheta_2$ of H where $(\vartheta_1 - \vartheta_2)^G = \chi_1 - \chi_2$ and χ_1 and χ_2 are under control. In these situations, information about $\mathrm{Irr}(G)$ can be obtained.

The earliest example of the use of this technique is in the proof of Frobenius' theorem. We shall give this before discussing any of the later refinements of these ideas.

In Theorem 6.34 we considered groups G having a normal subgroup N such that $\mathbf{C}_G(x) \subseteq N$ for all $1 \neq x \in N$. It follows immediately using Sylow's theorem and the fact that nontrivial p-groups have nontrivial centers that $(|N|, |G:N|) = 1$. By the Schur–Zassenhaus theorem we conclude that there exists $H \subseteq G$ with $NH = G$ and $N \cap H = 1$.

In this situation, let $g \in G - H$. Write $g = xn$ with $x \in H$ and $1 \neq n \in N$. If $y \in H \cap H^g$, then $y \in H^n$ and $y = h^n$ for some $h \in H$. Since $y \in H$, we have $[h, n] = h^{-1}h^n = h^{-1}y \in H$. Since $N \lhd G$, we have $[h, n] \in H \cap N = 1$ and $h \in \mathbf{C}(n) \subseteq N$. Thus $h = 1$ and $y = 1$. We conclude that $H \cap H^g = 1$.

(7.1) DEFINITION Let $H \subseteq G$, with $1 < H < G$. Assume that $H \cap H^g = 1$ whenever $g \in G - H$. Then H is a *Frobenius complement* in G. A group which contains a Frobenius complement is called a *Frobenius group*.

We have proved in the foregoing that groups satisfying the hypotheses of Theorem 6.34 with $1 < N < G$ are Frobenius groups. Frobenius' theorem is the converse of this.

(7.2) THEOREM (*Frobenius*) Let G be a Frobenius group with complement H. Then there exists $N \lhd G$ with $HN = G$ and $H \cap N = 1$.

The fact that the group N of Theorem 7.2 satisfies the condition that $\mathbf{C}_G(x) \subseteq N$ for all $1 \neq x \in N$ is not difficult to prove and is left to the reader.

Before beginning the proof of Theorem 7.2, we mention the curious fact that it is trivial to find N. What is hard is to prove that N is a subgroup.

(7.3) LEMMA Let H be a Frobenius complement in G. Let

$$N = \left(G - \bigcup_{x \in G} H^x\right) \cup \{1\}.$$

Then $|N| = |G:H|$. If $M \lhd G$ with $M \cap H = 1$, then $M \subseteq N$.

Proof Since $H = \mathbf{N}_G(H)$, there are $|G:H|$ distinct subgroups of the form H^x. These contain exactly $|G:H|(|H| - 1)$ nonidentity elements. The remaining elements of G constitute the set N. We have

$$|N| = |G| - |G:H|(|H| - 1) = |G| - |G| + |G:H| = |G:H|.$$

If $M \lhd G$ and $M \cap H = 1$, then $M \cap H^x = 1$ for all $x \in G$ and thus $M \subseteq N$. The proof is complete. ∎

We mention that except for some special cases, no proof of Theorem 7.2 is known which does not use characters.

(7.4) LEMMA Let H be a Frobenius complement in G. Let ϑ be a class function of H which satisfies $\vartheta(1) = 0$. Then $(\vartheta^G)_H = \vartheta$.

Proof Let $1 \neq h \in H$. Then

$$\vartheta^G(h) = (1/|H|) \sum_{x \in G} \vartheta^\circ(xhx^{-1}).$$

If $\vartheta^\circ(xhx^{-1}) \neq 0$, then $1 \neq xhx^{-1} \in H \cap H^{x^{-1}}$ and $x \in H$. Then $\vartheta^\circ(xhx^{-1}) = \vartheta(h)$. We have

$$\vartheta^G(h) = (1/|H|) \sum_{x \in H} \vartheta(h) = \vartheta(h).$$

Since $\vartheta^G(1) = |G:H|\vartheta(1) = 0$, the proof is complete. ∎

The proof of Theorem 7.2 may be motivated as follows. Assuming the theorem is true, let $\chi \in \mathrm{Irr}(G)$ with $N \subseteq \ker \chi$. Then $\chi_H \in \mathrm{Irr}(H)$. Now given $\chi_H = \varphi \in \mathrm{Irr}(H)$ we try to find $\chi \in \mathrm{Irr}(G)$. We do this for each $\varphi \in \mathrm{Irr}(H)$ and check that $\bigcap \ker \chi$ is the desired normal subgroup.

Proof of Theorem 7.2 Let $1_H \neq \varphi \in \mathrm{Irr}(H)$ and write $\vartheta = \varphi - \varphi(1)1_H$ so that $\vartheta(1) = 0$. Now $[\vartheta^G, \vartheta^G] = [\vartheta, (\vartheta^G)_H] = [\vartheta, \vartheta]$ by Lemma 7.4. Thus $[\vartheta^G, \vartheta^G] = 1 + \varphi(1)^2$. Now $[\vartheta^G, 1_G] = [\vartheta, 1_H] = -\varphi(1)$. We may therefore write $\vartheta^G = \varphi^* - \varphi(1)1_G$, where φ^* is a class function of G, $[\varphi^*, 1_G] = 0$, and $1 + \varphi(1)^2 = [\varphi^*, \varphi^*] + \varphi(1)^2$, so that $[\varphi^*, \varphi^*] = 1$. Since ϑ is a difference of characters, so is ϑ^G and hence φ^* is a difference of characters also. Since $[\varphi^*, \varphi^*] = 1$, it follows that $\pm\varphi^* \in \mathrm{Irr}(G)$. Furthermore, if $h \in H$, then

$$\varphi^*(h) = \vartheta^G(h) + \varphi(1) = \vartheta(h) + \varphi(1) = \varphi(h).$$

In particular, $\varphi^*(1) > 0$ and thus $\varphi^* \in \mathrm{Irr}(G)$.

For every nonprincipal $\varphi \in \mathrm{Irr}(H)$, we have now chosen an extension, $\varphi^* \in \mathrm{Irr}(G)$. Let $M = \bigcap_\varphi \ker \varphi^*$. If $x \in M \cap H$, then $\varphi(x) = \varphi^*(x) = \varphi^*(1) = \varphi(1)$ for all $\varphi \in \mathrm{Irr}(H)$ and thus $x = 1$. By Lemma 7.3, $M \subseteq N$.

Conversely, if $g \in G$ lies in no conjugate of H, then

$$\varphi^*(g) - \varphi(1) = \vartheta^G(g) = 0$$

and $g \in \ker \varphi^*$. It follows that $M = N$ and hence the normal subgroup M satisfies $|M| = |G:H|$. We have $|MH| = |M||H| = |G:H||H| = |G|$ and the result follows. ∎

The normal subgroup whose existence is guaranteed by Theorem 7.2 is called the *Frobenius kernel* of G. By Lemma 7.3, it is uniquely determined by H. We mention without proof that in fact a Frobenius group has a unique conjugacy class of complements and a unique kernel.

An entirely equivalent version of Frobenius' theorem may be stated as follows.

(7.5) COROLLARY Let G be a transitive permutation group on Ω with character χ. Assume $\chi(g) \leq 1$ for all $g \in G$ with $g \neq 1$. Then the set $\{g \in G \mid \chi(g) = 0\} \cup \{1\}$ is a transitive normal subgroup.

Proof Let $\alpha \in \Omega$ and assume $|\Omega| \geq 2$. If there exists any $g \in G, g \neq 1$ with $\chi(g) \neq 0$, then G_α is a Frobenius complement. By Theorem 7.2. and Lemma 7.3, $\{g \in G \mid \chi(g) = 0\} \cup \{1\} = N$ is the Frobenius kernel. It is transitive since $NG_\alpha = G$. ∎

We shall now discuss some of the ways that the ideas in Frobenius' proof have been extended and used in other contexts. We need to introduce some terminology.

(7.6) DEFINITION Let $X \subseteq G$ be a subset. Then X is a *T.I. set* (trivial intersection set) if for every $g \in G$, either $X^g = X$ or $X^g \cap X \subseteq \{1\}$.

(7.7) LEMMA Let X be a T.I. set in G and let φ and ϑ be class functions on $N = \mathbf{N}_G(X)$. Assume that φ and ϑ vanish on $N - X$ and that $\vartheta(1) = 0$. Then $\vartheta^G(x) = \vartheta(x)$ for all $x \in X$ and $[\vartheta^G, \varphi^G] = [\vartheta, \varphi]$.

Proof Let $x \in X$. Then $\vartheta^G(x) = (1/|N|)\sum_{y \in G} \vartheta^\circ(yxy^{-1})$. If $\vartheta^\circ(yxy^{-1}) \neq 0$, then $yxy^{-1} \in X \cap X^{y^{-1}}$ and $yxy^{-1} \neq 1$. It follows that $y \in N$ and $\vartheta^\circ(yxy^{-1}) = \vartheta(x)$. The first statement now follows.

We have $[\vartheta^G, \varphi^G] = [(\vartheta^G)_N, \varphi]$. Since φ vanishes on $N - X$ and $(\vartheta^G)_N - \vartheta$ vanishes on X, we conclude that $[(\vartheta^G)_N - \vartheta, \varphi] = 0$ and hence $[\vartheta^G, \vartheta^G] = [\vartheta, \varphi]$ as claimed. ∎

Because of the requirement that $\vartheta(1) = 0$, the preceding lemma cannot be applied if ϑ is a character. We usually take ϑ to be a difference of two characters. Such a difference is called a *generalized character*. Note that ϑ is a generalized character of G iff $[\vartheta, \chi] \in \mathbb{Z}$ for all $\chi \in \mathrm{Irr}(G)$. Also, the set of generalized characters of G is a ring.

We are now ready to consider groups whose Sylow 2-subgroup P is generalized quaternion. If $|P| \geq 16$, we shall prove the theorem of Brauer and Suzuki which asserts (among other things) that such a group cannot be simple. This theorem is also true if $|P| = 8$ but it is more difficult to prove in that case.

We shall use the following facts about a generalized quaternion group P:

(a) P has a cyclic subgroup of index 2;
(b) $|P : P'| = 4$;
(c) $|\mathbf{Z}(P)| = 2$;
(d) noncyclic subgroups of P are themselves generalized quaternion;
(e) P contains a unique involution.

We shall need to use the result of Problem 3.9 in the proof.

(7.8) THEOREM (*Brauer–Suzuki*) Let $P \in \mathrm{Syl}_2(G)$ be generalized quaternion with $|P| \geq 16$. Then there exists $N \lhd G$ with $|N|$ odd and such that G/N has a normal subgroup of order 2.

Note that possibly $N = 1$ in 7.8. We first prove the following weaker version. The full result will then follow easily.

(7.9) THEOREM Let $P \in \mathrm{Syl}_2(G)$ be generalized quaternion with $|P| \geq 16$. Then there exists $M \lhd G$ such that $|M|$ is even and G/M is nonabelian.

Proof Let $H \subseteq P$ be cyclic with $|P : H| = 2$. We have $P' \subseteq H$ and $|P'| = |P|/4 \geq 4$, so that $P' > \mathbf{Z}(P)$ and hence $\mathbf{C}_P(P') = H$. Let $C = \mathbf{C}_G(P')$ and $N = \mathbf{N}_G(P')$. Now $P \in \mathrm{Syl}_2(N)$ and $C \lhd N$ so that $H = P \cap C \in \mathrm{Syl}_2(C)$. Since a Sylow 2-subgroup of C is cyclic, it follows (for instance from Burnside's transfer theorem) that C has a normal 2-complement K and $K \lhd N$.

Now N/C is isomorphic to a subgroup of Aut(P'). Since P' is a cyclic 2-group, it follows that N/C is a 2-group. We therefore conclude that $N = KP$. Since $C = KH$ we have $|N : C| = 2$.

Let $U \subseteq P'$ with $|P' : U| = 2$. Let $X = C - UK$. We claim that X is a T.I. set and $N = \mathbf{N}(X)$. Now $C/K \cong H$ is cyclic and UK/K is its unique subgroup of order equal to $|U|$. It follows for $y \in C$ that $y \in X$ iff $o(yK)$ in C/K exceeds $|U|$. We conclude that $y \in X$ iff $(2|U|)|o(y)$.

Now $|P'| = 2|U|$ and $P' \lhd C$. Since P' contains all elements of C of order $2|U|$, it follows that $P' \subseteq \langle x \rangle$ for all $x \in X$. If $x \in X \cap X^g$, then $P' \subseteq \langle x \rangle$ and $(P')^g \subseteq \langle x \rangle$. Since $|P'| = |(P')^g|$, we have $P' = (P')^g$ and $g \in N$. Since clearly $N \subseteq \mathbf{N}(X)$, it now follows, as claimed, that X is a T.I. set and $N = \mathbf{N}(X)$. Since $|P'| \geq 4$, we also have $4|o(x)$ for every $x \in X$.

Now C/UK is cyclic of order 4. Let λ be a linear character of C with $\ker \lambda = UK$. Let $\vartheta = \lambda^N - (1_C)^N$. Since $\ker \lambda^N = UK \subseteq \ker(1_C)^N$, we conclude that ϑ vanishes on UK and in particular $\vartheta(1) = 0$. Clearly, ϑ vanishes on $N - C$ and hence ϑ vanishes on $N - X$. We may therefore apply Lemma 7.7 with $\varphi = \vartheta$ to conclude that $[\vartheta^G, \vartheta^G] = [\vartheta, \vartheta]$.

To compute $[\vartheta, \vartheta]$, observe that $(1_C)^N = 1_N + \mu$ where $\ker \mu = C$. We claim that $\lambda^N \in \text{Irr}(N)$. Otherwise, λ^N is a sum of linear characters and $P' \subseteq N' \subseteq \ker \lambda^N = UK$, which is not the case. Thus $\vartheta = \lambda^N - \mu - 1_N$ and $[\vartheta, \vartheta] = 3$.

Now $[\vartheta^G, \vartheta^G] = 3$ and $[\vartheta^G, 1_G] = [\vartheta, 1_N] = -1$. It follows that

$$\vartheta^G = \pm\chi_1 \pm \chi_2 - 1_G,$$

where $\chi_1, \chi_2 \in \text{Irr}(G)$ are not principal. Since $\vartheta^G(1) = \vartheta(1) = 0$, we conclude that the signs above are opposite and without loss we may write

$$\vartheta^G = \chi_1 - \chi_2 - 1_G.$$

Since $\vartheta^G(g) = 0$ unless g is conjugate to an element of X, we conclude

(1) $$\chi_1(g) - \chi_2(g) = 1 \qquad \text{if} \quad 4 \nmid o(g).$$

Since P has a unique involution, it follows that G has a unique conjugacy class of involutions. Let $\mathscr{K}_1$ be this class. Define the class function φ on G by

$$\varphi(g) = |\{(x, y) | x, y \in \mathscr{K}_1, xy = g\}|.$$

If $\varphi(g) \neq 0$, then $g = xy$ for involutions x and y and hence $g^x = yx = g^{-1}$. If $o(g)$ is even, let σ be the involution in $\langle g \rangle$. Then $\langle x, \sigma \rangle$ is an abelian, noncyclic group of order 4. Since P has no such subgroup, neither does G by Sylow's theorem. We conclude that $\varphi(g) = 0$ if $o(g)$ is even. Thus $\varphi(g)(\chi_1(g) - \chi_2(g) - 1) = 0$ for all $g \in G$. Therefore

(2) $$[\varphi, \chi_1 - \chi_2 - 1_G] = 0.$$

In $\mathbb{C}[G]$, let K_ν be the class sum corresponding to the conjugacy class $\mathscr{K}_\nu$. Then in the notation of Problem 3.9,

$$K_1 K_1 = \sum a_{11\nu} K_\nu,$$

and if $g \in \mathscr{K}_\nu$, then $\varphi(g) = a_{11\nu}$.

By Problem 3.9 we have for $g \in \mathscr{K}_\nu$ and $x \in \mathscr{K}_1$ that

$$\varphi(g) = a_{11\nu} = (|\mathscr{K}_1|^2/|G|) \sum_{\chi \in \mathrm{Irr}(G)} \chi(x)^2 \overline{\chi(g)}/\chi(1).$$

Since $\chi(x)$ and $\varphi(g)$ are real, we may rewrite this equation as

$$(|G|/|\mathscr{K}_1|^2)\varphi = \sum_{\chi \in \mathrm{Irr}(G)} (\chi(x)^2/\chi(1))\chi$$

and

$$(|G|/|\mathscr{K}_1|^2)[\varphi, \chi] = \chi(x)^2/\chi(1)$$

for all $\chi \in \mathrm{Irr}(G)$.

We conclude from Equation (2) that

$$\chi_1(x)^2/\chi_1(1) - \chi_2(x)^2/\chi_2(1) = 1.$$

From Equation (1) we have $\chi_2(x) = \chi_1(x) - 1$ since $4 \nmid o(x)$ and also $\chi_2(1) = \chi_1(1) - 1$. Substitution into the preceding equation yields

$$\chi_1(x)^2/\chi_1(1) - (\chi_1(x) - 1)^2/(\chi_1(1) - 1) = 1.$$

Simplifying this, we obtain

$$(\chi_1(x) - \chi_1(1))^2 = 0$$

and we conclude that $x \in \ker \chi_1$.

Since $o(x) = 2$, we have $|\ker \chi_1|$ is even. Now, $\chi_1(1) = 1 + \chi_2(1) \geq 2$ and thus $G/\ker \chi_1$ is nonabelian. The proof is now complete with $M = \ker \chi_1$. ∎

Proof of Theorem 7.8 Let U be the (normal) subgroup generated by all of the involutions in G. If U has a cyclic Sylow 2-subgroup, then U has a normal 2-complement N and $N \lhd G$. In this case, $U/N \lhd G/N$ and U/N is a cyclic 2-group. The result follows.

Assume then that the Sylow 2-subgroups of U are noncyclic. We derive a contradiction. Since $8 \mid |U|$, we may choose V, with $U \subseteq V \subseteq UP$ such that $|V : U| \leq 2$ and $16 \mid |V|$. Now Theorem 7.9 applies to V and we may choose $M \lhd V$ with V/M nonabelian and $|M|$ even.

Since a Sylow 2-subgroup of V contains a unique involution, it follows that all involutions of V are conjugate. Since $M \lhd V$ contains an involution, it follows that M contains all of the involutions of V and hence $U \subseteq M$. Thus $|V : M| \leq |V : U| \leq 2$ and this contradicts V/M being nonabelian and completes the proof. ∎

By the Brauer–Fowler theorem (Corollary 4.12), there are at most finitely many nonisomorphic simple groups which contain an involution whose centralizer is isomorphic to some given group C. Much work has been done in recent years to find all of the simple groups corresponding to various specific C. A key step is to find all possible orders of these simple groups and this often involves character theoretic techniques related to those in this chapter. As an illustration of this we prove the following.

(7.10) THEOREM Let $G = G'$ and suppose $\tau \in G$ is an involution with $\mathbf{C}_G(\tau)$ dihedral of order 8. Then $|G| = 168$ or 360.

We need a lemma.

(7.11) LEMMA (*Thompson*) Let $S \in \mathrm{Syl}_2(G)$ and suppose $M \subseteq S$ with $|S : M| = 2$. Let $\tau \in S$ be an involution which is not conjugate in G to any element of M. Then $\tau \notin G'$.

Proof Let G act by right multiplication on $\Omega = \{Mg \mid g \in G\}$. Then $|\Omega| = 2|G:S|$. Now if $Mg\tau = Mg$, then $g\tau g^{-1} \in M$, which is not the case. Thus τ has no fixed points on Ω and since $|G:S|$ is odd, it follows that τ induces an odd permutation on Ω. Therefore, there exists $A \lhd G$ with $|G:A| = 2$ and $\tau \notin A$. The result follows. ∎

Proof of Theorem 7.10 Let $D = \mathbf{C}_G(\tau)$ and $D \subseteq S \in \mathrm{Syl}_2(G)$. Then $\mathbf{Z}(S) \subseteq \mathbf{Z}(D) = \langle \tau \rangle$ and hence $S \subseteq \mathbf{C}(\tau) = D$. Thus $D \in \mathrm{Syl}_2(G)$. Let $M \subseteq D$ be cyclic of order 4 so that τ is the unique involution in M. Since $G = G'$, Lemma 7.11 guarantees that every involution in G is conjugate to τ.

Now M is a T.I. set in G since if $M \cap M^x \neq 1$, then $\tau \in M^x$ and thus $\tau = \tau^x$ and $x \in D$. Since $M \lhd D$ we have $M = M^x$. This argument also shows that $D = \mathbf{N}_G(M)$.

Let λ be a faithful linear character of M and let $\vartheta = (1_M - \lambda)^D$. Since λ^D is irreducible, it follows that $[\vartheta, \vartheta] = 3$. Also, $\vartheta(1) = 0$ and ϑ vanishes on $D - M$. Thus by Lemma 7.7 we have $[\vartheta^G, \vartheta^G] = 3$ and also $(\vartheta^G)_M = \vartheta$. Since $\vartheta^G(1) = \vartheta(1) = 0$, it follows that we may write $\vartheta^G = 1_G + \chi - \psi$, where $\chi, \psi \in \mathrm{Irr}(G)$. Calculation in D yields $4 = \vartheta(\tau) = \vartheta^G(\tau)$ and we have

$$(1) \qquad 0 = 1 + \chi(1) - \psi(1), \qquad 4 = 1 + \chi(\tau) - \psi(\tau).$$

Let $\mathscr{K}$ denote the (unique) conjugacy class of involutions in G and define the class function φ by

$$\varphi(g) = |\{(x, y) \mid x, y \in \mathscr{K}, xy = g\}|.$$

If $xy = g$ for involutions x and y, then $g^x = g^{-1}$ and conversely, if $x \in \mathscr{K}$, $x \neq g$, and $g^x = g^{-1}$, then $y = xg$ is an involution. It follows that $\varphi(g) = |\{x \in \mathscr{K} \mid x \neq g,\ g^x = g^{-1}\}|$. If $1 \neq g \in M$ and $g^x = g^{-1}$, then $\tau^x = \tau$ and $x \in D$. We conclude that $\varphi(g) = 4$ for $1 \neq g \in M$.

Reasoning as in the proof of Theorem 7.9 and using Problem 3.9, we have

$$\varphi = \frac{|\mathscr{K}|^2}{|G|} \sum_{\xi \in \mathrm{Irr}(G)} \frac{\xi(\tau)^2}{\xi(1)} \xi.$$

Since $|\mathscr{K}| = |G|/8$ this yields

$$[\vartheta^G, \varphi] = \frac{|G|}{2^6}\left[1 + \frac{\chi(\tau)^2}{\chi(1)} - \frac{\psi(\tau)^2}{\psi(1)}\right].$$

Also, $[\vartheta^G, \varphi] = [(1_M - \lambda), \varphi_M]$. Since φ has the constant value 4 on $M - \{1\}$, this yields $[\vartheta^G, \varphi] = (1 + i) + (2) + (1 - i) = 4$. We conclude that

$$2^8 = |G|\left[1 + \frac{\chi(\tau)^2}{\chi(1)} - \frac{\psi(\tau)^2}{\psi(1)}\right]. \tag{2}$$

Write $a = \chi(1)$ and $b = \chi(\tau)$ so that $\psi(1) = a + 1$ and $\psi(\tau) = b - 3$ by Equations (1).

By the second orthogonality relation we have

$$8 = |\mathbf{C}(\tau)| \geq 1 + \chi(\tau^2) + \psi(\tau)^2 = 1 + b^2 + (b - 3)^2.$$

Since $b \in \mathbb{Z}$, we conclude that $b = 1$ or 2.

Assume $b = 1$. Equation (2) yields $2^8 = |G|[1 + (1/a) - (4/(a + 1))]$ and

$$|G| = 2^8 a(a + 1)/(a - 1)^2.$$

Now $2 \mid a(a + 1)$ but $2^4 \nmid |G|$ and we conclude that $2^3 \mid (a - 1)$. Therefore $2^2 \nmid a(a + 1)$ and since $2^3 \,\|\, |G|$ we have $2^4 \nmid (a - 1)$. No odd prime divisor of $a - 1$ can divide $2^8 a(a + 1)$ and we conclude that $a - 1$ is a power of 2 and thus $a = 9$. This yields $|G| = 2^2 \times 9 \times 10 = 360$.

Now assume $b = 2$. Equation (2) yields $2^8 = |G|[1 + (4/a) - (1/(a + 1))]$ and

$$|G| = 2^8 a(a + 1)/(a + 2)^2.$$

Reasoning exactly as above, we conclude that $a + 2 = 8$ and $|G| = 2^2 \times 6 \times 7 = 168$. The proof is complete. ∎

We mention that $\mathrm{GL}(3, 2) \cong \mathrm{PSL}(2, 7)$ is the unique group G of order 168 with $G = G'$ and $A_6 \cong \mathrm{PSL}(2, 9)$ is the unique one of order 360.

We go back to the problem of obtaining information about the irreducible characters of a group from information about a subgroup. Let

$\mathscr{S} \subseteq \mathrm{Irr}(N)$. We introduce the notation $\mathbb{Z}[\mathscr{S}]$ for the set of $\mathbb{Z}$-linear combinations of elements of $\mathscr{S}$. (Thus $\mathbb{Z}[\mathrm{Irr}(N)]$ is the set of generalized characters of N.)

Suppose $N \subseteq G$ and that we can find a map $*: \mathbb{Z}[\mathscr{S}] \to \mathbb{Z}[\mathrm{Irr}(G)]$ such that $*$ is $\mathbb{Z}$-linear and that $[\varphi^*, \vartheta^*] = [\varphi, \vartheta]$ for all $\varphi, \vartheta \in \mathbb{Z}[\mathscr{S}]$. (Such a map is called a *linear isometry.*) In that case we have $[\chi^*, \chi^*] = 1$ for $\chi \in \mathscr{S}$ and thus $\pm\chi^* \in \mathrm{Irr}(G)$. Write $\varepsilon(\chi) = \pm 1$ so that $\varepsilon(\chi)\chi^* \in \mathrm{Irr}(G)$. The characters $\varepsilon(\chi)\chi^*$ are called the *exceptional characters* associated with $\mathscr{S}$ and $*$. They are in one-to-one correspondence with $\mathscr{S}$.

How might such a map $*$ be constructed? An easy example of a linear map $\mathbb{Z}[\mathscr{S}] \to \mathbb{Z}[\mathrm{Irr}(G)]$ is the induction map $\vartheta \to \vartheta^G$. This map, however, is rarely an isometry on $\mathbb{Z}[\mathscr{S}]$. By Lemma 7.7, we see that there are situations when induction is an isometry on $\mathbb{Z}[\mathscr{S}]^\circ = \{\vartheta \in \mathbb{Z}[\mathscr{S}] \mid \vartheta(1) = 0\}$. This occurs, for instance, if $N = \mathbf{N}(X)$, where X is a T.I. set and $\mathscr{S} = \{\chi \in \mathrm{Irr}(N) \mid \chi$ vanishes on $N - X\}$. The problem then is to extend a linear isometry from $\mathbb{Z}[\mathscr{S}]^\circ$ to all of $\mathbb{Z}[\mathscr{S}]$.

(7.12) DEFINITION Let $N \subseteq G$ and $\mathscr{S} \subseteq \mathrm{Irr}(N)$ with $|\mathscr{S}| \geq 2$. Suppose $\tau: \mathbb{Z}[\mathscr{S}]^\circ \to \mathbb{Z}[\mathrm{Irr}(G)]^\circ$ is a linear isometry. We say that $(\mathscr{S}, \tau)$ is *coherent* if τ can be extended to a linear isometry $*$ defined on $\mathbb{Z}[\mathscr{S}]$.

If τ is the induction map and $(\mathscr{S}, \tau)$ is coherent, we simply say $\mathscr{S}$ is coherent. It should be emphasized that even in this case, the map $*$ usually is not induction. The prototypical example of coherence is where N is a Frobenius complement in G. There, $\mathrm{Irr}(N)$ is coherent; and the proof of that fact is the essence of the proof of Frobenius' theorem.

If $(\mathscr{S}, \tau)$ is coherent, the map $*$ is not always uniquely defined. Nevertheless the set of exceptional characters $\varepsilon(\chi)\chi^*$ is uniquely determined by $(\mathscr{S}, \tau)$.

(7.13) LEMMA Let $(\mathscr{S}, \tau)$ be coherent and let $*$ be a linear isometry which extends τ. Then there exists $\varepsilon = \pm 1$ such that $\varepsilon\chi^* \in \mathrm{Irr}(G)$ for all $\chi \in \mathscr{S}$. The function $f: \mathscr{S} \to \mathrm{Irr}(G)$ defined by $f(\chi) = \varepsilon\chi^*$ is one-to-one. The image of f is $\{\psi \in \mathrm{Irr}(G) \mid [\vartheta^\tau, \psi] \neq 0$ for some $\vartheta \in \mathbb{Z}[\mathscr{S}]^\circ\}$.

Proof For $\chi \in \mathscr{S}$, $[\chi^*, \chi^*] = 1$, and we may choose $\varepsilon(\chi) = \pm 1$ so that $\varepsilon(\chi)\chi^* \in \mathrm{Irr}(G)$. We claim $\varepsilon(\chi) = \varepsilon(\xi)$ for all $\xi \in \mathscr{S}$. We have $\vartheta = \chi(1)\xi - \xi(1)\chi \in \mathbb{Z}[\mathscr{S}]^\circ$ and thus

$$\chi(1)\xi^* - \xi(1)\chi^* = \vartheta^\tau \in \mathbb{Z}[\mathrm{Irr}(G)]^\circ.$$

Evaluation at 1 yields $0 = \chi(1)\xi^*(1) - \xi(1)\chi^*(1)$ and hence $\xi^*(1)$ and $\chi^*(1)$ have the same sign. Thus $\varepsilon(\chi) = \varepsilon(\xi)$ as claimed.

The foregoing also shows that $[\vartheta^\tau, f(\chi)] \neq 0$ for some $\vartheta \in \mathbb{Z}[\mathscr{S}]^\circ$. Conversely, suppose $\psi \in \mathrm{Irr}(G)$ and $[\vartheta^\tau, \psi] \neq 0$, where $\vartheta \in \mathbb{Z}[\mathscr{S}]^\circ$. Then

$\vartheta = \sum_{\chi \in \mathscr{S}} a_\chi \chi$ and $\vartheta^\tau = \sum a_\chi \chi^*$. Therefore $0 \neq [\chi^*, \psi] = \varepsilon[f(\chi), \psi]$ for some $\chi \in \mathscr{S}$ and hence $f(\chi) = \psi$.

To show that f is one-to-one, it suffices to show that $*$ has trivial kernel. This follows since if $\vartheta^* = 0$, then $0 = [\vartheta^*, \vartheta^*] = [\vartheta, \vartheta]$ and hence $\vartheta = 0$. ∎

Suppose $N \subseteq G$, $\mathscr{S} \subseteq \mathrm{Irr}(N)$, $|\mathscr{S}| > 1$ and $\tau: \mathbb{Z}[\mathscr{S}]^\circ \to \mathbb{Z}[\mathrm{Irr}(G)]^\circ$ is a linear isometry. We seek conditions on $\mathscr{S}$ sufficient to guarantee that $(\mathscr{S}, \tau)$ is coherent. One such condition is that all $\chi \in \mathscr{S}$ have equal degrees. Although this is not hard to prove directly, we shall derive it as a corollary of the following more general result of Feit.

The main point of Feit's theorem may be summarized as follows. List the degrees $\chi(1)$ for $\chi \in \mathscr{S}$ in increasing order. Assume that the smallest of these divides all of the others. Then $\mathscr{S}$ is coherent if the degrees do not increase too rapidly.

(7.14) THEOREM (*Feit*) Let $N \subseteq G$, $\mathscr{S} \subseteq \mathrm{Irr}(N)$ and let $\tau: \mathbb{Z}[\mathscr{S}]^\circ \to \mathbb{Z}[\mathrm{Irr}(G)]^\circ$ be a linear isometry. Suppose $\mathscr{S} = \mathscr{S}_0 \cup \{\chi\}$, where $(\mathscr{S}_0, \tau)$ is coherent. Assume that there exists $\psi \in \mathscr{S}_0$ such that $\psi(1) \mid \chi(1)$ and

$$\chi(1) < \frac{1}{2\psi(1)} \sum_{\xi \in \mathscr{S}_0} \xi(1)^2.$$

Then $(\mathscr{S}, \tau)$ is coherent.

Proof Let $*: \mathbb{Z}[\mathscr{S}_0] \to \mathbb{Z}[\mathrm{Irr}(G)]$ be a linear isometry which extends τ on $\mathbb{Z}[\mathscr{S}_0]^\circ$. We shall define χ^* so that the map $*$ can be extended to all of $\mathbb{Z}[\mathscr{S}]$ by linearity.

Let $\chi(1) = d\psi(1)$ so that $\chi - d\psi \in \mathbb{Z}[\mathscr{S}]^\circ$. Define Δ by

(1) $$(\chi - d\psi)^\tau = \Delta - d\psi^* + \sum_{\xi \in \mathscr{S}_0} b_\xi \xi^*,$$

where $[\Delta, \xi^*] = 0$ for all $\xi \in \mathscr{S}_0$. Of course, Δ is a (possibly 0) generalized character of G and all $b_\xi \in \mathbb{Z}$. We shall show that $[\Delta, \Delta] = 1$ and all $b_\xi = 0$.

Now

$$1 + d^2 = [(\chi - d\psi), (\chi - d\psi)] = [(\chi - d\psi)^\tau, (\chi - d\psi)^\tau]$$

and thus

$$1 + d^2 = [\Delta, \Delta] + \sum_{\xi \neq \psi} b_\xi^2 + (b_\psi - d)^2.$$

Therefore

(2) $$[\Delta, \Delta] + \sum_\xi b_\xi^2 = 1 + 2\, d b_\psi.$$

Furthermore, since τ maps into $\mathbb{Z}[\mathrm{Irr}(G)]^\circ$, we have $(\chi - d\psi)^\tau(1) = 0$ and thus

(3) $$0 = \sum_\xi b_\xi \xi^*(1) - d\psi^*(1) + \Delta(1).$$

For $\xi \in \mathscr{S}_0$ let $\Xi = \psi(1)\xi - \xi(1)\psi \in \mathbb{Z}[\mathscr{S}_0]^\circ$. Then

$$[\Xi^\tau, (\chi - d\psi)^\tau] = [\Xi, (\chi - d\psi)] = d\xi(1).$$

Therefore

$$\begin{aligned} d\xi(1) &= [\Xi^\tau, \textstyle\sum b_\eta \eta^*] - d[\Xi^\tau, \psi^*] + [\Xi^\tau, \Delta] \\ &= \psi(1)b_\xi - \xi(1)b_\psi + d\xi(1) + 0 \end{aligned}$$

since $\Xi^\tau = \psi(1)\xi^* - \xi(1)\psi^*$. We conclude that

$$\psi(1)b_\xi = \xi(1)b_\psi.$$

Now define $\mu = b_\psi/\psi(1)$. We have then

(4) $$b_\xi = \mu\xi(1)$$

for all $\xi \in \mathscr{S}_0$.

Also, since $\Xi^\tau(1) = 0$, we have

$$\psi(1)\xi^*(1) = \xi(1)\psi^*(1)$$

and we may write

(5) $$\xi^*(1) = \alpha\xi(1)$$

for all $\xi \in \mathscr{S}_0$ and fixed $\alpha \neq 0$.

By hypothesis we have

(6) $$2d\psi(1)^2 < \sum_\xi \xi(1)^2.$$

Suppose $\mu \neq 0$. Then by (4) we obtain

$$2\, db_\psi{}^2 = \mu^2(2\, d\psi(1)^2) < \mu^2 \sum \xi(1)^2 = \sum b_\xi{}^2$$

and therefore

$$1 + 2\, db_\psi{}^2 \leq \sum b_\xi{}^2.$$

Now (2) yields

$$[\Delta, \Delta] + 2\, db_\psi{}^2 \leq 2\, db_\psi.$$

Since b_ψ is an integer $\neq 0$ (since $\mu \neq 0$) and $[\Delta, \Delta] \geq 0$, we obtain $\Delta = 0$ and $b_\psi = 1$. Thus $\Delta(1) = 0$ and (3), (5), and (4) yield

$$d\psi(1) = \sum b_\xi \xi(1) = \mu \sum \xi(1)^2.$$

Since $b_\psi = 1$, $\mu = 1/\psi(1)$ and thus

$$d\psi(1)^2 = \sum \xi(1)^2.$$

This contradicts (6) and proves $\mu = 0$.

It now follows that all $b_\xi = 0$ and (1) yields

$$(\chi - d\psi)^\tau = \Delta - d\psi^*.$$

From (2) we have $[\Delta, \Delta] = 1$ and we extend $*$ to $\mathbb{Z}[\mathscr{S}]$ linearly by defining $\chi^* = \Delta$.

We now show that $*$ is a linear isometry on $\mathbb{Z}[\mathscr{S}]$ and that it extends τ. If $\vartheta \in \mathbb{Z}[\mathscr{S}]^\circ$, with $[\vartheta, \chi] = a$, we can write

$$\vartheta = a(\chi - d\psi) + \varphi,$$

where $\varphi \in \mathbb{Z}[\mathscr{S}_0]^\circ$. Then

$$\vartheta^* = a(\chi - d\psi)^* + \varphi^* = a(\chi - d\psi)^\tau + \varphi^\tau = \vartheta^\tau$$

and thus $*$ extends τ. To show that $*$ is an isometry, it suffices to show $[\mu^*, \nu^*] = [\mu, \nu]$ for $\mu, \nu \in \mathscr{S}$. If $\mu, \nu \in \mathscr{S}_0$, we already know this. Since $[\Delta, \xi^*] = 0$ for $\xi \in \mathscr{S}_0$ and $[\Delta, \Delta] = 1$, the result now follows. ∎

(7.15) COROLLARY Let $N \subseteq G$ and $\mathscr{S} \subseteq \mathrm{Irr}(N)$ with $|\mathscr{S}| \geq 2$. Suppose $\tau: \mathbb{Z}[\mathscr{S}]^\circ \to \mathbb{Z}[\mathrm{Irr}(G)]$ is a linear isometry. Suppose that all $\chi \in \mathscr{S}$ have equal degrees. Then $(\mathscr{S}, \tau)$ is coherent.

Proof Use induction on $n = |\mathscr{S}|$. Let $\chi_1, \chi_2 \in \mathscr{S}$ be distinct. Then

$$[(\chi_1 - \chi_2)^\tau, (\chi_1 - \chi_2)^\tau] = [(\chi_1 - \chi_2), (\chi_1 - \chi_2)] = 2$$

and since $(\chi_1 - \chi_2)^\tau(1) = 0$, we may write $(\chi_1 - \chi_2)^\tau = \alpha - \beta$ where $\alpha, \beta \in \mathrm{Irr}(G)$ are distinct. If $n = 2$, define $\chi_1^* = \alpha$ and $\chi_2^* = \beta$ and extend by linearity to $\mathbb{Z}[\mathscr{S}]$. In this case $\mathbb{Z}[\mathscr{S}]^\circ = \{a(\chi_1 - \chi_2) | a \in \mathbb{Z}\}$ and thus $*$ agrees with τ on $\mathbb{Z}[\mathscr{S}]^\circ$.

If $n = 3$, let $\chi_3 \in \mathscr{S} - \{\chi_1, \chi_2\}$. Reasoning as above, $(\chi_1 - \chi_3)^\tau = \mu - \nu$ where $\mu, \nu \in \mathrm{Irr}(G)$ and $\mu \neq \nu$. Also

$$[(\chi_1 - \chi_2)^\tau, (\chi_1 - \chi_3)^\tau] = 1$$

and we conclude that either $\mu = \alpha$ or $\nu = \beta$ but not both. If $\mu = \alpha$, define $\chi_1^* = \alpha$, $\chi_2^* = \beta$, and $\chi_3^* = \nu$. If $\nu = \beta$, define $\chi_1^* = -\beta$, $\chi_2^* = -\alpha$, and $\chi_3^* = -\mu$. In either case, extend $*$ linearly to $\mathbb{Z}[\mathscr{S}]$ and check that $*$ agrees with τ on $\chi_1 - \chi_2$ and on $\chi_1 - \chi_3$ and that $*$ is an isometry. Since

$$\mathbb{Z}[\mathscr{S}]^\circ = \{a(\chi_1 - \chi_2) + b(\chi_1 - \chi_3) | a, b \in \mathbb{Z}\},$$

the result follows in this case.

Suppose $n > 3$. Write $\mathscr{S} = \mathscr{S}_0 \cup \{\chi\}$, where $3 \leq |\mathscr{S}_0| = n - 1$. By the inductive hypothesis, $(\mathscr{S}_0, \tau)$ is coherent. Choose $\psi \in \mathscr{S}_0$ and observe that

$\psi(1) \mid \chi(1)$ since $\psi(1) = \chi(1)$. Now

$$\frac{1}{2\psi(1)} \sum_{\xi \in \mathscr{S}_0} \xi(1)^2 = \frac{n-1}{2} \chi(1) > \chi(1)$$

and thus $(\mathscr{S}, \tau)$ is coherent by Theorem 7.14. ∎

The most common applications of coherence are in the situation of a "tamely imbedded" subgroup as defined in the paper of Feit and Thompson on the solvability of groups of odd order. For the purposes of this book, we give a considerably more restrictive definition which, nevertheless, is important in applications.

(7.16) DEFINITION Let $K \subseteq G$. Assume

(a) K is a T.I. set in G;
(b) $K < \mathbf{N}_G(K)$;
(c) $\mathbf{C}_G(k) \subseteq K$ for all $1 \neq k \in K$.

Then K is a *T.I.F.N.* subgroup of G.

The "F.N." in the foregoing definition stands for "Frobenius normalizer." Note that if $N = \mathbf{N}_G(K)$, where K is T.I.F.N. in G, then N is a Frobenius group and K is the Frobenius kernel. By Theorem 6.34, the irreducible characters of N are thus of two types, those with kernel containing K and those induced from nonprincipal $\varphi \in \operatorname{Irr}(K)$. If $\mathscr{S} = \{\chi \in \operatorname{Irr}(N) \mid K \nsubseteq \ker \chi\}$, then the object of what follows is to prove that $\mathscr{S}$ is coherent. Results of Feit and Sibley do, in fact, prove this in most cases. Note that if $K \subseteq G$ satisfies condition (a) of Definition 7.16, then (c) is equivalent to $\mathbf{C}_N(k) \subseteq K$ for all $1 \neq k \in K$, where $N = \mathbf{N}_G(K)$.

An important theorem of Thompson (not involving characters) asserts that Frobenius kernels are necessarily nilpotent. The nilpotence of T.I.F.N. subgroups will be assumed in what follows. (The reader may simply consider this to be an additional hypothesis in Definition 7.16.)

A more elementary fact about a T.I.F.N. subgroup K of G is that $(|G:K|, |K|) = 1$, that is, K is a Hall subgroup. To see this, let $p \mid |K|$, let $P \in \operatorname{Syl}_p(K)$ and suppose $P \subseteq S \in \operatorname{Syl}_p(G)$. Then $\mathbf{Z}(S) \subseteq \mathbf{C}(P) \subseteq K$ by part (c) of 7.16. Now $\mathbf{Z}(S) \neq 1$ and so $S \subseteq \mathbf{C}(\mathbf{Z}(S)) \subseteq K$, again by 7.16(c). Thus $p \nmid |G:K|$.

We mention some situations where T.I.F.N. subgroups arise. If $P \in \operatorname{Syl}_p(G)$, $|P| = p$ and $P = \mathbf{C}_G(P) < \mathbf{N}_G(P)$, then P is T.I.F.N. This happens, for instance in permutation groups of degree p and in the groups $\operatorname{PSL}(2, p)$.

A doubly transitive permutation group is a *Zassenhaus group* if some nonidentity element fixes two points but none fixes three. Let G be a Zassenhaus group on a set Ω and let $\alpha, \beta \in \Omega$ with $\alpha \neq \beta$. Then G_α is a Frobenius group with complement $G_{\alpha\beta}$. (We are assuming $|\Omega| > 2$ to avoid triviality.) Let K be the Frobenius kernel of G_α. The nonidentity elements of K are exactly those elements of G which fix α and no other point. It follows easily that K is T.I.F.N. in G and $G_\alpha = \mathbf{N}_G(K)$.

(7.17) LEMMA Let K be T.I.F.N. in G and let $N = \mathbf{N}_G(K)$ and $\mathscr{S} = \{\chi \in \mathrm{Irr}(N) \mid K \nsubseteq \ker \chi\}$. Then induction defines a linear isometry

$$\mathbb{Z}[\mathscr{S}]^\circ \to \mathbb{Z}[\mathrm{Irr}(G)]^\circ.$$

Also, if $\vartheta \in \mathbb{Z}[\mathscr{S}]^\circ$, then $(\vartheta^G)_N = \vartheta$.

Proof If $\chi \in \mathscr{S}$, then $\chi = \xi^N$ for some $\xi \in \mathrm{Irr}(K)$ by Theorem 6.34. Thus χ vanishes on $N - K$. If $\varphi, \vartheta \in \mathbb{Z}[\mathscr{S}]^\circ$, it follows that φ, ϑ vanish on $N - K$ and Lemma 7.7 yields $[\varphi^G, \vartheta^G] = [\varphi, \vartheta]$ and the first assertion is proved.

Also, $(\vartheta^G)_K = \vartheta_K$ by Lemma 7.7. Since ϑ vanishes $N - K$, it suffices to show that ϑ^G vanishes on $N - K$ in order to prove that $(\vartheta^G)_N = \vartheta$. However, no element of $N - K$ can be G-conjugate to an element of K since if $x \in N - K$ and $o(x) \mid |K|$, then $K < \langle K, x\rangle$ and $\langle K, x\rangle = K\langle x\rangle$ so that $|\langle K, x\rangle : K|$ is not relatively prime to $|K|$. This contradicts the fact that $(|G : K|, |K|) = 1$. It now follows from the definition of ϑ^G and the fact that ϑ vanishes on $N - K$ that ϑ^G also vanishes on $N - K$ and the proof is complete. ∎

(7.18) COROLLARY (*Brauer–Suzuki*) Let K be an abelian T.I.F.N. subgroup of G. Let $N = \mathbf{N}_G(K)$ and $e = |N : K|$. Let $\mathscr{S} = \{\chi \in \mathrm{Irr}(N) \mid K \nsubseteq \ker \chi\}$. Then either

(a) $|\mathscr{S}| = 1$, $|K| = 1 + e$ and K is an elementary abelian p-group or

(b) $\mathscr{S}$ is coherent and there exists a one-to-one function $f: \mathscr{S} \to \mathrm{Irr}(G)$ and $\varepsilon = \pm 1$ such that $(\chi - \xi)^G = \varepsilon(f(\chi) - f(\xi))$ for all $\chi, \xi \in \mathscr{S}$.

Proof All of the orbits of nonprincipal irreducible characters of K under the action of N have size e. Thus $\chi(1) = e$ for all $\chi \in \mathscr{S}$ and $|\mathscr{S}| = (|K| - 1)/e$. If $|\mathscr{S}| \geq 2$, then Corollary 7.15 yields that $\mathscr{S}$ is coherent. In this case, (b) follows from Lemma 7.13.

If $|\mathscr{S}| = 1$, then $|K| - 1 = e$ and the nonidentity elements of K are all conjugate in N. It follows that they all have the same prime order and (a) follows. ∎

The theorems of Feit and Sibley generalize Corollary 7.18 by dropping the assumption that K is abelian. If $|K| - 1 \neq e$, they prove that either $\mathscr{S}$ is coherent or K is a 2-group. Before proceeding with the proofs, we discuss some implications of coherence in the T.I.F.N. situation.

(7.19) LEMMA Let $K \subseteq G$ be T.I.F.N. and let $N = \mathbf{N}_G(K)$. Suppose $\mathscr{X} \subseteq \{\chi \in \mathrm{Irr}(N) \mid K \nsubseteq \ker \chi\}$ and that $\mathscr{X}$ is coherent. Let $*$ denote an isometry $\mathbb{Z}[\mathscr{X}] \to \mathbb{Z}[\mathrm{Irr}(G)]$ such that $*$ extends induction on $\mathbb{Z}[\mathscr{X}]^\circ$. Let $\alpha = \sum_{\chi \in \mathscr{X}} \chi(1)\chi$. Then

(a) If $\psi \in \mathrm{Irr}(G)$ and $\psi \notin \{\pm\chi^* \mid \chi \in \mathscr{X}\}$, then $\psi_N = a\alpha + \vartheta$, where a is rational and $[\vartheta, \chi] = 0$ for all $\chi \in \mathscr{X}$.

(b) $((\chi^*)_N - \chi)/\chi(1)$ is independent of $\chi \in \mathscr{X}$.

(c) $(\chi^*)_N - \chi = a\alpha + \vartheta$ for $\chi \in \mathscr{X}$, where a is rational and $[\vartheta, \xi] = 0$ for all $\xi \in \mathscr{X}$.

Proof (a) Let $\chi, \xi \in \mathscr{X}$. Then

$$\begin{aligned}[\psi_N, \chi(1)\xi - \xi(1)\chi] &= [\psi, (\chi(1)\xi - \xi(1)\chi)^G] \\ &= [\psi, \chi(1)\xi^* - \xi(1)\chi^*] = 0.\end{aligned}$$

Thus $\chi(1)[\psi_N, \xi] = \xi(1)[\psi_N, \chi]$ and hence $[\psi_N, \chi]/\chi(1) = a$ is independent of $\chi \in \mathscr{X}$. Therefore, $\psi_N = a\alpha + \vartheta$, where $[\vartheta, \chi] = 0$ for all $\chi \in \mathscr{X}$. Thus (a) is proved.

(b) Let $\chi, \xi \in \mathscr{X}$. Then $\Delta = \chi(1)\xi - \xi(1)\chi \in \mathbb{Z}[\mathscr{X}]^\circ$ and so $(\Delta^G)_N = \Delta$ by Lemma 7.17. However, $\Delta^G = \Delta^* = \chi(1)\xi^* - \xi(1)\chi^*$ and thus

$$\chi(1)\xi - \xi(1)\chi = \chi(1)(\xi^*)_N - \xi(1)(\chi^*)_N$$

and

$$\xi(1)((\chi^*)_N - \chi) = \chi(1)((\xi^*)_N - \xi).$$

Thus (b) follows.

(c) Let $\xi, \eta \in \mathscr{X}$ and write $\Delta = \xi(1)\eta - \eta(1)\xi$. Then

$$[(\chi^*)_N, \Delta] = [\chi^*, \Delta^G] = [\chi^*, \Delta^*] = [\chi, \Delta]$$

so that $[(\chi^*)_N - \chi, \Delta] = 0$. It now follows as in (a) that $(\chi^*)_N - \chi = a\alpha + \vartheta$, where $a = [(\chi^*)_N - \chi, \xi]/\xi(1)$ and $[\xi, \vartheta] = 0$ for $\xi \in \mathscr{X}$. ∎

(7.20) THEOREM Let $K \subseteq G$ be T.I.F.N. and let $N = \mathbf{N}_G(K)$. Suppose $\mathscr{S} = \{\chi \in \mathrm{Irr}(N) \mid K \nsubseteq \ker \chi\}$ is coherent and let $\mathscr{E} \subseteq \mathrm{Irr}(G)$ be the corresponding set of exceptional characters. Let $*$ be an isometry $\mathbb{Z}[\mathscr{S}] \to \mathbb{Z}[\mathscr{E}]$ which extends induction on $\mathbb{Z}[\mathscr{S}]^\circ$ and let $\varepsilon = \pm 1$ so that $\varepsilon\chi^* \in \mathscr{E}$ for $\chi \in \mathscr{S}$. Then

(a) $\mathscr{E} = \{\psi \in \mathrm{Irr}(G) \mid \psi$ is not constant on $K - \{1\}\}$.

(b) If $\psi = \varepsilon\chi^* \in \mathscr{E}$, then $\psi_K - \varepsilon\chi_K$ is constant on $K - \{1\}$. This constant has the form $m\chi(1)/|N:K|$ for some $m \in \mathbb{Z}$ which is independent of the choice of $\chi \in \mathscr{S}$.

(c) If $g \in G$ is not conjugate to an element of $K - \{1\}$, then $\psi(g)/\psi(1)$ is independent of the choice of $\psi \in \mathscr{E}$.

(d) If $\chi \in \mathscr{S}$, then $\chi^*(1)/\chi(1)$ is independent of the choice of χ.

Proof We apply Lemma 7.19 with $\mathscr{X} = \mathscr{S}$. Then $\alpha = \rho_N - \rho_{N/K}$ where ρ_N and $\rho_{N/K}$ are regular characters. In particular, α is constant on $K - \{1\}$. Also, if ϑ is a generalized character of N and $[\vartheta, \chi] = 0$, for all $\chi \in \mathscr{S}$, then K is in the kernel of every irreducible constituent of ϑ and hence ϑ_K is constant.

Now, if $\psi \notin \mathscr{E}$, then 7.19(a) yields $\psi_N = a\alpha + \vartheta$ and ψ is constant on $K - \{1\}$. If $\psi \in \mathscr{E}$, then 7.19(c) yields $\varepsilon\psi_N = \chi + a\alpha + \vartheta$ where $\varepsilon\psi = \chi^*$. However, $\chi \in \mathscr{S}$ cannot be constant on $K - \{1\}$, otherwise, by Problem 3.8, every nonprincipal irreducible character of K is a constituent of χ_K. This would force $|\mathscr{S}| = 1$ which is not the case. Thus ψ_N is not constant on $K - \{1\}$ and (a) is proved.

The first part of (b) is immediate from 7.19(c). To evaluate the constant, choose $\xi \in \mathscr{S}$ with $\xi(1) = |N:K|$, that is, $\xi = \lambda^N$ for some nonprincipal linear $\lambda \in \mathrm{Irr}(K)$. Now $(\xi^*)_K - \xi_K$ is a generalized character of K with constant value m on $K - \{1\}$. It follows that $m \in \mathbb{Z}$. If $\chi \in \mathscr{S}$ is arbitrary, 7.19(b) yields that $m'/\chi(1) = m/\xi(1)$ where m' is the constant value taken on by $(\chi^*)_K - \chi_K$ on $K - \{1\}$. Since $\xi(1) = |N:K|$, we have $m' = m\chi(1)/|N:K|$ and (b) follows.

Let $\chi, \xi \in \mathscr{S}$ so that $\xi(1)\chi^* - \chi(1)\xi^* = (\xi(1)\chi - \chi(1)\xi)^G$, which vanishes on elements $g \in G$ not conjugate to elements of $K - \{1\}$. Thus $\xi(1)\chi^*(g) - \chi(1)\xi^*(g) = 0$ and (c) follows.

Finally, (d) is immediate from 7.19(b). The proof is complete. ∎

We now begin work toward the theorems of Feit and Sibley.

(7.21) LEMMA Let N be a Frobenius group with Frobenius kernel K. Suppose $|N:K|$ is even. Then K is abelian.

Proof Let $t \in N$ be an involution. Then $t \notin K$ since $(|K|, N:K|) = 1$ and thus $x^t \neq x$ for $x \in K - \{1\}$. Now map $K \to K$ by $x \mapsto x^{-1}x^t$. This map is one-to-one since if $x^{-1}x^t = y^{-1}y^t$, then $yx^{-1} = (yx^{-1})^t$ and thus $yx^{-1} = 1$. Therefore the map is onto and every $x \in K$ has the form $u^{-1}u^t$. Thus $x^t = (u^{-1}u^t)^t = (u^t)^{-1}u = x^{-1}$. We now have $(xy)^t = (xy)^{-1} = y^{-1}x^{-1} = y^tx^t = (yx)^t$. Thus $xy = yx$ for $x, y \in K$ and the result follows. ∎

(7.22) LEMMA Let $P \in \mathrm{Syl}_p(G)$ with $p \neq 2$ and suppose P is T.I.F.N. in G and is nonabelian. Let $N = \mathbf{N}_G(P)$ and let $1 < Z \subseteq \mathbf{Z}(P)$ with $Z \lhd N$. Let $\mathscr{X} = \{\chi \in \mathrm{Irr}(N) \mid Z \nsubseteq \ker \chi\}$. Then $\mathscr{X}$ is coherent.

Note By Corollary 7.15, the lemma is also true if P is abelian except in the degenerate case that $|\mathscr{X}| = 1$. In that case $Z = P$ and $|P| = |N:P| + 1$.

Proof of Lemma 7.22 Let $e = |N:P|$. By Lemma 7.21, e is odd and thus $|N|$ is odd. Since $Z > 1$, $\mathscr{X} \neq \varnothing$. Let $\xi \in \mathscr{X}$ have minimum possible degree and let $\mathscr{X}_0 = \{\chi \in \mathscr{X} \mid \chi(1) = \xi(1)\}$. Since $|N|$ is odd, $\bar{\xi} \neq \xi$ and thus $|\mathscr{X}_0| \geq 2$. By Corollary 7.15, $\mathscr{X}_0$ is coherent. Now define sets $\mathscr{X}_i$ for $i > 0$ by $\mathscr{X}_i = \mathscr{X}_{i-1} \cup \{\chi_i\}$, where the χ_i are chosen from $\mathscr{X} - \mathscr{X}_0$ so that $\chi_i(1) \leq \chi_{i+1}(1)$. Thus $\mathscr{X}_0 \subseteq \mathscr{X}_1 \subseteq \cdots \subseteq \mathscr{X}_k = \mathscr{X}$. We use induction on i to show that $\mathscr{X}_i$ is coherent.

Now the degrees of the $\chi \in \mathscr{X}$ are all of the form ep^a and so $\xi(1) \mid \chi(1)$ for all $\chi \in \mathscr{X}$. It follows by Theorem 7.14 that in order to show that $\mathscr{X}_i$ is coherent, it suffices to show for $i \geq 1$ that

$$2\xi(1)\chi_i(1) < \sum_{\chi \in \mathscr{X}_{i-1}} \chi(1)^2.$$

We have

$$|N| = \sum_{\chi \in \mathrm{Irr}(N)} \chi(1)^2 = |N:Z| + \sum_{\chi \in \mathscr{X}} \chi(1)^2$$

and so $|P:Z|$ divides $\sum_{\chi \in \mathscr{X}} \chi(1)^2$. Thus

$$e^2|P:Z| \qquad \text{divides} \qquad \sum_{\chi \in \mathscr{X}} \chi(1)^2.$$

If $\chi \in \mathscr{X}$, then $\chi = \vartheta^N$ for some $\vartheta \in \mathrm{Irr}(P)$ and $\vartheta(1)^2 \leq |P:Z|$ by Corollary 2.30. Thus $\chi(1)^2 = e^2\vartheta(1)^2$ divides $e^2|P:Z|$ and hence $\chi_i(1)^2$ divides $\sum_{\chi \in \mathscr{X}} \chi(1)^2$. Also, $\chi_i(1)^2 \mid \chi_j(1)^2$ for $j \geq i$ and we conclude that $\chi_i(1)^2$ divides $\sum_{\chi \in \mathscr{X}_{i-1}} \chi(1)^2$. In particular, since $2\xi(1) < p\xi(1) \leq \chi_i(1)$, we have

$$2\xi(1)\chi_i(1) < \chi_i(1)^2 \leq \sum_{\chi \in \mathscr{X}_{i-1}} \chi(1)^2$$

and the result follows. ∎

(7.23) LEMMA Let $N \subseteq G$ and $\mathscr{S} \subseteq \mathrm{Irr}(N)$. Let $\alpha: \mathbb{Z}[\mathscr{S}]^\circ \to \mathbb{Z}[\mathrm{Irr}(G)]^\circ$ be a linear isometry and let $\mathscr{X}, \mathscr{Y} \subseteq \mathscr{S}$ with $\mathscr{X} \cap \mathscr{Y} = \varnothing$ and $(\mathscr{X}, \alpha)$ and $(\mathscr{Y}, \alpha)$ coherent. Let σ and τ be isometries on $\mathbb{Z}[\mathscr{X}]$ and $\mathbb{Z}[\mathscr{Y}]$, respectively, where σ and τ are as implied by coherence. Then

(a) $[\chi^\sigma, \eta^\tau] = 0$ for all $\chi \in \mathscr{X}$ and $\eta \in \mathscr{Y}$.
(b) Suppose $\chi_0 \in \mathscr{X}$ and $\eta_0 \in \mathscr{Y}$ with

$$(\chi_0(1)\eta_0 - \eta_0(1)\chi_0)^\alpha = \chi_0(1)\eta_0{}^\tau - \eta_0(1)\chi_0{}^\sigma.$$

Then $\mathscr{X} \cup \mathscr{Y}$ is coherent.

Proof For $\mu, \nu \in \mathscr{S}$, write $\Delta(\mu, \nu) = \mu(1)\nu - \nu(1)\mu \in \mathbb{Z}[\mathscr{S}]^\circ$. Let $\delta, \varepsilon = \pm 1$ so that $\delta\chi^\sigma$ and $\varepsilon\eta^\tau \in \mathrm{Irr}(G)$ for $\chi \in \mathscr{X}$ and $\eta \in \mathscr{Y}$. To prove (a), suppose $[\chi^\sigma, \eta^\tau] \neq 0$. Then $\delta\chi^\sigma = \vartheta = \varepsilon\eta^\tau$ for some $\vartheta \in \mathrm{Irr}(G)$. Pick $\chi_1 \in \mathscr{X} - \{\chi\}$ and $\eta_1 \in \mathscr{Y} - \{\eta\}$ and write $\psi = \delta\chi_1{}^\sigma$ and $\xi = \varepsilon\eta_1{}^\tau$. Then $\psi \neq \vartheta \neq \xi$ and we have

$$0 = [\Delta(\chi, \chi_1), \Delta(\eta, \eta_1)] = [\Delta(\chi, \chi_1)^\alpha, \Delta(\eta, \eta_1)^\alpha]$$
$$= \varepsilon\delta[(\chi(1)\psi - \chi_1(1)\theta), (\eta(1)\xi - \eta_1(1)\theta)]$$
$$= \varepsilon\delta(\chi_1(1)\eta_1(1) + \chi(1)\eta(1)[\psi, \xi]).$$

Since $[\psi, \xi] \geq 0$, this is impossible, and (a) is proved.

Now fix $\chi_0 \in \mathscr{X}$ and $\eta_0 \in \mathscr{Y}$ and assume $\Delta(\chi_0, \eta_0)^\alpha = \chi_0(1)\eta_0{}^\tau - \eta_0(1)\chi_0{}^\sigma$. Define $*$ on $\mathscr{X} \cup \mathscr{Y}$ by $\chi^* = \chi^\sigma$ for $\chi \in \mathscr{X}$ and $\eta^* = \eta^\tau$ for $\eta \in \mathscr{Y}$ and extend $*$ by linearity to $\mathbb{Q}[\mathscr{X} \cup \mathscr{Y}]$. It follows from (a) and the fact that σ and τ are isometries, that $*$ is an isometry and $*: \mathbb{Z}[\mathscr{X} \cup \mathscr{Y}] \to \mathbb{Z}[\mathrm{Irr}(G)]$. It thus suffices to show that $*$ agrees with the linear extension of α to $\mathbb{Q}[\mathscr{X} \cup \mathscr{Y}]^\circ$. However, $*$ agrees with α on $\mathbb{Q}[\mathscr{X}]^\circ$, $\mathbb{Q}[\mathscr{Y}]^\circ$ and $\Delta(\chi_0, \eta_0)$. A dimension count shows that these span $\mathbb{Q}[\mathscr{X} \cup \mathscr{Y}]^\circ$ over $\mathbb{Q}$ and the result follows. ▌

(7.24) THEOREM (*Sibley*) Let $P \in \mathrm{Syl}_p(G)$ be T.I.F.N. in G with $p \neq 2$. Let $N = \mathbf{N}_G(P)$ and $\mathscr{S} = \{\chi \in \mathrm{Irr}(N) \mid P \nsubseteq \ker \chi\}$. Let $e = |N : P|$. Then one of the following occurs.

(a) $|\mathscr{S}| = 1$, P is elementary abelian, and $|P| = e + 1$.
(b) $\mathscr{S}$ is coherent.

Proof If P is abelian, this is included in Corollary 7.18. Assume that P is nonabelian so that e is odd by Lemma 7.21, and thus $|N|$ is odd. Let $Z = P' \cap \mathbf{Z}(P) > 1$ and let $\mathscr{Y} = \{\eta \in \mathscr{S} \mid Z \nsubseteq \ker \eta\}$. Let $\mathscr{X} = \{\chi \in \mathscr{S} \mid P' \subseteq \ker \chi\}$. Then $\mathscr{Y}$ is coherent by Lemma 7.22 and every $\chi \in \mathscr{X}$ satisfies $\chi(1) = e$. Since $\chi \neq \bar{\chi}$ for $\chi \in \mathscr{X}$, we have $\mathscr{X}$ coherent by Corollary 7.15. Since $Z \subseteq P'$, we have $\mathscr{X} \cap \mathscr{Y} = \varnothing$. Most of the proof is devoted to showing that $\mathscr{X} \cup \mathscr{Y}$ is coherent.

Let $*: \mathbb{Z}[\mathscr{X}] \to \mathbb{Z}[\mathrm{Irr}(G)]$ be a linear isometry which extends induction on $\mathbb{Z}[\mathscr{X}]^\circ$. Similarly, let τ be an isometry on $\mathbb{Z}[\mathscr{Y}]$ which extends induction. Since $\mathscr{X} \cap \mathscr{Y} = \varnothing$, Lemma 7.23 yields $[\chi^*, \eta^\tau] = 0$ for $\chi \in \mathscr{X}$ and $\eta \in \mathscr{Y}$.

Let $\alpha = (1/e) \sum_{\eta \in \mathscr{Y}} \eta(1)\eta$. Since $e \mid \eta(1)$ for $\eta \in \mathscr{Y}$, α is a character of N. For $\chi \in \mathscr{X}$, $\chi^* \neq \pm\eta^\tau$ for any $\eta \in \mathscr{Y}$ and Lemma 7.19(a) yields

$$(\chi^*)_N = \vartheta_\chi + a_\chi \alpha,$$

where ϑ_χ is a generalized character of N such that $[\vartheta_\chi, \eta] = 0$ for $\eta \in \mathscr{Y}$ and $a_\chi \in \mathbb{Q}$.

Since all $\chi(1)$ are equal for $\chi \in \mathscr{X}$, Lemma 7.19(b) yields that

$$(\chi^*)_N - \chi = \vartheta_\chi - \chi + a_\chi \alpha$$

is independent of $\chi \in \mathscr{X}$. It follows that $\vartheta_\chi - \chi = \Delta$ and $a_\chi = a$ are independent of $\chi \in \mathscr{X}$ and

(1) $$(\chi^*)_N = \chi + \Delta + a\alpha.$$

The next several paragraphs are devoted to proving that $a \in \mathbb{Z}$.

Since Z is contained in the kernel of $\chi \in \mathscr{X}$ and of every irreducible constituent of Δ, we have for $z \in Z$ that $\chi(z) = \chi(1)$ and $\Delta(z) = \Delta(1)$ and hence Equation (1) yields

$$\chi^*(1) - \chi^*(z) = a(\alpha(1) - \alpha(z)).$$

We also have $\rho_N = \rho_{N/Z} + e\alpha$ where ρ_N and $\rho_{N/Z}$ are regular characters. For $z \in Z - \{1\}$ we thus have

$$|N| = \rho_N(1) - \rho_N(z) = e(\alpha(1) - \alpha(z))$$

and hence

(2) $$\chi^*(1) - \chi^*(z) = a|N|/e = a|P|.$$

We now compare $\chi^*(1)$ and $\chi^*(z)$ in a different way. Let $\mathscr{K}_0, \mathscr{K}_1, \ldots$ be the conjugacy classes of G, numbered so that

(a) $\mathscr{K}_0 = \{1\}$;
(b) $\mathscr{K}_i \cap Z \neq \varnothing$ iff $i \leq r$;
(c) $\mathscr{K}_i \cap P \neq \varnothing$ iff $i \leq s$.

Note that since P is a T.I. set with $N = \mathbf{N}(P)$, we have $\mathscr{K}_i \cap P$ is a conjugacy class of N for $i \leq s$ and $\mathscr{K}_i \cap P \subseteq Z \lhd N$ for $i \leq r$.

Write $K_i \in \mathbb{C}[G]$ for the class sum corresponding to $\mathscr{K}_i$ and $K_i K_j = \sum a_{ij\mu} K_\mu$, where $a_{ij\mu} \in \mathbb{Z}$. Now fix $i, j \leq r$. Let $\psi \in \operatorname{Irr}(G)$ and let $\omega(K_\mu) = \psi(x)|\mathscr{K}_\mu|/\psi(1)$ for $x \in \mathscr{K}_\mu$ as in Chapter 3. Let $R \subseteq \mathbb{C}$ be the ring of algebraic integers so that $\omega(K_\mu) \in R$. Write $\psi(1) = mp^t$ with $p \nmid m$, and let $q = |P|/p^t$. We have

$$\omega(K_i)\omega(K_j) = \sum_\mu a_{ij\mu}\omega(K_\mu).$$

For $\mu > s$, we claim that $\omega(K_\mu) \in qR$. From the T.I.F.N. property and the fact that $\mathscr{K}_\mu \cap P = \varnothing$, it follows that $|P|$ divides $|\mathscr{K}_\mu|$ and thus $m(\omega(K_\mu)/q) \in R$. Since also $q(\omega(K_\mu)/q) \in R$ and $(m, q) = 1$, it follows that $\omega(K_\mu)/q \in R$ as claimed. Thus

$$\omega(K_i)\omega(K_j) \equiv \sum_{\mu=0}^{s} a_{ij\mu}\omega(K_\mu) \bmod qR.$$

Now let $r < \mu \leq s$ and $x \in \mathscr{K}_\mu \cap P$. Let $C = \mathbf{C}_P(x)$ and let $\Omega = \{(u, v) | u \in \mathscr{K}_i,\ v \in \mathscr{K}_j,\ uv = x\}$. Then $a_{ij\mu} = |\Omega|$ and C acts on Ω by $(u, v)^c = (u^c, v^c)$. If $c \in C - \{1\}$ and $(u, v)^c = (u, v)$, then $u, v \in \mathbf{C}_G(c) \subseteq P$ and so $u \in \mathscr{K}_i \cap P \subseteq Z$ and similarly $v \in Z$. However $uv = x \in \mathscr{K}_\mu$ and $\mathscr{K}_\mu \cap Z = \varnothing$. This contradiction shows that all orbits of C on Ω have size $|C|$ and thus $|C|$ divides $a_{ij\mu}$.

We have $\mathbf{C}_G(x) \subseteq P$ and so $\mathbf{C}_G(x) = C$ and $|P:C|$ divides $|\mathscr{K}_\mu|$. It now follows that $m(a_{ij\mu}\omega(K_\mu)/q) \in R$ and since $q(a_{ij\mu}\omega(K_\mu)/q) \in R$, we have $a_{ij\mu}\omega(K_\mu) \in qR$ and

$$\omega(K_i)\omega(K_j) \equiv \sum_{\mu=0}^{r} a_{ij\mu}\omega(K_\mu) \bmod qR.$$

Now suppose $\psi \in \mathrm{Irr}(G)$ is such that ψ is constant on $Z - \{1\}$. Then all $\omega(K_\mu)$ are equal (to ω, say) for $1 \le \mu \le r$ and $\omega(K_0) = 1$. Also write $a_{ij} = \sum_{\mu=1}^{r} a_{ij\mu}$, so that

$$\omega^2 \equiv a_{ij0} + a_{ij}\omega \bmod qR.$$

Since $|N|$ is odd, the nonidentity elements of Z are not conjugate in N (and hence not in G either) to their inverses. Thus $a_{110} = 0$. Assume $\mathscr{K}_2$ is the class of inverses of $\mathscr{K}_1$ so that $a_{120} = |\mathscr{K}_1| = |G|/|P|$. We thus have

(3) $$a_{11}\omega \equiv \omega^2 \equiv |G|/|P| + a_{12}\omega \bmod qR.$$

In the special case that $\psi = 1_G$, we have $\omega = |G|/|P|$ and $q = |P|$ so that

$$a_{11}|G|/|P| \equiv |G|/|P| + a_{12}|G|/|P| \bmod |P|$$

and $a_{11} \equiv 1 + a_{12} \bmod |P|$. Now (3) yields

$$(1 + a_{12})\omega \equiv |G|/|P| + a_{12}\omega \bmod qR$$

and

(4) $$\omega \equiv |G|/|P| \bmod qR.$$

We apply this to the character $\psi = \varepsilon\chi^* \in \mathrm{Irr}(G)$ with $\varepsilon = \pm 1$ and $\chi \in \mathscr{X}$. Note that ψ is constant on $Z - \{1\}$ by (2) and thus (4) applies.

We have for $z \in Z - \{1\}$ that

$$\chi^*(z)|G:P|/\chi^*(1) = \omega \equiv |G:P| \bmod qR.$$

Since $|P|$ divides $q\chi^*(1)$, this yields

$$\chi^*(z)|G:P| \equiv \chi^*(1)|G:P| \bmod |P|R.$$

Now (2) yields

$$|G:P| \cdot |P|a = |G:P|(\chi^*(1) - \chi^*(z)) \in |P|R$$

and hence $|G:P|a \in R$. Since also $|P|a = \chi^*(1) - \chi^*(z) \in R$ and $(|G:P|, |P|) = 1$, we have $a \in R \cap \mathbb{Q} = \mathbb{Z}$ as desired.

Now let $\chi, \chi_1 \in \mathscr{X}$ and $\eta \in \mathscr{Y}$. Note that $\chi(1) = e$ divides $\eta(1)$ and we write $c = \eta(1)/e$. Let $\varphi = c\chi - \eta \in \mathbb{Z}[\mathscr{S}]^\circ$. We have

$$[\varphi^G, \chi_1{}^*] = [\varphi, (\chi_1{}^*)_N] = [\varphi, \chi_1 + \Delta + a\alpha].$$

Since $[\chi, \alpha] = 0 = [\eta, \chi_1 + \Delta]$, this yields

$$[\varphi^G, \chi_1{}^*] = c[\chi, \chi_1 + \Delta] - a[\eta, \alpha].$$

However, $[\eta, \alpha] = c$ by definition of α and since $a \in \mathbb{Z}$, we have $c \mid [\varphi^G, \chi_1{}^*]$. This calculation also shows that

$$[\varphi^G, \chi^*] = c + [\varphi^G, \chi_1{}^*]$$

for $\chi_1 \neq \chi$. This yields

$$\varphi^G = c(1 + b)\chi^* + cb \sum_{\xi \in \mathscr{X};\, \xi \neq \chi} \xi^* - \Gamma,$$

where $b \in \mathbb{Z}$ and $[\Gamma, \xi^*] = 0$ for all $\xi \in \mathscr{X}$.

We also have

$$[\varphi^G, \varphi^G] = [\varphi, \varphi] = 1 + c^2$$

and thus there are two possibilities:

(i) $b = 0$ and $[\Gamma, \Gamma] = 1$; or
(ii) $b = -1$, $|\mathscr{X}| = 2$ and $[\Gamma, \Gamma] = 1$.

In situation (ii), we can replace $*$ by $**$ where $\chi_1^{**} = -\chi_2{}^*$ and $\chi_2^{**} = -\chi_1{}^*$ for $\mathscr{X} = \{\chi_1, \chi_2\}$. The result of this change is to put us into situation (i). We thus assume that $(c\chi - \eta)^G = c\chi^* - \Gamma$.

Now let $\eta_1 \in \mathscr{Y} - \{\eta\}$ so that

$$\begin{aligned}\eta_1(1) = [\varphi, \eta(1)\eta_1 - \eta_1(1)\eta] &= [\varphi^G, \eta(1)\eta_1{}^\tau - \eta_1(1)\eta^\tau] \\ &= [\Gamma, \eta_1(1)\eta^\tau - \eta(1)\eta_1{}^\tau].\end{aligned}$$

Thus again we have two possibilities:

(i) $\Gamma = \eta^\tau$; or
(ii) $\Gamma = -\eta_1{}^\tau$ and $\eta(1) = \eta_1(1)$.

If $\Gamma \neq \eta^\tau$, then $\eta_1{}^\tau = -\Gamma$ for every $\eta_1 \in \mathscr{Y} - \{\eta\}$ and thus $|\mathscr{Y}| = 2$ and $\mathscr{Y} = \{\eta, \eta_1\}$ with $\eta(1) = \eta_1(1)$. In this situation, we can redefine τ and thus we may assume that (i) occurs. Thus

$$(c\chi - \eta)^G = c\chi^* - \eta^\tau$$

and hence $\mathscr{X} \cup \mathscr{Y}$ is coherent by Lemma 7.23(b).

To complete the proof, we observe that $\sum_{\eta \in \mathscr{Y}} \eta(1)^2 = |N| - |N:Z|$ and so is divisible by $|P:Z|$ and hence by $e^2|P:Z|$. Let $\psi \in \mathscr{S} - (\mathscr{X} \cup \mathscr{Y})$. Then as in Lemma 7.22, we have $\psi = \vartheta^N$, with $\vartheta \in \mathrm{Irr}(P)$ and $\vartheta(1)^2 \leq |P:Z|$. Thus

$$\psi(1)^2 = e^2\vartheta(1)^2 \leq e^2|P:Z| \leq \sum_{\xi \in \mathscr{Y}} \xi(1)^2.$$

Let $\chi \in \mathscr{X}$ so that $\chi(1) = e$ and $\psi(1) \geq p\chi(1) > 2\chi(1)$ since $\psi \notin \mathscr{X}$ and so $\vartheta(1) \geq p$. Thus

$$2\chi(1)\psi(1) < \psi(1)^2 \leq \sum_{\xi \in \mathscr{X} \cup \mathscr{Y}} \xi(1)^2.$$

Since $\mathscr{X} \cup \mathscr{Y}$ is coherent and $\chi(1) | \psi(1)$ for all $\psi \in \mathscr{S}$, repeated application of Theorem 7.14 yields that $\mathscr{S}$ is coherent and the proof is complete. ∎

(7.25) THEOREM (*Feit–Sibley*) Let $K \subseteq G$ be T.I.F.N. with $N = \mathbf{N}_G(K)$, $e = |N:K|$ and $\mathscr{S} = \{\chi \in \mathrm{Irr}(N) | K \nsubseteq \ker \chi\}$. Then one of the following occurs.

(a) $|\mathscr{S}| = 1$, $|K| = e + 1$ and K is an elementary abelian p-group.
(b) $K \in \mathrm{Syl}_2(G)$ and $|K:K'| < 4e^2$.
(c) $\mathscr{S}$ is coherent.

Proof If K is abelian, either (a) or (c) occurs by Corollary 7.18. Assume then, that K is nonabelian and so e is odd. If $|K:K'| = e + 1$, then K/K' is an elementary abelian p-group and thus K is a p-group. Since $e + 1$ is even $p = 2$ and (b) follows. Assume therefore, that $|\{\chi \in \mathscr{S} | K' \subseteq \ker \chi\}| \geq 2$. This set is thus coherent by Corollary 7.15. Let $L \lhd N$ with $L \subseteq K'$ minimal such that $\mathscr{X} = \{\chi \in \mathscr{S} | L \subseteq \ker \chi\}$ is coherent and assume that $\mathscr{S}$ is not coherent so that $L > 1$.

Let $M \lhd N$ be such that L/M is a chief factor of N. Since

$$\{\chi \in \mathscr{S} | M \subseteq \ker \chi\}$$

is not coherent, repeated application of Theorem 7.14 yields $\psi \in \mathrm{Irr}(N)$ such that

$$2e\psi(1) \geq \sum_{\chi \in \mathscr{X}} \chi(1)^2 = |N:L| - |N:K| = e(|K:L| - 1)$$

and $M \subseteq \ker \psi$. Let $Z/M = \mathbf{Z}(K/M)$. Then $Z \cap L > M$ and hence $L \subseteq Z$. Also $\psi = \vartheta^N$ for some $\vartheta \in \mathrm{Irr}(K)$ and $\vartheta(1)^2 \leq |K:Z|$ so that $\psi(1)^2 \leq e^2|K:Z|$. Thus

$$4e^4|K:Z| \geq 4e^2\psi(1)^2 \geq e^2(|K:L| - 1)^2.$$

Now write $a = |K:L|$ and $b = |Z:L|$.We have $b|K:Z| = |K:L| = a$ and

$$4e^2a = 4e^2b|K:Z| \geq b(|K:L| - 1)^2 = b(a - 1)^2.$$

If $4e^2 \leq b(a - 2)$, this would yield $ba(a - 2) \geq b(a - 1)^2$, which is not the case. Thus $b(a - 2) < 4e^2$.

If $Z = L$, then $\mathbf{Z}(K/M) = L/M$ is a p-group for some prime and thus K/M is a p-group. Since $M \subseteq K'$, it follows that K is a p-group. By Theorem 7.24, we have $p = 2$. Also, $a - 2 < 4e^2$ so that

$$|K:K'| \leq |K:L| = a \leq 4e^2 + 1.$$

Since $|K:K'|$ is a power of 2 and $e > 1$ is odd, we have $|K:K'| < 4e^2$ and (b) holds.

If $Z > L$, then we must have $e|(|Z:L| - 1)$ and $b \geq e + 1$. Also, K/M is nonabelian and thus $Z < K$ and $e|(|K:Z| - 1)$. Thus

$$a \geq (e + 1)b \geq (e + 1)^2 \geq e^2 + 2.$$

Thus $4e^2 > b(a - 2) \geq (e + 1)e^2 \geq 4e^2$ since $e \geq 3$. This contradiction proves the theorem. ∎

We remark that situation (b) can actually occur with $\mathscr{S}$ not coherent. The inequality $|K:K'| < 4e^2$ can be sharpened. In fact, the groups in which (b) occurs have been classified.

Problems

(7.1) Let $N \lhd G$, $H \subseteq G$, with $NH = G$ and $N \cap H = 1$. Show that the following are equivalent:

(a) $\mathbf{C}_G(n) \subseteq N$ for all $1 \neq n \in N$;
(b) $\mathbf{C}_H(n) = 1$ for all $1 \neq n \in N$;
(c) $\mathbf{C}_G(h) \subseteq H$ for all $1 \neq h \in H$;
(d) Every $x \in G - N$ is conjugate to an element of H;
(e) If $1 \neq h \in H$, then h is conjugate to every element of Nh.
(f) H is a Frobenius complement in G.

Note Problem 7.1 does not involve characters. It is included to acquaint the reader with some elementary properties of Frobenius groups. Much deeper information is known.

(7.2) (*Wielandt*) Let $H \subseteq G$ with $M \lhd H$ and suppose that $H \cap H^x \subseteq M$ whenever $x \notin H$. Show that there exists $N \lhd G$ with $NH = G$ and $N \cap H = M$.

Hint Note that $H - M$ is a T.I. set. Mimic the proof of Frobenius' theorem.

(7.3) Let $H \subseteq G$ and $\xi \in \text{Irr}(H)$. Suppose $(\xi - \xi(1)1_H)^G = \vartheta$ and $[\vartheta, \vartheta] = 1 + \xi(1)^2$. Show that there exists $N \lhd G$ with $N \cap H = \ker \xi$ and every $x \in G - N$ conjugate to some element of H.

(7.4) Let $H < G$ and suppose induction to G is an isometry on $\mathbb{Z}[\text{Irr}(H)]^\circ$. Show that H is a Frobenius complement in G.

(7.5) Let $N \subseteq G$ and $\mathscr{S} \subseteq \text{Irr}(N)$. Assume that induction to G is an isometry on $\mathbb{Z}[\mathscr{S}]^\circ$ and $|\mathscr{S}| \geq 2$. Suppose $\mathscr{S}_0 = \{\xi \in \mathscr{S} \mid \xi \text{ is extendible to } G\} \neq \varnothing$.

Let $d = \text{g.c.d.}(\{\xi(1) \mid \xi \in \mathscr{S}_0\})$ and assume $d \mid \xi(1)$ for all $\xi \in \mathscr{S}$. Show that $\mathscr{S}_0 = \mathscr{S}$, $\mathscr{S}$ is coherent, and that every $\xi \in \mathscr{S}$ has a *unique* extension to G.

(7.6) Let $P \in \text{Syl}_p(G)$ with $|P| = p$, $\mathbf{C}_G(P) = P$, and $|\mathbf{N}_G(P) : P| = e$. Show that G has at most $e + (p-1)/e$ irreducible characters with degree not divisible by p.

Hints Note that P is T.I.F.N. in G if $e > 1$. If $e < p - 1$, let $\mathscr{E}$ be the corresponding set of exceptional characters of G. Show $\sum_{\chi \in \mathscr{E}} |\chi(x)|^2 \geq p - e$, where $1 \neq x \in P$. If $\chi(x) = 0$, then $p \mid \chi(1)$.

Note It is a consequence of some of R. Brauer's deep results in block theory that in the situation of Problem 7.6, the upper bound is always attained.

(7.7) In the situation of the preceding problem, assume that G has exactly $e + (p-1)/e$ irreducible characters of p'-degree. Let $P = \langle x \rangle$. If $\chi \in \text{Irr}(G)$ is not exceptional (in particular if $e = p - 1$), show that $\chi(x) = -1, 0$, or 1 and that $\chi(1) \equiv \chi(x) \bmod p$. If $e < p - 1$ and $\mathscr{E}$ is the corresponding set of exceptional characters, let $\sigma = \sum_{\chi \in \mathscr{E}} \chi(x)$. Show that $\sigma = \pm 1$ and $\chi(1) \equiv -e\sigma \bmod p$ for $\chi \in \mathscr{E}$.

Note The results of Problem 7.7 together with the equation

$$\sum_{\chi \in \text{Irr}(G)} \chi(1)\chi(x) = 0$$

and the facts that $\chi(1) \mid |G|$ and $\sum \chi(1)^2 = |G|$ provide a powerful tool for computing the degrees of the irreducible characters of a group which satisfies the hypotheses of Problem 7.6.

(7.8) In the situation of Lemma 7.13, assume that $|\mathscr{S}| > 2$. Prove that $*$, ε, and f are uniquely determined by $(\mathscr{S}, \tau)$.

(7.9) Let $K \subseteq G$ be T.I.F.N. Let χ be a (possibly reducible) character of G which is constant on $K - \{1\}$. Assume that $\mathscr{S} = \{\psi \in \text{Irr}(\mathbf{N}_G(K)) \mid K \nsubseteq \ker \psi\}$ is coherent. Show that the multiplicity with which each exceptional character appears in χ is proportional to its degree.

(7.10) Let K be T.I.F.N. in G and assume that

$$\mathscr{S} = \{\psi \in \text{Irr}(\mathbf{N}(K)) \mid K \nsubseteq \ker \psi\}$$

is coherent. Let $\mathscr{T}$ be the corresponding set of exceptional characters of G. Let $M = \bigcap_{\chi \in \mathscr{T}} \ker \chi$. Show that $M \cap K = 1$.

(7.11) In the situation of Problem 7.10, show that either $M = \ker \chi$ for all $\chi \in \mathscr{T}$ or else $MK \lhd G$.

Hint Suppose $\chi \in \mathcal{T}$ and $U = \ker \chi > M$. Then U/M is a π-group where π is the set of prime divisors of $|K|$ and so $U \subseteq MK$. Use the Frattini argument.

(7.12) Let K be T.I.F.N. in G and let $e = |\mathbf{N}(K):K|$. Suppose that G has a faithful irreducible character of degree $<2e$. Show that K is abelian.

(7.13) In the situation of Problem 7.12, assume that $K \ntriangleleft G$. Show that K is an elementary abelian p-group.

Hint Let λ be a linear character of K with $|K:\ker \lambda| = p$. Show that there exists faithful $\chi \in \mathrm{Irr}(G)$ with $\chi(1) < 2e$ and $[\chi_K, \lambda] \neq 0$. Use Problem 7.11.

(7.14) Let G be a simple Zassenhaus group in which the stabilizer of two points has even order e. Let the degree of G (the number of points permuted) be $k + 1$. Show that k is a prime power.

Hint Count involutions to prove that G has a nonprincipal irreducible character of degree $<2e$.

Note Problem 7.14 is true without the assumption that e is even. This is a theorem of Feit and the case where e is odd is proved in his book.

(7.15) Let $N \subseteq G$ and $\mathcal{S} \subseteq \mathrm{Irr}(N)$. Suppose that $\tau: \mathbb{Z}[\mathcal{S}]^\circ \to \mathbb{Z}[\mathrm{Irr}(G)]^\circ$ is a linear isometry. Assume that $|\mathcal{S}| \geq 3$ and that for every $\chi, \psi \in \mathcal{S}$, we have $(\chi(1)\psi - \psi(1)\chi)^\tau = a\vartheta - b\varphi$ for some $\varphi, \vartheta \in \mathrm{Irr}(G)$ with $a, b \in \mathbb{Z}$. Show that $(\mathcal{S}, \tau)$ is coherent.

Hint Suppose $\chi, \psi, \eta \in \mathcal{S}$ are distinct. Write

$$(\chi(1)\psi - \psi(1)\chi)^\tau = a\vartheta - b\varphi, \qquad (\eta(1)\psi - \psi(1)\eta)^\tau = c\mu - d\nu$$

with $a, b, c, d > 0$. Then exactly one of $\vartheta = \mu$ or $\varphi = \nu$ holds.

(7.16) Let $N \subseteq G$, $\mathcal{S} \subseteq \mathrm{Irr}(N)$ and suppose $\tau: \mathbb{Z}[\mathcal{S}]^\circ \to \mathbb{Z}[\mathrm{Irr}(G)]^\circ$ is a linear isometry. Let $\mathcal{S}_1, \mathcal{S}_2 \subseteq \mathcal{S}$ such that $\mathcal{S}_1 \cap \mathcal{S}_2 \neq \varnothing$. Assume either that $|\mathcal{S}_1| \geq 3$ or that $\mathcal{S}_1 = \{\psi, \chi\}$ with $\psi(1) = \chi(1)$. Show that $(\mathcal{S}_1 \cup \mathcal{S}_2, \tau)$ is coherent.

Hint Let $f_i: \mathcal{S}_i \to \mathrm{Irr}(G)$ and $\varepsilon_i = \pm 1$ be such that $\varepsilon_i f_i$ defines an appropriate extension of τ on $\mathbb{Z}[\mathcal{S}_i]$. If $|\mathcal{S}_i| \geq 3$ for $i = 1, 2$, use the hint for Problem 7.15 to show that $\varepsilon_1 = \varepsilon_2$ and $f_1(\chi) = f_2(\chi)$ for $\chi \in \mathcal{S}_1 \cap \mathcal{S}_2$.

(7.17) Let $K \subseteq G$ be T.I.F.N. with $N = \mathbf{N}_G(K)$ and

$$\mathcal{S} = \{\chi \in \mathrm{Irr}(N) \mid K \nsubseteq \ker \chi\}.$$

Suppose that ψ is an exceptional character of G corresponding to χ. Show that the values of ψ and χ generate the same field over $\mathbb{Q}$.

(7.18) Let $K_1, K_2 \subseteq G$ be T.I.F.N. with $N_i = \mathbf{N}_G(K_i)$ and

$$\mathscr{S}_i = \{\chi \in \mathrm{Irr}(N_i) \mid K_i \not\subseteq \ker \chi\}$$

and assume that the $\mathscr{S}_i$ are coherent with $\mathscr{E}_i$ the corresponding sets of exceptional characters. Suppose $\mathscr{E}_1 \cap \mathscr{E}_2 \neq \varnothing$. Show that $\mathscr{E}_1 = \mathscr{E}_2$ and K_1 is conjugate to K_2.

Hint To show that $K_2 = K_1{}^g$, it suffices to show that $x^g \in K_2$ for some $x \in K_1$, $x \neq 1$. Use the fact that the K_i are nilpotent to prove this.

(7.19) Suppose $|G|$ is odd and that $\mathbf{C}_G(x)$ is abelian for every $1 \neq x \in G$. Assume that G is neither abelian nor a Frobenius group and show that every nonprincipal $\chi \in \mathrm{Irr}(G)$ is exceptional for some T.I.F.N. subgroup.

Note A group with all $\mathbf{C}(x)$ abelian for $1 \neq x \in G$ is called a CA-group. Problem 7.19 is a step in M. Suzuki's determination of all simple CA-groups.

(7.20) Let $K \subseteq G$ be T.I.F.N. with $|K|$ even. Let $N = \mathbf{N}_G(K) < G$ and $\mathscr{S} = \{\psi \in \mathrm{Irr}(N) \mid K \not\subseteq \ker \psi\}$. Assume that $\mathscr{S}$ is coherent. By carrying out the following steps, show that G has a nontrivial normal subgroup of odd order.

(a) Show that G has a unique conjugacy class of involutions and that N contains exactly $e = |N : K|$ involutions.

(b) Let $\psi \in \mathscr{S}$ and let ψ^* be the corresponding exceptional character of G. For $1 \neq x \in K$ we have $\psi^*(x) = \varepsilon\psi(x) + a_\psi$, where $a_\psi \in \mathbb{Z}$ is independent of x. Show that

$$\nu_2(\psi^*) = \varepsilon\nu_2(\psi) + (\psi^*(1) - \varepsilon\psi(1) - a_\psi)/|K|$$

where ν_2 is as in Theorem 4.5.

(c) Show that $(\psi^*)_K = \psi_K + a_\psi 1_K$.

(d) Show that there exists $\psi \in \mathscr{S}$ with $\ker(\psi^*) \neq 1$.

(e) Complete the proof.

Hints (a) See the hint for Problem 4.11.

(b) Let $\mathscr{A}_1 = \{x \in G \mid x \neq 1, x^2 = 1\}$;

$\mathscr{A}_2 = \{x \in G \mid x^2 \neq 1, x \text{ is conjugate to an element of } K\}$;

$\mathscr{A}_3 = G - (\mathscr{A}_1 \cup \mathscr{A}_2)$.

Compute $\sum \psi^*(x^2)$ separately on each of $\mathscr{A}_1$, $\mathscr{A}_2$, and $\mathscr{A}_3$. Note that $\sum_{x \in \mathscr{A}_3} \psi^*(x^2) = \sum_{x \in \mathscr{A}_3} \psi^*(x)$.

(c) Let $m_\psi = (\psi^*(1) - \varepsilon\psi(1) - a_\psi)/|K|$.

Use (b) to study the behavior of m_ψ as ψ varies over $\mathscr{S}$ and conclude that $m_\psi = 0$. Use the fact that m_ψ is proportional to $\psi(1)$.

(d) Note that K is not elementary abelian.

Notes A corollary of Problem 7.20 is that a simple Zassenhaus group of odd degree has degree $1 + 2^a$. This is another special case of Feit's theorem.

There do exist simple groups with T.I.F.N. subgroups of even order. If $G = \mathrm{SL}(2, 2^n)$, with $n \geq 2$, then the Sylow 2-subgroup K of G is T.I.F.N. It is elementary abelian of order 2^n. In this case, coherence fails because $|\mathbf{N}(K):K| = 2^n - 1$ and the set $\mathscr{S}$ contains only one character.

If $G = Sz(2^n)$ with odd $n > 1$ [the Suzuki simple group of order

$$(2^n - 1)(2^{2n})(2^{2n} + 1)],$$

then again the Sylow 2-subgroup is T.I.F.N. Here, case (b) of Theorem 7.25 holds.

(7.21) (*Sibley*) In the situation of Theorem 7.20, let $e = |N:K|$ and $k = |K|$. If m is as in 7.20 (b), show that

$$-e/(\sqrt{k} + 1) < m < e/(\sqrt{k} - 1).$$

In particular, if K is not an elementary abelian p-group, show that $m = 0$.

Hints Let $1 \neq x \in K$ and use the inequality

$$|\mathbf{C}_G(x)| > \sum_{\psi \in \mathscr{E}} \psi(x)\overline{\psi(x)},$$

where $\mathscr{E}$ is the set of exceptional characters. Derive that

$$2m + e > m^2(k - 1)/e.$$

(7.22) Let $K \subseteq G$ be T.I.F.N. with $N = \mathbf{N}(K)$ and

$$\mathscr{S} = \{\chi \in \mathrm{Irr}(N) \mid K \not\subseteq \ker \chi\}$$

coherent. Let $\psi \in \mathrm{Irr}(G)$ with $K \not\subseteq \ker \psi$.

(a) If ψ is nonexceptional, show that $|K| \leq \psi(1) + 1 - [\psi_K, 1_K]$.
(b) If ψ is exceptional, show that $|K| < \psi(1)^2$.

Hints (b) Reduce to the case that $\psi_N = \chi + \Lambda$, where $\chi \in \mathscr{S}$ and either $\Lambda = 0$ or $K \subseteq \ker \Lambda$. If $\Lambda \neq 0$, appeal to Problem 7.21. If $\Lambda = 0$, pick an irreducible constituent, $\eta \neq 1_G$ of $\psi\overline{\psi}$ and apply part (a) to η.

8 Brauer's theorem

There are many ways in which class functions of a group can be constructed. For instance, we could define $\vartheta(g) = |\mathbf{N}_G(\langle g \rangle)|$ or $\vartheta(g) = 1$ if g^2 is conjugate to some fixed $x \in G$, and $\vartheta(g) = 0$ otherwise. It is perhaps too much to expect that such arbitrarily defined functions will turn out to be characters. (For instance, in the second example given, $\vartheta(1) = 0$ if $x \neq 1$.) One might hope, however, that a well chosen ϑ will be a generalized character, that is, $\vartheta \in \mathbb{Z}[\mathrm{Irr}(G)]$, or at least that it is an R-linear combination of irreducible characters for some specified ring R with $\mathbb{Z} \subseteq R \subseteq \mathbb{C}$. (We denote the set of these *R-generalized characters* by $R[\mathrm{Irr}(G)]$.)

How can one decide if a given class function ϑ is an R-generalized character? Of course, if one knows $\mathrm{Irr}(G)$, the answer is easy: simply check that $[\vartheta, \chi] \in R$ for all $\chi \in \mathrm{Irr}(G)$. A more typical situation is that one has enough information about some family $\mathscr{H}$ of subgroups of G so that it can be shown that $\vartheta_H \in R[\mathrm{Irr}(H)]$ for every $H \in \mathscr{H}$. The main point of Brauer's theorem is that for suitable families $\mathscr{H}$ the last relationship is sufficient to guarantee that $\vartheta \in R[\mathrm{Irr}(G)]$. For instance, this will be true if $\mathscr{H}$ is the collection of all nilpotent subgroups of G.

(8.1) DEFINITION Let R be a ring with $\mathbb{Z} \subseteq R \subseteq \mathbb{C}$ and let $\mathscr{H}$ be a family of subgroups of G.

(a) $\mathscr{R}_R(G, \mathscr{H})$ is the set of class functions ϑ of G such that $\vartheta_H \in R[\mathrm{Irr}(H)]$ for all $H \in \mathscr{H}$.

(b) $\mathscr{I}_R(G, \mathscr{H})$ is the set of R-linear combinations of characters ψ^G for $\psi \in \mathrm{Irr}(H)$, $H \in \mathscr{H}$.

If $R = \mathbb{Z}$, we delete the subscripts and write $\mathscr{R}(G, \mathscr{H})$ and $\mathscr{I}(G, \mathscr{H})$.

(8.2) LEMMA Let $\mathscr{H}$ be a collection of subgroups of G and let R be a ring with $\mathbb{Z} \subseteq R \subseteq \mathbb{C}$. Then

(a) $\mathscr{I}(G, \mathscr{H}) \subseteq \mathscr{I}_R(G, \mathscr{H}) \subseteq R[\mathrm{Irr}(G)] \subseteq \mathscr{R}_R(G, \mathscr{H})$.
(b) $\mathscr{R}_R(G, \mathscr{H})$ is a ring in which $\mathscr{I}_R(G, \mathscr{H})$ is an ideal.

(Addition and multiplication are pointwise.)

Proof The containments in (a) are all obvious from the definitions. To prove (b) observe that $\mathbb{Z}[\mathrm{Irr}(H)]$ is a ring for every $H \in \mathscr{H}$. Since $\mathbb{Z} \subseteq R$, it follows that $R[\mathrm{Irr}(H)]$ is a ring. If $\varphi, \vartheta \in \mathscr{R}_R(G, \mathscr{H})$, then

$$(\varphi\vartheta)_H = \varphi_H \vartheta_H \in R[\mathrm{Irr}(H)]$$

and hence $\varphi\vartheta \in \mathscr{R}_R(G, \mathscr{H})$, which is, therefore, a ring.

To prove that $\mathscr{I}_R(G, \mathscr{H})$ is an ideal, we use the fact that if α is a class function of $H \subseteq G$ and β is a class function of G, then $(\alpha\beta_H)^G = \alpha^G\beta$. This is immediate from the definition of induction (and appears as Problem 5.3).

Let $\varphi \in \mathscr{I}_R(G, \mathscr{H})$ and $\vartheta \in \mathscr{R}_R(G, \mathscr{H})$. Then $\varphi = \sum_{H \in \mathscr{H}} (\psi_{(H)})^G$ with $\psi_{(H)} \in R[\mathrm{Irr}(H)]$. Now

$$\varphi\vartheta = \sum_H (\psi_{(H)})^G\vartheta = \sum_H (\psi_{(H)}\vartheta_H)^G.$$

Since $\psi_{(H)}\vartheta_H \in R[\mathrm{Irr}(H)]$, it follows that $\varphi\vartheta \in \mathscr{I}_R(G, \mathscr{H})$ and the proof is complete. ∎

It follows from Lemma 8.2 that in order to prove that $\mathscr{I}_R(G, \mathscr{H}) = R[\mathrm{Irr}(G)]$, it suffices to show that $1_G \in \mathscr{I}(G, \mathscr{H})$. Furthermore, if this can be done for some family $\mathscr{H}$, it follows that

$$\mathscr{I}_R(G, \mathscr{H}) = R[\mathrm{Irr}(G)] = \mathscr{R}_R(G, \mathscr{H})$$

for all R with $\mathbb{Z} \subseteq R \subseteq \mathbb{C}$.

(8.3) DEFINITION (*Brauer*) A group E is *p-elementary* (where p is a prime) if E is the direct product of a cyclic group and a p-group. We say that E is *elementary* if it is p-elementary for some prime.

(8.4) THEOREM (*Brauer*) Let $\mathbb{Z} \subseteq R \subseteq \mathbb{C}$, where R is a ring.

(a) A class function ϑ of G is an R-generalized character iff $\vartheta_E \in R[\mathrm{Irr}(E)]$ for every elementary $E \subseteq G$.

(b) Every $\chi \in \mathrm{Irr}(G)$ is a $\mathbb{Z}$-linear combination of characters of the form λ^G for linear characters λ of elementary subgroups of G.

Let $\mathscr{E}$ be the set of elementary subgroups of G. Statement (a) of Brauer's theorem is exactly the assertion that $R[\mathrm{Irr}(G)] = \mathscr{R}_R(G, \mathscr{E})$. This part of the result is often called the "characterization of characters."

Every $E \in \mathscr{E}$ is nilpotent and hence is an M-group by Corollary 6.14. If $\varphi \in \operatorname{Irr}(E)$, then $\varphi = \lambda^E$, where λ is a linear character of some $E_0 \subseteq E$. Since $E_0 \in \mathscr{E}$ and $\varphi^G = \lambda^G$, it follows that $\mathscr{I}(G, \mathscr{E})$ is exactly the set of $\mathbb{Z}$-linear combinations of λ^G for linear $\lambda \in \operatorname{Irr}(E)$ with $E \in \mathscr{E}$. Therefore, statement (b) of Brauer's theorem amounts to the assertion that $\mathbb{Z}[\operatorname{Irr}(G)] = \mathscr{I}(G, \mathscr{E})$. This is often called the "theorem on induced characters."

By Lemma 8.2, both parts of Brauer's Theorem 8.4 will be proved when we show that $1_G \in \mathscr{I}(G, \mathscr{E})$. We work toward this goal. The proof given here is based on an idea of B. Banaschewski. It is suggested that the reader also consult the Brauer–Tate paper for another approach.

(8.5) LEMMA (*Banaschewski*) Let S be a nonempty finite set and let R be a ring of $\mathbb{Z}$-valued functions defined on S (with pointwise addition and multiplication). If the function 1_S with constant value 1 does not lie in R, then there exists $x \in S$ and a prime p, such that p divides $f(x)$ for every $f \in R$.

Proof For each $x \in S$, let $I_x = \{f(x) \mid f \in R\}$. Then I_x is an additive subgroup of $\mathbb{Z}$. If for some $x \in S$, we have $I_x < \mathbb{Z}$, then $I_x \subseteq (p)$ for some prime and the result follows. Assume then, that $I_x = \mathbb{Z}$ for every $x \in S$. For each x, we may therefore choose $f_x \in R$ with $f_x(x) = 1$. Thus $f_x - 1_S$ vanishes at x and $\prod_{x \in S} (f_x - 1_S) = 0$. Expanding this product yields an expression for 1_S as a linear combination of products of the functions f_x. Thus $1_S \in R$. ∎

To obtain a ring to which we can apply Lemma 8.5, we consider permutation characters of G, that is, characters of the form $(1_H)^G$ for subgroups $H \subseteq G$.

(8.6) LEMMA Let $H, K \subseteq G$. Then $(1_H)^G(1_K)^G = \sum a_U(1_U)^G$ for subgroups $U \subseteq H$ and integers $a_U \geq 0$.

Proof Write $\vartheta = (1_K)^G$ so that $(1_H)^G(1_K)^G = \vartheta(1_H)^G = (\vartheta_H)^G$ by Problem 5.3. By Lemma 5.14, ϑ is the permutation character of G acting on the set of right cosets of K. It follows that ϑ_H is a permutation character of H and hence $\vartheta_H = \sum a_U(1_U)^H$ for subgroups $U \subseteq H$ and integers $a_U \geq 0$. Since $((1_U)^H)^G = (1_U)^G$, the result follows. ∎

(8.7) COROLLARY The set of $\mathbb{Z}$-linear combinations of characters of G of the form $(1_H)^G$ is a ring $P(G)$. Let $\mathscr{H}$ be a collection of subgroups of G with the property that if $K \subseteq H \in \mathscr{H}$, then $K \in \mathscr{H}$. Let $P(G, \mathscr{H})$ denote the set of $\mathbb{Z}$-linear combinations of characters of the form $(1_H)^G$ with $H \in \mathscr{H}$. Then $P(G, \mathscr{H})$ is an ideal of $P(G)$.

To apply Corollary 8.7, we define a class of groups more general than elementary groups.

(8.8) DEFINITION A group H is *p-quasi-elementary* if H has a cyclic normal p-complement for the prime p. We say that H is *quasi-elementary* if it is p-quasi-elementary for some p.

Note that subgroups of p-quasi-elementary groups are themselves p-quasi-elementary. Thus in the notation of Corollary 8.7, $P(G, \mathscr{H})$ is closed under multiplication, where $\mathscr{H}$ is either the set of all quasi-elementary subgroups of G or is the set of all p-quasi-elementary subgroups of G for some fixed prime p.

(8.9) LEMMA Let $x \in G$ and let p be a prime. Then there exists a p-quasi-elementary subgroup, $H \subseteq G$, such that $(1_H)^G(x)$ is not divisible by p.

Proof Let C be the p-complement in the group $\langle x \rangle$. (Possibly $C = 1$ or $C = \langle x \rangle$.) Let $N = \mathbf{N}_G(C)$, so that $x \in N$. Since $\langle x \rangle / C$ is a p-group, we may choose $H/C \in \mathrm{Syl}_p(N/C)$ with $\langle x \rangle \subseteq H$. Then C is a normal p-complement for H and H is p-quasi-elementary.

Now $(1_H)^G$ is the permutation character of G acting on the right cosets of H, and so $(1_H)^G(x) = |\{Hy \mid Hyx = Hy, y \in G\}|$. If $Hyx = Hy$, we have $x^{y^{-1}} \in H$ and hence $C^{y^{-1}} \subseteq H$. However, C is the unique p-complement in H and hence $C^{y^{-1}} = C$ and $y \in N$. We therefore need to count the number of fixed points in the action of $\langle x \rangle$ on the cosets of H in N.

Since $C \lhd N$, and $C \subseteq H$, we see that C is in the kernel of the action of N on the cosets of H in N. Since $\langle x \rangle / C$ is a p-group, it follows that the number of nonfixed cosets is divisible by p and hence $(1_H)^G(x) \equiv |N : H| \bmod p$. By the choice of H, $p \nmid |N : H|$ and the proof is complete. ∎

(8.10) THEOREM (*L. Solomon*) Let $\mathscr{H}$ be the set of quasi-elementary subgroups of G and $\mathscr{H}_p$ the set of p-quasi-elementary subgroups for some prime p. Then

(a) $1_G \in P(G, \mathscr{H})$.
(b) $m1_G \in P(G, \mathscr{H}_p)$ for some $m \in \mathbb{Z}$ with $p \nmid m$.

Proof By Corollary 8.7, $P(G, \mathscr{H})$ is a ring of $\mathbb{Z}$-valued functions on G. If $1_G \notin P(G, \mathscr{H})$, then by Lemma 8.5, there exists $x \in G$ and a prime p with $p \mid \varphi(x)$ for all $\varphi \in P(G, \mathscr{H})$. This contradicts Lemma 8.9 and so (a) is proved.

For (b) let $R = \{\varphi + np1_G \mid \varphi \in P(G, \mathscr{H}_p), n \in \mathbb{Z}\}$. Then R is a ring. If there exists $x \in G$ and a prime q, with $q \mid \varphi(x)$ for all $\varphi \in R$, then since $p1_G \in R$, we have $q = p$ and thus we have a contradiction to Lemma 8.9. By Lemma 8.5, we conclude that $1_G \in R$ and hence $(1 - np)1_G \in P(G, \mathscr{H}_p)$ for some $n \in \mathbb{Z}$. This completes the proof. ∎

Only part (a) of Theorem 8.10 is needed to prove Brauer's theorem; however, part (b) is useful for certain refinements of the result. Recall that we need $1_G \in \mathscr{I}(G, \mathscr{E})$. The method of proof is essentially to use 8.10 to reduce the problem to quasi-elementary groups. We need a lemma.

(8.11) LEMMA Suppose $G = CP$ where $C \lhd G$, P is a p-group and $p \nmid |C|$. Let λ be a linear character of C which is invariant in G and suppose $\mathbf{C}_C(P) \subseteq \ker \lambda$. Then $\lambda = 1_C$.

Proof Let $K = \ker \lambda$. Since λ is linear, it takes on distinct constant values on the distinct cosets of K in C. Since λ is invariant under the action of P, it follows that P normalizes each coset Kx for $x \in C$.

In the conjugation action of P on Kx, the number of moved points is divisible by p. Since $p \nmid |K|$, we have $p \nmid |Kx|$ and thus $Kx \cap \mathbf{C}_C(P) \neq \varnothing$. Since $\mathbf{C}_C(P) \subseteq K$ by hypothesis, we conclude that $Kx = K$ and thus $K = C$ and $\lambda = 1_C$ as claimed. ∎

Proof of Theorem 8.4 As was remarked following the statement of the theorem, it suffices to prove that $1_G \in \mathscr{I}(G, \mathscr{E})$. We use induction on $|G|$. We may therefore suppose that $1_H \in \mathscr{I}(H, \mathscr{E}_H)$ whenever $H < G$, where $\mathscr{E}_H$ is the set of elementary subgroups of H. Thus $\mathbb{Z}[\mathrm{Irr}(H)] = \mathscr{I}(H, \mathscr{E}_H)$ for these H by Lemma 8.2. By transitivity of induction (Problem 5.1), it follows for $\varphi \in \mathscr{I}(H, \mathscr{E}_H)$, that $\varphi^G \in \mathscr{I}(G, \mathscr{E}_H) \subseteq \mathscr{I}(G, \mathscr{E})$. Thus for all $H < G$ and $\varphi \in \mathrm{Irr}(H)$, we have $\varphi^G \in \mathscr{I}(G, \mathscr{E})$ and it will suffice to show that 1_G is a $\mathbb{Z}$-linear combination of characters induced from proper subgroups.

By Solomon's Theorem 8.10(a) we are done if G is not quasi-elementary and so we may assume that G has the cyclic normal p-complement C. Let $P \in \mathrm{Syl}_p(G)$ and $Z = \mathbf{C}_C(P)$. Since we may clearly assume that G is not elementary, we have $Z < C$ and $E = PZ < G$.

Write $(1_E)^G = 1_G + \Xi$, where Ξ is a possibly reducible character of G. We shall show that every irreducible constituent of Ξ is induced from a proper subgroup and thus $1_G = (1_E)^G - \Xi \in \mathscr{I}(G, \mathscr{E})$ as desired.

Let χ be an irreducible constituent of Ξ. Now $CE = G$ and $C \cap E = Z$ so that

$$1_C + \Xi_C = ((1_E)^G)_C = (1_Z)^C$$

by Problem 5.2. Thus

$$1 = [(1_Z)^C, 1_C] = [1_C + \Xi_C, 1_C]$$

and hence $[\Xi_C, 1_C] = 0$. Thus $[\chi_C, 1_C] = 0$.

Let λ be an irreducible constituent of χ_C. Then $\lambda \neq 1_C$. However, $Z \lhd G$ and so $Z \subseteq \ker((1_E)^G)$ and hence $Z \subseteq \ker \chi$. We therefore have $Z \subseteq \ker \lambda$ and thus by Lemma 8.11, λ is not invariant in G. Let $T = I_G(\lambda) < G$. By Theorem 6.11, $\chi = \psi^G$ for some $\psi \in \mathrm{Irr}(T)$. The result now follows. ∎

The following is a useful special case of part (a) of Brauer's theorem.

(8.12) COROLLARY Let χ be a class function of G. Suppose

(a) χ_E is a generalized character for every elementary $E \subseteq G$;
(b) $[\chi, \chi] = 1$;
(c) $\chi(1) \geq 0$.

Then $\chi \in \text{Irr}(G)$.

Proof Using (a) and Theorem 8.4(a), we have $\chi = \sum a_\xi \xi$ for $\xi \in \text{Irr}(G)$ and $a_\xi \in \mathbb{Z}$. From (b) we conclude that at most one $a_\xi \neq 0$ and $\chi = \pm\xi$ for some $\xi \in \text{Irr}(G)$. Then (c) yields the result. ∎

Before going on to reap some of the harvest of applications of Brauer's theorem, we digress briefly to derive the "local" version of the result. In the Brauer–Tate proof, this is obtained as an intermediate result.

(8.13) COROLLARY Let $\mathscr{E}_p$ be the set of p-elementary subgroups of G for some fixed prime p. Then

$$m1_G \in \mathscr{I}(G, \mathscr{E}_p)$$

for some $m \in \mathbb{Z}$ with $p \nmid m$.

Proof By Theorem 8.10(b), there exists $m \in \mathbb{Z}$ with $p \nmid m$ such that $m1_G$ is a $\mathbb{Z}$-linear combination of characters of the form $(1_H)^G$ for p-quasi-elementary $H \subseteq G$. Let $\mathscr{E}_H$ denote the set of elementary subgroups of H. If H is p-quasi-elementary, then $\mathscr{E}_H \subseteq \mathscr{E}_p$ and by transitivity of induction, we have

$$(1_H)^G \in \mathscr{I}(G, \mathscr{E}_H) \subseteq \mathscr{I}(G, \mathscr{E}_p).$$

The result now follows. ∎

We shall use Brauer's theorem to remove the solvability hypothesis from Theorem 6.25 on extending characters. To do this we need part (c) of the following result. [Part (b) will be used in Chapter 13.]

(8.14) LEMMA Let $N \lhd G$ and let $\chi \in \text{Irr}(G)$ with $\chi_N = \vartheta \in \text{Irr}(N)$. Let $\psi \in \text{Irr}(G)$. For each coset Ng of N in G compute

$$\eta(Ng) = (1/|N|) \sum_{x \in Ng} \psi(x)\overline{\chi(x)}.$$

We have

(a) If $[\psi_N, \vartheta] \neq 0$, then $\eta \in \text{Irr}(G/N)$.
(b) If $[\psi_N, \vartheta] = 0$, then $\eta(Ng) = 0$ for all g.
(c) If $\chi = \psi$, then $\eta(Ng) = 1$ for all g.

Proof Conjugates of Ng in G/N all have the form $Ng^h = h^{-1}(Ng)h$ for $h \in G$. It follows that η is a class function on G/N. We may thus write $\eta = \sum b_\xi \xi$, where ξ runs over $\mathrm{Irr}(G/N)$ and $b_\xi \in \mathbb{C}$. Now

$$b_\xi = [\eta, \xi] = \frac{1}{|G:N|} \sum_{Ng \in G/N} \eta(Ng)\overline{\xi(Ng)}$$
$$= \frac{1}{|G|} \sum_{x \in G} \psi(x)\overline{\chi(x)}\overline{\xi(x)},$$

where in the last expression, we view $\xi \in \mathrm{Irr}(G)$ and use the fact that as such, ξ is constant on cosets of N.

We thus have $b_\xi = [\psi, \chi\xi]$. Now by Corollary 6.17, $\chi\xi \in \mathrm{Irr}(G)$ and the characters $\chi\xi$ are distinct for distinct $\xi \in \mathrm{Irr}(G/N)$. Therefore $\eta = \sum b_\xi \xi$ satisfies $\eta = 0$ unless $\psi = \chi\xi$ for some $\xi \in \mathrm{Irr}(G/N)$ in which case $\eta = \xi$. By Corollary 6.17, $\psi = \chi\xi$ for some ξ iff $[\psi_N, \vartheta] \neq 0$. The result now follows. ∎

(8.15) THEOREM (*Gallagher*) Let $N \lhd G$ with $\vartheta \in \mathrm{Irr}(N)$ invariant in G. Suppose $(\vartheta(1), |G:N|) = 1$. Then ϑ is extendible to G iff $\det \vartheta$ is extendible.

Proof The "only if" part of the assertion is trivial. Assume $\lambda = \det \vartheta$ is extendible to $\mu \in \mathrm{Irr}(G)$. If $N \subseteq H \subseteq G$ with H/N solvable, then ϑ is extendible to H by Theorem 6.25. By Lemma 6.24, there is a unique extension $\chi_{(H)}$, such that $\det(\chi_{(H)}) = \mu_H$. Define the function χ on G as follows. If $g \in G$, let $H = \langle N, g \rangle$ so that H/N is cyclic. Set $\chi(g) = \chi_{(H)}(g)$. We shall use Corollary 8.12 to show that $\chi \in \mathrm{Irr}(G)$.

First we establish that χ is a class function. If $g, x \in G$, let $H = \langle N, g \rangle$, so that $H^x = \langle N, g^x \rangle$. Define ψ on H^x by $\psi(h^x) = \chi_{(H)}(h)$ for $h \in H$. Since the map $h \mapsto h^x$ is an isomorphism and $\chi_{(H)} \in \mathrm{Irr}(H)$, we have $\psi \in \mathrm{Irr}(H^x)$. Also, $(\det \psi)(h^x) = (\det(\chi_{(H)}))(h) = \mu(h) = \mu(h^x)$, so that $\det \psi = \mu_{H^x}$. Furthermore, if $n \in N$, we have $\psi(n) = \chi_{(H)}(n^{x^{-1}}) = \vartheta(n^{x^{-1}}) = \vartheta^x(n) = \vartheta(n)$. It now follows from the uniqueness of $\chi_{(H^x)}$, that $\chi_{(H^x)} = \psi$ and hence

$$\chi(g^x) = \chi_{(H^x)}(g^x) = \psi(g^x) = \chi_{(H)}(g) = \chi(g)$$

and χ is a class function.

Next, let $E \subseteq G$ be elementary and set $H = NE$ so that $H/N \cong E/(N \cap E)$ is nilpotent and $\chi_{(H)}$ is defined. We shall show that the restriction $\chi_H = \chi_{(H)}$ and thus $\chi_E = (\chi_{(H)})_E$ which is a character of E. If $g \in H$, then $\langle N, g \rangle = K \subseteq H$. Clearly, $(\chi_{(H)})_K$ is an extension of ϑ to K and $\det((\chi_{(H)})_K) = \mu_K$ so that $(\chi_{(H)})_K = \chi_{(K)}$. Therefore, $\chi(g) = \chi_{(K)}(g) = \chi_{(H)}(g)$, hence $\chi_H = \chi_{(H)}$ as claimed, and χ_E is a character.

Next, we compute $[\chi, \chi]$. For each coset Ng of N in G, we have

$$\sum_{x \in Ng} |\chi(x)|^2 = \sum_{x \in Ng} |\chi_{(H)}(x)|^2,$$

where $H = \langle N, g\rangle = \langle N, x\rangle$ for all $x \in Ng$. Since $(\chi_{(H)})_N = \vartheta \in \mathrm{Irr}(N)$, Lemma 8.14(c) yields $\sum_{x\in Ng} |\chi_{(H)}(x)|^2 = |N|$ and it follows that $\sum_{x\in G} |\chi(x)|^2 = |G:N|\cdot|N| = |G|$ and $[\chi, \chi] = 1$.

Finally, $\chi(1) = \vartheta(1) > 1$ and hence $\chi \in \mathrm{Irr}(G)$ by Corollary 8.12. Clearly, $\chi_N = \vartheta$ and the proof is complete. ∎

(8.16) COROLLARY Let $N \lhd G$ and $\vartheta \in \mathrm{Irr}(N)$ with ϑ invariant in G. Suppose $(|G:N|, o(\vartheta)\vartheta(1)) = 1$. Then ϑ has a unique extension, $\chi \in \mathrm{Irr}(G)$ with $(|G:N|, o(\chi)) = 1$. In fact, $o(\chi) = o(\vartheta)$. In particular, this holds if $(|G:N|, |N|) = 1$.

Proof This is immediate from Corollary 6.27 and Theorem 8.15. (See the proof of Corollary 6.28.) ∎

We have already seen several conditions sufficient to guarantee that a character value is zero. The following is a powerful one.

(8.17) THEOREM Let $\chi \in \mathrm{Irr}(G)$ and suppose $p \nmid (|G|/\chi(1))$ for some prime p. Then $\chi(g) = 0$ whenever $p \mid o(g)$.

Proof Define ϑ on G by $\vartheta(g) = \chi(g)$ if $p \nmid o(g)$ and $\vartheta(g) = 0$ if $p \mid o(g)$. We shall use Brauer's theorem to show that ϑ is a generalized character.

Let $E \subseteq G$ be elementary. Since E is nilpotent, we may write $E = P \times Q$, where P is a p-group and $p \nmid |Q|$. If $x \in E$ and $p \nmid o(x)$, we have $x \in Q$ and so ϑ vanishes on $E - Q$ and $\vartheta_Q = \chi_Q$.

Let $\psi \in \mathrm{Irr}(E)$. We have

$$|E|[\vartheta_E, \psi] = \sum_{x\in Q} \chi(x)\overline{\psi(x)} = |Q|[\chi_Q, \psi_Q].$$

Since $|E| = |P||Q|$, we conclude that $|P|[\vartheta_E, \psi] \in \mathbb{Z}$.

Now let $\omega = \omega_\chi$ be the algebra homomorphism $\mathbf{Z}(\mathbb{C}[G]) \to \mathbb{C}$ associated with χ so that $\omega(K) = \chi(g)|\mathscr{K}|/\chi(1)$, where $\mathscr{K}$ is the conjugacy class containing g and $K = \Sigma\mathscr{K} \in \mathbb{C}[G]$. We shall write $\omega(g)$ for the algebraic integer $\omega(K)$, so that

$$\chi(g) = \chi(1)\omega(g)/|\mathscr{K}| = \chi(1)\omega(g)|\mathbf{C}_G(g)|/|G|$$

Now

$$|E|[\vartheta_E, \psi] = \sum_{x\in Q} \chi(x)\overline{\psi(x)} = \frac{\chi(1)}{|G|} \sum_{x\in Q} \omega(x)\overline{\psi(x)}|\mathbf{C}_G(x)|.$$

Since $P \subseteq \mathbf{C}_G(x)$ for $x \in Q$, we have

$$\frac{|G||Q|}{\chi(1)}[\vartheta_E, \psi] = \frac{|G||E|}{\chi(1)|P|}[\vartheta_E, \psi] = \sum_{x\in Q} \omega(x)\overline{\psi(x)}|\mathbf{C}_G(x):P|,$$

which is an algebraic integer. Since $|P|[\vartheta_E, \psi] \in \mathbb{Z}$, we have $[\vartheta_E, \psi] \in \mathbb{Q}$ and thus $(|G||Q|/\chi(1))[\vartheta_E, \psi] \in \mathbb{Z}$. Since $|G||Q|/\chi(1) \in \mathbb{Z}$ and is relatively prime to $|P|$, we conclude that $[\vartheta_E, \psi] \in \mathbb{Z}$ and hence ϑ_E is a generalized character of E.

Brauer's theorem now yields that ϑ is a generalized character of G. In particular $[\vartheta, \chi] \in \mathbb{Z}$. Now $[\vartheta, \chi] = (1/|G|) \sum \{|\chi(g)|^2 \mid p \nmid o(g)\}$, and so $0 < [\vartheta, \chi] \le [\chi, \chi] = 1$. We conclude that $[\vartheta, \chi] = [\chi, \chi]$ and

$$0 = [(\chi - \vartheta), \chi] = \frac{1}{|G|} \sum \{|\chi(g)|^2 \mid p \mid o(g)\}.$$

The result now follows. ∎

A character $\chi \in \mathrm{Irr}(G)$ is said to have *p-defect zero* if p does not divide $|G|/\chi(1)$.

In Chapter 2 we discussed the question of what information can be obtained about a group from a knowledge of its character table. Since D_8 and Q_8 have the same table, one cannot determine the orders of the elements in the various classes. Nevertheless, it is possible to determine the sets of prime divisors of these orders. To do this, we shall apply Theorem 8.4(a) in a situation where $R \neq \mathbb{Z}$.

If $g \in G$, we shall use the notation $\pi(g)$ to denote the set of prime divisors of $o(g)$.

(8.18) LEMMA Let $g \in G$ and let π be a set of primes. Then there exist unique $x, y \in G$ with

(a) $g = xy = yx$;
(b) $\pi(x) \subseteq \pi$ and $\pi(y) \cap \pi = \varnothing$.

Furthermore, $x, y \in \langle g \rangle$.

Proof Write $o(g) = mn$ such that every prime divisor of m is in π and no prime divisor of n is in π. Then $(m, n) = 1$ and we have $km + ln = 1$ for some $k, l \in \mathbb{Z}$. Let $x = g^{ln}$ and $y = g^{km}$ so that $g = xy = yx$ and $x, y \in \langle g \rangle$. Since $x^m = 1 = y^n$, (b) follows.

Now suppose $g = uv = vu$ with $\pi(u) \subseteq \pi$ and $\pi(v) \cap \pi = \varnothing$. Then $u \subseteq \mathbf{C}(g) \subseteq \mathbf{C}(x)$ and hence $\pi(xu^{-1}) \subseteq \pi(x) \cup \pi(u) \subseteq \pi$. Similarly,

$$\pi(y^{-1}v) \subseteq \pi(v) \cup \pi(y),$$

so that $\pi(y^{-1}v) \cap \pi(xu^{-1}) = \varnothing$. Since $y^{-1}v = xu^{-1}$, we have $xu^{-1} = 1 = y^{-1}v$ and uniqueness follows. ∎

If g, π, x, and y are as in Lemma 8.18, we write $x = g_\pi$ and $y = g_{\pi'}$. If $\pi = \{p\}$, we write $x = g_p$ and $y = g_{p'}$.

(8.19) LEMMA Let $|G| = mn$ with $(m, n) = 1$ and let π be the set of prime divisors of m. Let $g \in G$, with $g = g_{\pi'}$, and define ϑ on G by

$$\vartheta(x) = \begin{cases} n & \text{if } x_{\pi'} \text{ is conjugate to } g, \\ 0 & \text{if } x_{\pi'} \text{ is not conjugate to } g. \end{cases}$$

Let R be the ring of algebraic integers in $\mathbb{Q}[\varepsilon]$, where ε is a primitive nth root of 1. Then ϑ is an R-generalized character.

Proof Let $E \subseteq G$ be elementary. It suffices to show that $\vartheta_E \in R[\mathrm{Irr}(E)]$. Since E is nilpotent, we may write $E = H \times K$, where H is a π-group and K is a π'-group. Then for $x \in E$, we have $x = uv$, $u \in H$, $v \in K$ and thus $u = x_\pi$ and $v = x_{\pi'}$.

If g is not conjugate to any element of K, then $\vartheta_E = 0$ and there is nothing to prove. Suppose then, that $g_1, g_2, \ldots, g_r$ are the distinct elements of K which are conjugate to g. Then the elements $x \in E$ with $x_{\pi'}$ conjugate to g are exactly the elements of the cosets Hg_i, $1 \le i \le r$.

Let $\psi \in \mathrm{Irr}(E)$. By Theorem 4.21, we have $\psi = \xi \times \eta$ for some $\xi \in \mathrm{Irr}(H)$ and $\eta \in \mathrm{Irr}(K)$. We have

$$|E|[\vartheta_E, \psi] = \sum_{i=1}^{r} \sum_{u \in H} \vartheta(ug_i)\overline{\psi(ug_i)} = n \sum_{i=1}^{r} \overline{\eta(g_i)} \sum_{u \in H} \overline{\xi(u)}.$$

Since $\sum_{u \in H} \overline{\xi(u)} = 0$ unless $\xi = 1_H$, we have either $[\vartheta_E, \psi] = 0$ or

$$|E|[\vartheta_E, \psi] = n|H| \sum_{i=1}^{r} \overline{\eta(g_i)}.$$

Now $|E| = |K||H|$ and $|K|\,|\,n$ and thus $[\vartheta_E, \psi]$ is a multiple of $\sum \overline{\eta(g_i)}$ which lies in R. The result now follows. ∎

(8.20) THEOREM Let R be the ring of algebraic integers in $\mathbb{Q}[\varepsilon]$, where ε is a primitive $|G|$th root of 1. Let $p \in \mathbb{Z}$ be a prime and let I be a maximal ideal of R with $p \in I$. Let $x, y \in G$. Then $x_{p'}$ and $y_{p'}$ are conjugate in G iff $\chi(x) \equiv \chi(y) \bmod I$ for every $\chi \in \mathrm{Irr}(G)$.

Proof Write $|G| = p^a n$, with $p \nmid n$, and define ϑ on G by $\vartheta(g) = n$ if $g_{p'}$ is conjugate to $x_{p'}$ and $\vartheta(g) = 0$, otherwise. By Lemma 8.19, ϑ is an R-generalized character of G.

Suppose $\chi(x) \equiv \chi(y) \bmod I$ for every $\chi \in \mathrm{Irr}(G)$. It follows that $\vartheta(x) \equiv \vartheta(y) \bmod I$. Now $\vartheta(x) = n$. If $\vartheta(y) = 0$, then $n \equiv 0 \bmod I$ and $n \in I$. Since $p \in I$ and $p \nmid n$, it follows that $1 = (p, n) \in I$ and this is a contradiction. Thus $\vartheta(y) = n$ and hence $y_{p'}$ is conjugate to $x_{p'}$ as desired.

Conversely, we may suppose $x_{p'} = g = y_{p'}$ and we show that $\chi(x) \equiv \chi(g) \equiv \chi(y) \bmod I$. Write $\chi_{\langle x \rangle} = \sum \lambda_i$, where the λ_i are linear characters. Since $g \in \langle x \rangle$, it suffices to show that $\lambda(g) \equiv \lambda(x) \bmod I$ for linear λ.

Now $x = gh$, where $h = x_p \in \langle x \rangle$ and $\lambda(x) = \lambda(g)\lambda(h)$. Let $\lambda(h) = \delta$ so that $\delta^{p^m} = 1$ for some m. For $\alpha \in R$, let α^* denote its image in R/I which is a field of characteristic p. We have $0^* = (\delta^*)^{p^m} - 1^* = (\delta^* - 1^*)^{p^m}$ and thus $\delta^* - 1^* = 0^*$ and $\delta \equiv 1 \bmod I$. It follows that $\lambda(x) \equiv \lambda(g) \bmod I$ and the result follows. ∎

Note that by Theorem 8.20, it follows that if $x_{p'}$ and $y_{p'}$ are conjugate and $\chi \in \mathrm{Irr}(G)$, then $\chi(x) - \chi(y)$ lies in *every* maximal ideal of R which contains p.

(8.21) THEOREM (*G. Higman*) Let $\mathscr{K}_1, \mathscr{K}_2, \ldots, \mathscr{K}_k$ be the conjugacy classes of G. Then the character table of G determines the sets of primes $\pi_\mu = \pi(g_\mu)$ for $g_\mu \in \mathscr{K}_\mu$.

Proof Let $\mathrm{Irr}(G) = \{\chi_i | 1 \le i \le k\}$. We are given the complex numbers $a_{i\mu} = \chi_i(g_\mu)$ for $g_\mu \in \mathscr{K}_\mu$. If $1 \in \mathscr{K}_\nu$, then ν is the unique integer $1 \le \nu \le k$, such that $a_{i\nu} \in \mathbb{Z}$ and $a_{i\nu} \ge |a_{i\mu}|$ for all i, μ and hence we know $\pi_\nu = \varnothing$. For notational simplicity, we now assume $1 \in \mathscr{K}_1$.

Next we compute $\sum_i (a_{i1})^2 = |G|$ and let R be the ring of algebraic integers in $\mathbb{Q}[\varepsilon]$, where ε is a primitive $|G|$th root of 1. For each prime, $p \mid |G|$, we choose a maximal ideal of R, $I_p \supseteq pR$. Construct the equivalence relation $\sim_p$ on $\{\mu | 1 \le \mu \le k\}$ by setting $\mu \sim_p \nu$ if $a_{i\mu} \equiv a_{i\nu} \bmod I_p$ for all i, $1 \le i \le k$. By Theorem 8.20, $\mu \sim_p \nu$ iff the elements of $\mathscr{K}_\mu$ and $\mathscr{K}_\nu$ have conjugate p'-parts. In particular, if $\mu \sim_p \nu$, then $\pi_\mu \cup \{p\} = \pi_\nu \cup \{p\}$. Furthermore, given μ, there exists ν, with $p \notin \pi_\nu$, such that $\mu \sim_p \nu$. To see this, take $g \in \mathscr{K}_\mu$ and choose ν so that $g_{p'} \in \mathscr{K}_\nu$.

Let π be a set of prime divisors of G. Using induction on $|\pi|$, we show how to construct the set $\mathscr{S}_\pi = \{\mu | \pi_\mu = \pi\}$. We have $\mathscr{S}_\varnothing = \{1\}$. If $\pi \ne \varnothing$, write $\pi = \pi_0 \cup \{p\}$, where $p \notin \pi_0$. By the preceding remarks, it follows that

$$\mathscr{S}_\pi = \{\mu | \mu \notin \mathscr{S}_{\pi_0} \quad \text{and} \quad \mu \sim_p \nu \quad \text{for some} \quad \nu \in \mathscr{S}_{\pi_0}\}.$$

The proof is now complete. ∎

Transfer theory is a tool commonly used for producing normal subgroups of a group, especially normal p-complements or normal π-complements. Many of these transfer theorems can also be proved using characters, and in particular using Brauer's theorem. An example of a typical transfer theorem is the following: Let H be a Hall subgroup of G (that is, $(|H|, |G:H|) = 1$) and suppose that H is nilpotent and that if $x, y \in H$ are any two elements which are conjugate in G, then x and y are already conjugate in H. Then H has a normal complement in G.

In the above situation, let π be the set of prime divisors of $|H|$. Using the assumption that H is nilpotent, it is not too hard to prove that if $U \subseteq G$ is a nilpotent π-subgroup, then U is conjugate to a subgroup of H. Because of this, the following result is a generalization of the above transfer theorem.

(8.22) THEOREM (*Brauer–Suzuki*) Let H be a Hall subgroup of G and suppose that whenever two elements of H are conjugate in G, then they are already conjugate in H. Assume for every elementary subgroup $E \subseteq G$, that if $|E|\,|\,|H|$, then E is conjugate to a subgroup of H. Then there exists $K \triangleleft G$ with $HK = G$ and $H \cap K = 1$.

Proof Let π be the set of prime divisors of $|H|$, so that by hypothesis, every elementary π-subgroup of G is conjugate to a subgroup of H. Let ϑ be a class function of H. We define a class function $\tilde{\vartheta}$ on G as follows. For $g \in G$, $\langle g_\pi \rangle$ is an elementary π-group and hence g_π is conjugate to some element $x \in H$. Set $\tilde{\vartheta}(g) = \vartheta(x)$ and observe that this is well defined since if g_π is also conjugate to $y \in H$, then x and y are conjugate in H by hypothesis and so $\vartheta(x) = \vartheta(y)$.

For $x \in H$, let $\mu(x)$ denote the number of elements $g \in G$ with g_π conjugate to x. Let $x_1, x_2, \ldots, x_t$ be representatives for the conjugacy classes of H. If ϑ and φ are class functions of H, we have

$$(*) \qquad [\tilde{\vartheta}, \tilde{\varphi}] = \frac{1}{|G|} \sum_{i=1}^{t} \vartheta(x_i)\overline{\varphi(x_i)}\mu(x_i).$$

Now let R be a ring with $\mathbb{Z} \subseteq R \subseteq \mathbb{C}$ and suppose that ϑ is an R-generalized character of H. We claim that $\tilde{\vartheta}$ is an R-generalized character of G. To prove this, let $E \subseteq G$ be elementary. We show that $(\tilde{\vartheta})_E$ is an R-generalized character of E. Write $E = U \times V$, where U is a π-group and V is a π'-group. We may assume that $U \subseteq H$.

If $u \in U$ and $v \in V$, then $(uv)_\pi = u$ and hence $\tilde{\vartheta}(uv) = \vartheta(u)$. Since $U \subseteq H$, we may write $\vartheta_U = \sum a_\psi \psi$ for $\psi \in \mathrm{Irr}(U)$ and $a_\psi \in R$. It follows that $(\tilde{\vartheta})_E = \sum_\psi a_\psi(\psi \times 1_V)$, which is an R-generalized character of E. Thus $\tilde{\vartheta}$ is an R-generalized character of G by Brauer's theorem as claimed.

Now let R be the ring of algebraic integers in $\mathbb{Q}[\varepsilon]$, where ε is a primitive $|H|$th root of 1. For $1 \le i \le t$, define $\vartheta_i = \sum_{\varphi \in \mathrm{Irr}(H)} \overline{\varphi(x_i)}\varphi$, so that $\vartheta_i(x_j) = 0$ if $i \ne j$ and $\vartheta_i(x_i) = |\mathbf{C}_H(x_i)|$ by the second orthogonality relation. Also, ϑ_i is an R-generalized character of H and hence $[\tilde{\vartheta}_i, 1_G] \in R$ and

$$[\tilde{\vartheta}_i, 1_G] = \frac{1}{|G|} \sum_{j=1}^{t} \vartheta_i(x_j)\mu(x_j) = \frac{\mu(x_i)|\mathbf{C}_H(x_i)|}{|G|}.$$

We conclude that $\mu(x_i)|\mathbf{C}_H(x_i)|/|G|$ is a positive rational number which is an algebraic integer. It follows that it is a positive rational integer and thus

$$\mu(x_i)|\mathbf{C}_H(x_i)|/|G| \ge 1$$

and $\mu(x_i) \geq |G|/|\mathbf{C}_H(x_i)|$ for all i. We now have

$$|G| = \sum_{i=1}^{t} \mu(x_i) \geq \frac{|G|}{|H|} \sum \frac{|H|}{|\mathbf{C}_H(x_i)|} = \frac{|G|}{|H|} \cdot |H| = |G|.$$

Therefore, equality holds throughout and $\mu(x_i) = |G|/|\mathbf{C}_H(x_i)|$. Thus Equation (*) yields

$$[\tilde{\vartheta}, \tilde{\varphi}] = \frac{1}{|H|} \sum_{i=1}^{t} \frac{\vartheta(x_i)\overline{\varphi(x_i)}|H|}{|\mathbf{C}_H(x_i)|} = [\vartheta, \varphi].$$

Now suppose $\chi \in \mathrm{Irr}(H)$. By taking $R = \mathbb{Z}$, we conclude that $\tilde{\chi}$ is a generalized character of G. Since $[\tilde{\chi}, \tilde{\chi}] = [\chi, \chi] = 1$ and $\tilde{\chi}(1) = \chi(1) > 0$, it follows that $\tilde{\chi} \in \mathrm{Irr}(G)$.

Let $K = \bigcap \{\ker(\tilde{\chi}) \mid \chi \in \mathrm{Irr}(H)\}$. Then $K \lhd G$. If $u \in H \cap K$, then $\chi(u) = \tilde{\chi}(u) = \tilde{\chi}(1) = \chi(1)$ for all $\chi \in \mathrm{Irr}(H)$ and thus $u = 1$. On the other hand, if $p \notin \pi$, $P \in \mathrm{Syl}_p(G)$, and $v \in P$, then $v_\pi = 1$ and $\tilde{\chi}(v) = \chi(1) = \tilde{\chi}(1)$ for all $\chi \in \mathrm{Irr}(H)$, so that $v \in K$. Thus $P \subseteq K$ and hence $|K| \geq |G : H|$. It follows that $HK = G$. ∎

To demonstrate the power of Theorem 8.22, we show how easily Burnside's transfer theorem follows from it.

(8.23) COROLLARY Let $P \in \mathrm{Syl}_p(G)$ and suppose $P \subseteq \mathbf{Z}(\mathbf{N}(P))$. Then there exists $K \lhd G$ with $KP = G$ and $K \cap P = 1$.

Proof If $E \subseteq G$ is a p-group, then E is conjugate to some subgroup of P by Sylow's theorems. It therefore suffices to assume $x, y \in P$ are conjugate in G and show that x and y are conjugate in P.

We have then, $y = x^g$ for some $g \in G$. Since P is abelian, $P \subseteq \mathbf{C}_G(y)$ and $P^g \subseteq \mathbf{C}_G(x^g) = \mathbf{C}_G(y)$. Thus $P^{gc} = P$ for some $c \in \mathbf{C}_G(x^g)$ and hence $gc \in \mathbf{N}(P) \subseteq \mathbf{C}(x)$ by hypothesis. Now $y = x^g = (x^g)^c = x^{gc} = x$ and the proof is complete. ∎

We also remark that Theorem 8.22 provides an alternate proof of Frobenius' Theorem 7.2. It is routine to check that a Frobenius complement necessarily satisfies the hypotheses of Theorem 8.22.

Our next results depend on Corollary 8.13 where we restrict attention to p-elementary subgroups for some fixed p.

(8.24) THEOREM (*Dade*) Let $N \lhd G$ with G/N a p-group. Let $\vartheta \in \mathrm{Irr}(N)$ be invariant in G. Then there exists a p-elementary subgroup $E \subseteq G$, and

$\varphi \in \mathrm{Irr}(E \cap N)$, such that

(a) $NE = G$;
(b) φ is invariant in E;
(c) $[\vartheta_{E \cap N}, \varphi]$ is prime to p.

Proof By Corollary 8.13, we have

$$(*) \qquad m1_G = \sum_{\psi} a_\psi \psi^G,$$

where ψ runs over irreducible characters of p-elementary subgroups, $a_\psi \in \mathbb{Z}$ and $p \nmid m$. If E is p-elementary and $\psi \in \mathrm{Irr}(E)$, we compute $[\beta, (\psi^G)_N]$ for G-invariant class functions β of N. Now $(\beta^G)_N = |G:N|\beta$ and $(\beta^{EN})_N = |EN:N|\beta$. Since β^G and β^{EN} both vanish on $EN - N$, we have $(\beta^G)_{EN} = |G:EN|\beta^{EN}$. Now

$$[\beta, (\psi^G)_N] = [\beta^G, \psi^G] = [(\beta^G)_{EN}, \psi^{EN}],$$

where the latter equality follows since $\psi^G = (\psi^{EN})^G$. This yields

$$\begin{aligned}[\beta, (\psi^G)_N] = |G:EN|[\beta^{EN}, \psi^{EN}] &= |G:EN|[\beta, (\psi^{EN})_N] \\ &= |G:EN|[\beta, (\psi_{E \cap N})^N],\end{aligned}$$

where we have used Problem 5.2 to obtain the last equality.

We apply this with $\beta = \vartheta\bar{\vartheta}$. For each ψ in Equation (*), write $\psi \in \mathrm{Irr}(E_\psi)$ for p-elementary $E_\psi \subseteq G$. We obtain

$$\begin{aligned} m = [\vartheta\bar{\vartheta}, m1_N] &= \sum_\psi a_\psi[\vartheta\bar{\vartheta}, (\psi^G)_N] \\ &= \sum_\psi a_\psi |G:NE_\psi|[\vartheta\bar{\vartheta}, (\psi_{E_\psi \cap N})^N].\end{aligned}$$

Since $p \nmid m$, there exists ψ such that p does not divide $|G:NE|[\vartheta\bar{\vartheta}, (\psi_{E \cap N})^N]$, where we have written $E = E_\psi$. Since G/N is a p-group and $p \nmid |G:NE|$, we conclude that $NE = G$, and (a) follows.

Now write $L = E \cap N$ so that

$$[\vartheta\bar{\vartheta}, (\psi_L)^N] = [\vartheta_L \bar{\vartheta}_L, \psi_L] = [\vartheta_L, \psi_L \vartheta_L]$$

is prime to p. Since ϑ is invariant in G we have ϑ_L is invariant in E and we may write $\vartheta_L = \sum_\Delta e_\Delta \Delta$, where Δ runs over sums of orbits of the action of E on $\mathrm{Irr}(L)$ and $e_\Delta \in \mathbb{Z}$. Write $\gamma = \psi_L \vartheta_L$ so that γ is invariant in E. We have

$$[\vartheta_L, \psi_L \vartheta_L] = \sum_\Delta e_\Delta [\Delta, \gamma]$$

is prime to p. Choose Δ such that $p \nmid e_\Delta[\Delta, \gamma]$ and write $\Delta = \varphi_1 + \cdots + \varphi_t$, where the $\varphi_i \in \mathrm{Irr}(L)$ and are conjugate under E.

Since γ is invariant under E, all $[\varphi_i, \gamma]$ are equal and $[\Delta, \gamma] = t[\varphi_1, \gamma]$. However t is a divisor of $|E:L| = |G:N|$ and so is a power of p. Since $p\nmid[\Delta, \gamma]$, we have $t = 1$ and $\varphi_1 = \varphi$ is invariant in E. This is (b). Finally, $[\vartheta_L, \varphi] = e_\Delta$ is prime to p and the proof is complete. ∎

An interesting special case is when $N = G$.

(8.25) COROLLARY Let $\chi \in \mathrm{Irr}(G)$ and let p be a fixed prime. Then there exists p-elementary $E \subseteq G$ and $\varphi \in \mathrm{Irr}(E)$ with $[\chi_E, \varphi] \not\equiv 0 \bmod p$.

Examples exist which show that in Corollary 8.25, E cannot always be taken to be a p-group.

Dade's Theorem 8.24 has a nice application to the question of character extendibility.

(8.26) THEOREM Let $N \lhd G$ with G/N a p-group. Let $P \in \mathrm{Syl}_p(G)$ and assume that $P \cap N \subseteq \mathbf{Z}(P)$ and that every linear character of $P \cap N$ is extendible to P. Then every G-invariant irreducible character of N is extendible to G.

Proof Let $\vartheta \in \mathrm{Irr}(N)$ be invariant in G. By Theorem 8.24, choose a p-elementary subgroup $E \subseteq G$, with $EN = G$, and $\varphi \in \mathrm{Irr}(E \cap N)$ such that $p\nmid[\vartheta_{E \cap N}, \varphi]$.

We have $E = Q \times C$, where Q is a p-group and C is cyclic. We may assume that C is a p'-group so that $E \cap N = (Q \cap N) \times C$ and we write $\varphi = \alpha \times \lambda$ where $\alpha \in \mathrm{Irr}(Q \cap N)$ and $\lambda \in \mathrm{Irr}(C)$. We may assume $Q \subseteq P$ so that $Q \cap N \subseteq P \cap N$, which is abelian. It follows that α is extendible to $P \cap N$ and thence to P by hypothesis. Therefore α has an extension $\hat{\alpha} \in \mathrm{Irr}(Q)$ and $\hat{\varphi} = \hat{\alpha} \times \lambda$ is an extension of φ to E.

Now since $EN = G$, we have $(\hat{\varphi}^G)_N = (\hat{\varphi}_{E \cap N})^N = \varphi^N$ and hence

$$[(\hat{\varphi}^G)_N, \vartheta] = [\varphi^N, \vartheta] = [\varphi, \vartheta_{E \cap N}],$$

which is prime to p. Therefore, there exists an irreducible constituent χ of $\hat{\varphi}^G$ with $p\nmid[\chi_N, \vartheta]$. Since ϑ is invariant in G, it follows that $\chi_N = e\vartheta$ with $p\nmid e$. Now e divides $|G:N|$ by Problem 6.7 and since G/N is a p-group it follows that $e = 1$ and χ extends ϑ. (Note that Problem 6.7 follows immediately from Corollary 6.19 by induction.) ∎

We remark that in the notation of Theorem 8.26, the hypothesis on P will automatically be satisfied if P is abelian. More generally, it suffices for $P' \cap N = 1$.

It has been mentioned that if $N \lhd G$ and $\vartheta \in \mathrm{Irr}(N)$ is invariant in G, then it suffices to show that ϑ is extendible to the inverse images in G of the Sylow subgroups of G/N in order to prove that ϑ is extendible to G. It is for this reason that it is useful to obtain results like Theorem 8.26 which give sufficient conditions for extendibility when the factor group is a p-group.

Problems

(8.1) Let $\mathbb{Z} \subseteq R \subseteq \mathbb{C}$, where R is a ring and the additive group $(\mathbb{Z}, +)$ is a direct summand of $(R, +)$. Let $\mathscr{H}$ be a family of subgroups of G. Show that $\mathscr{I}_R(G, \mathscr{H}) \cap \mathbb{Z}[\mathrm{Irr}(G)] = \mathscr{I}(G, \mathscr{H})$.

Hint Write $R = \mathbb{Z} \dotplus S$. If $\vartheta = \sum a_\psi \psi^G$, where all $a_\psi \in S$, then $[\vartheta, \chi] \in S$ for all $\chi \in \mathrm{Irr}(G)$.

Note The Brauer–Tate proof of Brauer's theorem uses this fact applied to $R = \mathbb{Z}[\varepsilon]$, where ε is a root of unity.

(8.2) Let $\mathscr{A}$ be the collection of abelian subgroups of the group G. Show that the following statements are equivalent:

(a) $1_G \in \mathscr{I}(G, \mathscr{A})$;
(b) $\mathscr{R}(G, \mathscr{A}) = \mathbb{Z}[\mathrm{Irr}(G)]$;
(c) every Sylow subgroup of G is abelian.

Hint If a Sylow p-subgroup of G is nonabelian, consider the function $\vartheta = (1/p)\rho$, where ρ is the regular character of G.

(8.3) Let χ be a character of G. For $g \in G$ and prime p, write $g = g_p g_{p'}$ in the notation introduced following Lemma 8.18. Define the class function χ_p by $\chi_p(g) = \chi(g_{p'})$. Show that $(\chi_p)_N$ is a character for every nilpotent $N \subseteq G$. Show that if G is not nilpotent then χ_p is not a character of G for some $\chi \in \mathrm{Irr}(G)$ and some prime p.

Note In fact, χ_p is a character for every $\chi \in \mathrm{Irr}(G)$ iff G has a normal Sylow p-subgroup. Of course, by Brauer's theorem, χ_p is always a generalized character.

(8.4) Let χ and ψ be characters of G and suppose $|G| = mn$, where $(m, n) = 1$. Let

$$\alpha = \sum \chi(g)\overline{\psi(g)},$$

where the sum is taken over those $g \in G$ such that $o(g) | m$. Show that $\alpha/m \in \mathbb{Z}$.

(8.5) Let $\chi \in \mathrm{Irr}(G)$ and suppose $|G| = mn$ with $(m, n) = 1$. Assume that $\chi(x) = 0$ for all $1 \neq x \in G$ such that $x^n = 1$. Suppose $y \in G$ and $y^m \neq 1$. Show that $\chi(y) = 0$.

(8.6) Let G and H be groups with classes $\mathscr{K}_i$ and $\mathscr{L}_i$ respectively and irreducible characters χ_i and ψ_i, respectively. Assume that whenever $g \in \mathscr{K}_i$ and $h \in \mathscr{L}_i$ we have $\chi_j(g) = \psi_j(h)$ for all i and j. (In short, G and H have identical character tables.) Let $x \in \mathbf{Z}(P)$ where $P \in \mathrm{Syl}_p(G)$. Suppose $x \in \mathscr{K}_i$. Let $y \in \mathscr{L}_i$. Show $o(x) = o(y)$.

Hint If $o(y) > o(x)$ and $z \in H$, with $\langle y \rangle = \langle z \rangle$ and $y^p = z^p$, show that $z \in \mathscr{L}_i$ and conclude that $p \,|\, |\mathscr{L}_i|$. See the hint for Problem 2.12.

(8.7) Suppose G and H have identical character tables and that G is solvable and has an abelian Sylow p-subgroup. Show that G and H have isomorphic Sylow p-subgroups.

Hint If G has no nontrivial normal p'-subgroup, then its Sylow p-subgroup is normal. Use Problem 8.6.

(8.8) (*Dade*) Let $N \lhd H \subseteq G$. Assume the following:

(a) $(|H:N|, |G:H|) = 1$.
(b) If x, $x^g \in H$ for some $g \in G$, then there exists $h \in H$ and $n \in N$ with $x^g = x^h n$.
(c) If $E \subseteq G$ is elementary and $(|E|, |G:H|) = 1$, then $E^g \subseteq H$ for some $g \in G$.

Show that there exists $M \lhd G$ such that $HM = G$ and $H \cap M = N$.

Hint Mimic the proof of Theorem 8.22. Consider only those class functions ϑ of H which are constant on cosets of N.

Note Many of the important "transfer" theorems follow from Problem 8.8, for instance, the *Focal Subgroup Theorem*:

Let $P \in \mathrm{Syl}_p(G)$ and let $N = \langle x^{-1}x^g \,|\, x \in P, x^g \in P, g \in G \rangle$. Then $\mathbf{A}^p(G) \cap P = N$.

Here, $\mathbf{A}^p(G)$ denotes the normal subgroup of G minimal such that the factor group is an abelian p-group. Problem 8.8 yields $M \lhd G$ with $MP = G$ and $M \cap P = N$. It follows that $M \supseteq \mathbf{A}^p(G)$. The reverse inclusion follows from $N \subseteq G' \subseteq \mathbf{A}^p(G)$.

(8.9) Show that a group G is quasi-elementary iff 1_G cannot be written in the form $\sum a_H(1_H)^G$ for proper subgroups H, with $a_H \in \mathbb{Z}$.

Hint For "only if": Assume $1_G = \sum a_H(1_H)^G$, where the H run over representatives for the conjugacy classes of proper subgroups. Let C be a cyclic normal p-complement for G. Choose H_0 minimal, such that $p \nmid (a_{H_0}|G:H_0|)$, and let λ be a faithful linear character of $C/(C \cap H_0)$. Consider the numbers $[(a_H(1_H)^G)_C, \lambda]$.

Note Problem 8.9 essentially says that Theorem 8.10(a) is the best possible.

(8.10) Show that the integers m in Theorem 8.10(b) and Corollary 8.13 can be taken to be the p'-part of $|G|$.

(8.11) Let $\chi \in \mathrm{Irr}(G)$ be a character of defect zero with respect to the prime p. Let $Q \subseteq G$, with $p \nmid |Q|$ and let $P \in \mathrm{Syl}_p(\mathbf{C}_G(Q))$. Show that $(1/|P|)\chi_Q$ is a character of Q.

(8.12) Let $P \lhd G$, where P is a p-group, and let $\chi \in \mathrm{Irr}(G/P)$ be of defect zero in G/P with respect to p. Let $\vartheta \in \mathrm{Irr}(P)$ be invariant in G. Define the function ψ on G by

$$\begin{aligned} \psi(g) &= 0 && \text{if} \quad g_p \notin P \\ \psi(g) &= \vartheta(g_p)\chi(g_{p'}) && \text{if} \quad g_p \in P. \end{aligned}$$

(a) Show that ψ is a generalized character of G.
(b) If $G = P\mathbf{C}_G(P)$, show that $\psi \in \mathrm{Irr}(G)$.

(8.13) Let $N \lhd G$ be a Hall π-subgroup. Define the function v on N by $v(x) = |\{y \in \mathbf{C}_G(x) \mid y \text{ is of } \pi'\text{-order}\}|$. Show that v is a generalized character of N.

Hint If ϑ is a G-invariant character of N (which is not necessarily irreducible), define $\tilde{\vartheta}$ on G by

$$\tilde{\vartheta}(g) = \vartheta(g_\pi).$$

Show that $\tilde{\vartheta}$ is a generalized character of G and compute $[\tilde{\vartheta}, 1_G]$.

9 Changing the field

So far we have restricted our attention almost exclusively to characters, representations, and modules over the complex numbers. In this chapter we digress from our study of the properties of $\text{Irr}(G)$ to consider irreducible group representations over arbitrary fields. (In prime characteristic, this falls far short of the general case, since not all representations are completely reducible.) In particular, if $F \subseteq E$ is a field extension we explore the connections between the irreducible E-representations and F-representations of a group. In prime characteristic, we shall see that this situation is (surprisingly) under somewhat better control than it is in characteristic zero, which will be more fully considered in Chapter 10.

Let $F \subseteq E$ and let $\mathfrak{X}$ be an F-representation of a group G. Then $\mathfrak{X}$ maps G into a group of nonsingular matrices over F that, of course, are also nonsingular over E. We may, therefore, view $\mathfrak{X}$ as an E-representation of G. As such we denote it by $\mathfrak{X}^E$. (The superscript thus merely indicates a change in point of view.) If $\mathfrak{X}_1$ and $\mathfrak{X}_2$ are similar F-representations, then ${\mathfrak{X}_1}^E$ and ${\mathfrak{X}_2}^E$ are similar and it follows that if $\mathfrak{X}$ corresponds to the $F[G]$-module V, then there is a uniquely defined (up to isomorphism) $E[G]$-module V^E that corresponds to $\mathfrak{X}^E$. (For those readers familiar with tensor products, we remark that $V^E \cong V \otimes_F E$.) We shall not, however, need to refer to V^E again, since it is usually easier to work with the representation $\mathfrak{X}^E$.

The F-representation $\mathfrak{X}$ may be extended by linearity to obtain a representation of $F[G]$, which we shall continue to call $\mathfrak{X}$. Under this convention, the $E[G]$-representation $\mathfrak{X}^E$ is an extension of the $F[G]$-representation $\mathfrak{X}$.

If $\mathfrak{X}^E$ is irreducible, then clearly so is $\mathfrak{X}$. However, $\mathfrak{X}^E$ may well reduce, even if $\mathfrak{X}$ is irreducible. To illustrate what can happen, we consider two examples, both for the field extension $\mathbb{R} \subseteq \mathbb{C}$. Let $G = \langle g \rangle$ be cyclic of order 3 and let $\mathfrak{X}$ be the $\mathbb{R}$-representation defined by

$$\mathfrak{X}(g) = \begin{pmatrix} 0 & 1 \\ -1 & -1 \end{pmatrix}.$$

Then $\mathfrak{X}^{\mathbb{C}}$ affords the character $\lambda + \bar{\lambda}$, where λ is a faithful linear character of G and so $\mathfrak{X}^{\mathbb{C}}$ is reducible. Since λ is not real valued, it is not afforded by any $\mathbb{R}$-representation and it follows that $\mathfrak{X}$ is irreducible over $\mathbb{R}$.

Now let $Q_8 = \langle a, b \rangle$ be the quaternion group of order 8, where $o(a) = 4 = o(b)$. Define

$$\mathfrak{X}(a) = \begin{pmatrix} 0 & 1 & 0 & 0 \\ -1 & 0 & 0 & 0 \\ 0 & 0 & 0 & -1 \\ 0 & 0 & 1 & 0 \end{pmatrix}, \qquad \mathfrak{X}(b) = \begin{pmatrix} 0 & 0 & 1 & 0 \\ 0 & 0 & 0 & 1 \\ -1 & 0 & 0 & 0 \\ 0 & -1 & 0 & 0 \end{pmatrix}.$$

It is not hard to check that this defines a representation of Q_8. (In fact this is the representation whose module is the quaternion algebra $\mathbb{H} = \mathbb{R} + \mathbb{R}i + \mathbb{R}j + \mathbb{R}k$ with respect to the basis $\{1, i, j, k\}$, where $a = i$ and $b = j$ act by right multiplication.) Now $\mathfrak{X}^{\mathbb{C}}$ affords the character 2χ, where $\chi \in \mathrm{Irr}(Q_8)$ and $\chi(1) = 2$. By Problem 2.5(b), χ is not afforded by any $\mathbb{R}$-representation of Q_8 and thus $\mathfrak{X}$ is irreducible over $\mathbb{R}$. (This can also be deduced from the fact that $\mathbb{H}$ is a division algebra.)

(9.1) DEFINITION Let $\mathfrak{X}$ be an F-representation of G. Then $\mathfrak{X}$ is *absolutely irreducible* if $\mathfrak{X}^E$ is irreducible for every field $E \supseteq F$.

(9.2) THEOREM Let $\mathfrak{X}$ be an irreducible F-representation of G with degree n. The following are equivalent.

(a) $\mathfrak{X}$ is absolutely irreducible.
(b) $\mathfrak{X}^E$ is irreducible for every finite degree extension $E \supseteq F$.
(c) The centralizer of $\mathfrak{X}(G)$ in the matrix ring $M_n(F)$ consists of scalar matrices.
(d) $\mathfrak{X}(F[G]) = M_n(F)$.

Proof That (a) implies (b) is trivial. Now assume (b) and let $M \in M_n(F)$, with $M\mathfrak{X}(g) = \mathfrak{X}(g)M$ for all $g \in G$. Let E be a finite degree extension of F, chosen so that M has an eigenvalue, $\lambda \in E$. Since $\mathfrak{X}^E$ defines an irreducible representation of $E[G]$ and $M - \lambda I$ is a singular matrix centralizing its image, it follows from Schur's Lemma 1.5 that $M - \lambda I = 0$. Thus (c) follows.

That (c) implies (d) is immediate from the Double Centralizer Theorem 1.16. Finally, assume (d) and let $L \supseteq F$. Since every $M \in M_n(L)$ is an L-linear combination of matrices in $M_n(F)$, it follows that $\mathfrak{X}(L[G]) = M_n(L)$. Thus for every L representation $\mathfrak{Y}$ similar to $\mathfrak{X}$, we have $\mathfrak{Y}(L[G]) = M_n(L)$ and so $\mathfrak{Y}$ cannot be in reduced form. Thus $\mathfrak{X}$ is irreducible over L and the proof is complete. ∎

(9.3) DEFINITION The field F is a *splitting field* for G if every irreducible F-representation of G is absolutely irreducible.

(9.4) COROLLARY If F is algebraically closed, then F is a splitting field for every group.

Proof Since F has no proper finite degree extension, condition (b) of Theorem 9.2 holds for every irreducible F-representation of G. ∎

Suppose F is a splitting field for G and $E \supseteq F$. Then every irreducible F-representation $\mathfrak{X}$ determines an irreducible E-representation $\mathfrak{X}^E$. If $\{\mathfrak{X}_i\}$ is a set of representatives for the similarity classes of irreducible F-representations of G, then $\{\mathfrak{X}_i^E\}$ is a complete set of representatives for the irreducible E-representations of G. In order to prove this, we need to discuss the "irreducible constituents" of a possibly reducible representation.

If V is an $F[G]$-module, then a *composition series* for V is a chain of submodules

$$V = V_0 > V_1 > \cdots > V_n = 0$$

such that each V_{i-1}/V_i is an irreducible module. The modules V_{i-1}/V_i are the *factors* of the series. The Jordan–Hölder theorem asserts that the factors of any two composition series for V are the same up to isomorphism (and counting multiplicities). An irreducible module W isomorphic to a factor of some (and hence all) composition series for V is called an *irreducible constituent* of V. If W is isomorphic to an irreducible submodule of V, then W is a constituent of V. We call it a *bottom* constituent in that case. Similarly any irreducible homomorphic image of V is a constituent. These are the *top* constituents of V. If V is completely reducible [e.g., if $\mathrm{char}(F) \nmid |G|$], then every irreducible constituent of V is both a top and a bottom constituent. (Caution: the converse is false.)

All of the preceding remarks may be translated into the language of representations. If $\mathfrak{X}$ is an F-representation of G corresponding to the $F[G]$-module V, then $\mathfrak{X}$ is similar to a representation $\mathfrak{Z}$ in triangular block form

$$\mathfrak{Z}(g) = \begin{pmatrix} \mathfrak{Z}_1(g) & & & * \\ & \mathfrak{Z}_2(g) & & \\ & & \ddots & \\ 0 & & & \mathfrak{Z}_n(g) \end{pmatrix},$$

where the irreducible representations $\mathfrak{Z}_i$ correspond to the factors V_{i-1}/V_i.

Representations similar to the $\mathfrak{Z}_i$ are the irreducible constituents of $\mathfrak{X}$; those similar to $\mathfrak{Z}_1$ being top constituents and those similar to $\mathfrak{Z}_n$ being bottom constituents. (Of course some of the other irreducible constituents of $\mathfrak{X}$ may also be top or bottom constituents since $\mathfrak{Z}$ is not necessarily the only representation in triangular block form to which $\mathfrak{X}$ is similar.) Note that the character afforded by a representation is the sum of the characters of all of the irreducible constituents (counting multiplicities). We emphasize that by the Jordan–Hölder theorem, the representation $\mathfrak{X}$ has only finitely many irreducible constituents (up to similarity), namely those which appear in any single triangular block representation similar to $\mathfrak{X}$.

(9.5) COROLLARY Let F be a field and G a group.

(a) Every irreducible F-representation of G is a top constituent of the regular F-representation $\mathfrak{R}$.

(b) There exist only finitely many similarity classes of irreducible F-representations of G.

(c) If $E \supseteq F$ and $\mathfrak{Y}$ is an irreducible E-representation of G, then $\mathfrak{Y}$ is a constituent of $\mathfrak{X}^E$ for some irreducible F-representation $\mathfrak{X}$.

Proof Statement (a) is immediate from Lemma 1.14 and (b) follows from (a) via the Jordan–Hölder theorem. Let $\mathfrak{Z}$ be any F-representation of G. The irreducible constituents of $\mathfrak{Z}^E$ may be found by taking the irreducible constituents $\mathfrak{X}_i$ of $\mathfrak{Z}$ and then finding the irreducible constituents of the $\mathfrak{X}_i^E$. Now (c) follows by applying this remark to $\mathfrak{Z} = \mathfrak{R}$. ∎

The following result is a useful tool for establishing the similarity of two F-representations of G and for other purposes.

(9.6) THEOREM Let $\mathfrak{X}$ be an irreducible representation of $F[G]$ and let $a \in F[G]$. Then there exists $b \in F[G]$ such that $\mathfrak{X}(b) = \mathfrak{X}(a)$ and $\mathfrak{Y}(b) = 0$ for every irreducible $F[G]$-representation $\mathfrak{Y}$ which is not similar to $\mathfrak{X}$.

Proof Let $\{\mathfrak{X}_i\}$ be a set of representatives for the similarity classes of irreducible $F[G]$-representations. Let $I_i = \{x \in F[G] \mid \mathfrak{X}_i(x) = 0\}$ so that I_i is an ideal of $F[G]$ and in the notation of Problem 1.4, $J(F[G]) = \bigcap I_i$. Let $A = F[G]/(\bigcap I_i)$, so that each $\mathfrak{X}_i$ may be viewed as a representation of the algebra A. As such, the $\mathfrak{X}_i$ are irreducible and pairwise nonsimilar. In particular, $J(A) = 0$ and by Problem 1.5, A is semisimple. By Theorem 1.15, A has minimal ideals M_i, such that $\mathfrak{X}_j(M_i) = 0$ if $j \neq i$ and $\mathfrak{X}_i(M_i) = \mathfrak{X}_i(A) = \mathfrak{X}_i(F[G])$. Now suppose $\mathfrak{X} = \mathfrak{X}_1$. Choose b in the inverse image in $F[G]$ of M_1 with $\mathfrak{X}(b) = \mathfrak{X}(a)$. The result follows. ∎

(9.7) COROLLARY Let $\mathfrak{X}$ and $\mathfrak{Y}$ be irreducible F-representations of G. Suppose $F \subseteq E$ and that $\mathfrak{X}^E$ and $\mathfrak{Y}^E$ have a common irreducible constituent. Then $\mathfrak{X}$ is similar to $\mathfrak{Y}$.

Proof Let $\mathfrak{Z}$ be an irreducible constituent of $\mathfrak{X}^E$ and of $\mathfrak{Y}^E$. If $\mathfrak{X}$ and $\mathfrak{Y}$ are not similar, view them as representations of $F[G]$ and choose $b \in F[G]$ with $\mathfrak{X}(b) = \mathfrak{X}(1)$ and $\mathfrak{Y}(b) = 0$. It follows that if $\mathfrak{U}$ is any $E[G]$-representation similar to $\mathfrak{X}^E$, then $\mathfrak{U}(b)$ is the identity matrix and hence $\mathfrak{Z}(b)$ is the identity matrix. A similar argument, working with $\mathfrak{Y}^E$, shows that $\mathfrak{Z}(b) = 0$ and this contradiction proves the result. ∎

Note that Corollary 9.7 asserts that the representation $\mathfrak{X}$ in Corollary 9.5(c) is unique up to similarity.

(9.8) COROLLARY Let F be a splitting field for G and let $\{\mathfrak{X}_i\}$ be a set of representatives for the similarity classes of irreducible F-representations. Suppose $E \supseteq F$. Then E is a splitting field and $\{\mathfrak{X}_i^E\}$ is a set of representatives for the irreducible E-representations of G.

Proof Since the $\mathfrak{X}_i$ are absolutely irreducible, the $\mathfrak{X}_i^E$ are absolutely irreducible. They are pairwise nonsimilar by Corollary 9.7. Finally, suppose $\mathfrak{Y}$ is any irreducible E-representation. By Corollary 9.5(c), $\mathfrak{Y}$ is a constituent of $\mathfrak{X}_i^E$ for some i. Since $\mathfrak{X}_i^E$ is irreducible, it is similar to $\mathfrak{Y}$ and the proof is complete. ∎

(9.9) THEOREM Let E be a splitting field for G and let $F \subseteq E$. Then F is a splitting field iff every irreducible E-representation of G is similar to $\mathfrak{Y}^E$ for some F-representation $\mathfrak{Y}$.

Proof Suppose F is a splitting field. Then by Corollary 9.8, every irreducible E-representation is as desired. Conversely, let $\mathfrak{Z}$ be any irreducible F-representation and let $\mathfrak{X}$ be an irreducible constituent of $\mathfrak{Z}^E$. By hypothesis, there exists an irreducible F-representation $\mathfrak{Y}$, such that $\mathfrak{Y}^E$ is similar to $\mathfrak{X}$ and hence $\mathfrak{Y}^E$ and $\mathfrak{Z}^E$ have an irreducible constituent in common. By Corollary 9.7, $\mathfrak{Y}$ is similar to $\mathfrak{Z}$, and thus $\mathfrak{Z}^E$ is absolutely irreducible. It follows that $\mathfrak{Z}$ is absolutely irreducible since the only F-matrices which centralize all $\mathfrak{Z}(g)$ are scalar. The proof is complete. ∎

(9.10) COROLLARY Let F be any field and G a group. Then some finite degree extension of F is a splitting field for G.

Proof Let $\bar{F}$ be the algebraic closure of F, so that $\bar{F}$ is a splitting field. Let $\{\mathfrak{X}_i\}$ be a set of representatives for the similarity classes of irreducible $\bar{F}$-representations. By Corollary 9.5(b), $|\{\mathfrak{X}_i\}| < \infty$ and hence only finitely many elements of $\bar{F}$ occur as entries in any of the matrices $\mathfrak{X}_i(g)$ for $g \in G$. Adjoin all of these elements to F so as to obtain the field E. Since $\bar{F}$ is algebraic over F, it follows that $|E : F| < \infty$. Since each $\mathfrak{X}_i$ may be viewed as an E-representation of G, it follows from Theorem 9.9 that E is a splitting field. ∎

(9.11) COROLLARY Let $\mathbb{Q} \subseteq F \subseteq \mathbb{C}$. Then F is a splitting field for G iff every $\chi \in \mathrm{Irr}(G)$ is afforded by some F-representation.

Proof If F is a splitting field and $\chi \in \mathrm{Irr}(G)$, choose a $\mathbb{C}$-representation $\mathfrak{X}$ that affords χ. By Theorem 9.9, $\mathfrak{X}$ is similar to $\mathfrak{Y}^{\mathbb{C}}$ for some F-representation $\mathfrak{Y}$. Then $\mathfrak{Y}$ affords χ.

Conversely, suppose that every $\chi \in \mathrm{Irr}(G)$ is afforded by some F-representation. Let $\mathfrak{X}$ be an irreducible $\mathbb{C}$-representation and let $\mathfrak{Y}$ be an F-representation which affords the same character. Then by the linear independence of $\mathrm{Irr}(G)$ and using the fact that $\mathbb{C}$ has characteristic zero, it follows that $\mathfrak{X}$ is the unique irreducible constituent of $\mathfrak{Y}^{\mathbb{C}}$ and occurs with multiplicity 1. Thus $\mathfrak{Y}^{\mathbb{C}}$ is similar to $\mathfrak{X}$. The result now follows from Theorem 9.9. ∎

We shall now discuss the character theory of a group over an arbitrary splitting field. The following result is actually true without the assumption that E is a splitting field. We shall prove the more general fact later.

(9.12) LEMMA Let E be a splitting field for G. Then the characters of nonsimilar irreducible E-representations of G are nonzero, distinct, and linearly independent over E.

Proof Let $\{\mathfrak{X}_i\}$ be a set of representatives for the similarity classes of irreducible E-representations, and let χ_i be the character afforded by $\mathfrak{X}_i$. View χ_i as being defined on all of $E[G]$. Since $\mathfrak{X}_i(E[G])$ is a full matrix ring over E, we may choose $a_i \in E[G]$ with $\chi_i(a_i) = 1$. By Theorem 9.6, we may assume that $\chi_j(a_i) = 0$ if $i \neq j$. The result is now immediate. ∎

If E is a splitting field for G, we shall use the notation $\mathrm{Irr}_E(G)$ to denote the set of characters of the (absolutely) irreducible E-representations of G. The point of the next result is that in some sense $\mathrm{Irr}_E(G)$ is not as dependent on the particular field E as the notation would indicate.

(9.13) LEMMA Let E be a splitting field for G and let $\chi \in \mathrm{Irr}_E(G)$. Suppose $K \subseteq E$ is a subfield which contains $\chi(g)$ for all $g \in G$ and let $F \supseteq K$ be another splitting field for G. Then $\chi \in \mathrm{Irr}_F(G)$.

Proof We may replace F by a K-isomorphic copy and assume that E, $F \subseteq L$ for some field L. By Corollary 9.8, L is a splitting field for G and $\mathrm{Irr}_F(G) = \mathrm{Irr}_L(G) = \mathrm{Irr}_E(G)$. The result follows. ∎

The next result is of great importance in studying representations in prime characteristic. Its proof depends on Wedderburn's theorem which asserts that finite division rings are commutative. As the quaternion group of order 8 shows, Theorem 9.14 would be false in characteristic zero.

(9.14) THEOREM Let $\mathfrak{X}$ be an absolutely irreducible E-representation of G, where E is of prime characteristic. Suppose that $\mathfrak{X}$ affords the character χ and that $\chi(g) \in F$ for all $g \in G$, where F is some subfield of E. Then $\mathfrak{X}$ is similar to $\mathfrak{Y}^E$ for some absolutely irreducible F-representation $\mathfrak{Y}$.

Proof First we consider the crucial special case that $|E| < \infty$. Working by induction on $|E|$, it is no loss to assume that F is a maximal subfield of E. Let A be the set of all F-linear combinations of the matrices $\mathfrak{X}(g)$ for $g \in G$. Then A is an F-subalgebra of the matrix ring $M_n(E)$, where n is the degree of $\mathfrak{X}$. If $a \in \mathbf{Z}(A)$, then a is a matrix over E which commutes with all $\mathfrak{X}(g)$. Since $\mathfrak{X}$ is absolutely irreducible, a is a scalar matrix and thus $F \cdot 1 \subseteq \mathbf{Z}(A) \subseteq E \cdot 1$. It follows that $\mathbf{Z}(A)$ is a finite integral domain, and hence is a field. Since F is a maximal sufield of E, we conclude that either $\mathbf{Z}(A) = F \cdot 1$ or $\mathbf{Z}(A) = E \cdot 1$.

If $\mathbf{Z}(A) = E \cdot 1$, then $A = E \cdot A = \mathfrak{X}(E[G])$. Since $\mathfrak{X}$ is absolutely irreducible, we conclude that $A = M_n(E)$ and thus if $\alpha \in E - F$, we can choose $a \in A$ with $\operatorname{tr}(a) = \alpha$. However, a is an F-linear combination of matrices of the form $\mathfrak{X}(g)$, each of which has trace in F. This contradiction shows that $\mathbf{Z}(A) = F \cdot 1$.

Next, we claim that the F-algebra A is algebra isomorphic to the full matrix algebra $M_k(F)$ for some integer k. To show this, we first establish that A is semisimple. By Problem 1.5, it suffices to show that A has no nonzero nilpotent ideals. If I is a nilpotent ideal of A, then $E \cdot I$ is a nilpotent ideal in $E \cdot A = \mathfrak{X}(E[G]) = M_n(E)$. It follows that $I = 0$. (Any nonzero matrix has a multiple which is not nilpotent.)

By Theorem 1.15, the semisimple F-algebra A is a direct sum of minimal ideals. Since $\dim_F(\mathbf{Z}(A)) = 1$, it follows that there is only one direct summand and hence in the notation of 1.15, $A = M(A) \cong A_M$ for some irreducible A-module M. Let $D = \mathbf{E}_A(M)$, so that D is a division ring by Schur's Lemma 1.5. Since $\dim_F(M) < \infty$ and $|F| < \infty$, we have $|D| < \infty$ and thus D is commutative by Wedderburn's theorem.

By the Double Centralizer Theorem 1.16, we have $A_M = \mathbf{E}_D(M)$ and thus $D \subseteq \mathbf{Z}(A_M) \cong F$. Since D contains the scalar multiplications, $1 \cdot F$, of F on M, we conclude that $D = 1 \cdot F$. Thus if $k = \dim_F(M)$, then $A \cong A_M = \mathbf{E}_D(M) = \mathbf{E}_F(M) \cong M_k(F)$ and we have an F-algebra isomorphism $\vartheta : A \to M_k(F)$. Thus $\mathfrak{Y} = \mathfrak{X}\vartheta : F[G] \to M_k(F)$ is a representation of $F[G]$. It is absolutely irreducible since it maps onto $M_k(F)$.

We claim that $\mathfrak{Y}^E$ is similar to $\mathfrak{X}$. To see this, let $\mathfrak{Z}$ be an irreducible F-representation such that $\mathfrak{X}$ is a constituent of $\mathfrak{Z}^E$. If $\mathfrak{Z}$ is not similar to $\mathfrak{Y}$, use Theorem 9.6 to choose $b \in F[G]$ such that $\mathfrak{Z}(b) = 0$ and $\mathfrak{Y}(b) \neq 0$. Since $\mathfrak{X}$ is a constituent of $\mathfrak{Z}^E$, we have $\mathfrak{X}(b) = 0$ and thus $\mathfrak{Y}(b) = (\mathfrak{X}(b))\vartheta = 0$. This contradiction shows that $\mathfrak{Y}$ is similar to $\mathfrak{Z}$ and thus $\mathfrak{X}$ is a constituent of $\mathfrak{Y}^E$. Since $\mathfrak{Y}$ is absolutely irreducible, the claim follows and the theorem is proved when E is finite.

We now consider the general case where E may be infinite. Since we may replace E by a larger field, it is no loss to assume that E is algebraically closed. Let $K \subseteq E$ be the prime field and let $L \supseteq K$ be a splitting field with $|L:K| < \infty$ (by Corollary 9.10). Since E is algebraically closed, we may assume $L \subseteq E$. Since L is a splitting field, $\mathfrak{X}$ is similar to $\mathfrak{Z}^E$ for some L-representation $\mathfrak{Z}$ and $\mathfrak{Z}$ affords χ which takes values in $L \cap F$.

Since $|L| < \infty$, the first part of the proof yields an absolutely irreducible $(L \cap F)$-representation $\mathfrak{Y}$, such that $\mathfrak{Y}^L$ is similar to $\mathfrak{Z}$ and hence $\mathfrak{Y}^E$ is similar to $\mathfrak{X}$. Now $\mathfrak{Y}^F$ is the desired F-representation and the proof is complete. ∎

Theorem 9.14 remains true if $\mathfrak{X}$ is irreducible but not absolutely irreducible. This will be proved in Corollary 9.23.

By the *exponent* of G we mean the least positive integer n, such that $g^n = 1$ for all $g \in G$. Clearly $n \mid |G|$.

(9.15) COROLLARY Let G have exponent n and assume the polynomial $x^n - 1$ splits into linear factors in the field F. If F has prime characteristic, then it is a splitting field for G.

Proof Let $E \supseteq F$ be a splitting field and let $\chi \in \mathrm{Irr}_E(G)$. Then $\chi(g)$ is a sum of nth roots of unity and hence $\chi(g) \in F$. The result follows by Theorems 9.14 and 9.9. ∎

An entirely different proof shows that Corollary 9.15 also holds in characteristic zero. This result of R. Brauer depends on his theorem on induced characters and will be proved in Chapter 10.

Let E be any field, σ an automorphism of E, and $\mathfrak{X}$ an E-representation of G. We can apply σ to every entry in the matrix $\mathfrak{X}(g)$ for every $g \in G$. What results is a new representation, denoted $\mathfrak{X}^\sigma$, which may or may not be similar to $\mathfrak{X}$. Similarly, if $\vartheta : G \to E$ is a function, we write $\vartheta^\sigma(g) = (\vartheta(g))^\sigma$. Suppose $\mathfrak{X}$ affords the E-character χ and that χ takes values in $F \subseteq E$. If $\tau \in \mathrm{Aut}(F)$, it is not obvious that χ^τ is an E-character of G.

(9.16) LEMMA Let E be a splitting field for G and let $\chi \in \mathrm{Irr}_E(G)$. Suppose $\chi(g) \in F \subseteq E$ for all $g \in G$ and let $\tau \in \mathrm{Aut}(F)$. Then $\chi^\tau \in \mathrm{Irr}_E(G)$.

Proof Let $\bar{E}$ be an algebraic closure for E and let $\bar{F} \subseteq \bar{E}$ be an algebraic closure for F. Then $\mathrm{Irr}_E(G) = \mathrm{Irr}_{\bar{E}}(G) = \mathrm{Irr}_{\bar{F}}(G)$ by Corollary 9.8. Since τ is extendible to an automorphism of $\bar{F}$, the result follows. ∎

Let $K \subseteq E$ be a field extension and suppose χ is an E-character of G. We write $K(\chi)$ to denote the subfield of E generated by K and the character values $\chi(g)$ for $g \in G$. Note that $K(\chi)$ is contained in a splitting field for a polynomial of the form $x^n - 1$ over K. Since this polynomial yields a Galois extension with an abelian Galois group, it follows that $K(\chi)$ is a finite degree Galois extension of K and the Galois group $\mathscr{G}(K(\chi)/K)$ is abelian.

Assume that E is a splitting field for G and that $F \subseteq E$. If $\chi, \psi \in \operatorname{Irr}_E(G)$, we say that χ and ψ are *Galois conjugate over* F if $F(\chi) = F(\psi)$ and there exists $\sigma \in \mathscr{G}(F(\chi)/F)$ such that $\chi^\sigma = \psi$. It is clear that this defines an equivalence relation on $\operatorname{Irr}_E(G)$.

(9.17) LEMMA Let E be a splitting field for G and let $\chi \in \operatorname{Irr}_E(G)$. Let $\mathscr{S}$ be the equivalence class of χ with respect to Galois conjugacy over F where $F \subseteq E$. We have

(a) If $K \subseteq E$, $K(\chi) = K$, and $\sigma \in \mathscr{G}(K/K \cap F)$, then $\chi^\sigma \in \mathscr{S}$.
(b) If $F_0 \subseteq F$ and $\psi \in \mathscr{S}$, then ψ is Galois conjugate to χ over F_0.
(c) $|\mathscr{S}| = |F(\chi):F|$.

Proof (a) By 9.16, $\chi^\sigma \in \operatorname{Irr}_E(G)$. Now $(K \cap F)(\chi) \supseteq K \cap F$ is a normal extension and so $(K \cap F)(\chi) \subseteq K$ is invariant under σ. It is thus no loss to assume that $K = (K \cap F)(\chi)$ and so $K \subseteq F(\chi)$ and no proper subfield of $F(\chi)$ contains both F and K. It follows by Galois theory that restiction maps $\mathscr{G}(F(\chi)/F)$ onto $\mathscr{G}(K/K \cap F)$ and so $\chi^\sigma = \chi^\tau$ for some $\tau \in \mathscr{G}(F(\chi)/F)$. Now $F(\chi^\tau) = F(\chi)$ and so $\chi^\sigma = \chi^\tau \in \mathscr{S}$.

(b) This is immediate by applying (a) to F_0 and taking $K = F(\chi)$.

(c) We have that $\mathscr{S}$ is the orbit of χ under $\mathscr{G}(F(\chi)/F)$. By definition of $F(\chi)$, the stabilizer of χ in this group is trivial and so $|\mathscr{S}| = |\mathscr{G}(F(\chi)/F)| = |F(\chi):F|$. ∎

Now let F be any field and let $\mathfrak{X}$ be an irreducible F-representation of G. Let $E \supseteq F$ be a splitting field for G. What does $\mathfrak{X}^E$ look like? We shall prove that it is completely reducible; that all irreducible constituents occur with equal multiplicity; that the characters of these constituents constitute a Galois conjugacy class over F and that the common multiplicity is 1 except possibly when F has characteristic zero.

(9.18) LEMMA Let $F \subseteq E$ with $|E:F| = n < \infty$ and let V be an irreducible $E[G]$-module corresponding to the E-representation $\mathfrak{X}$. Then V may be viewed as an $F[G]$-module and as such let it correspond to the F-representation $\mathfrak{Z}$. Then

(a) $\deg \mathfrak{Z} = n \deg \mathfrak{X}$.
(b) $\mathfrak{Z}$ has a unique (up to similarity) irreducible constituent. It is the F-representation $\mathfrak{Y}$ such that $\mathfrak{X}$ is a constituent of $\mathfrak{Y}^E$.
(c) If $\mathfrak{X}$ affords the E-character χ and $F(\chi) = F$, then $\mathfrak{Z}$ affords $n\chi$.

Proof We may certainly view V as an F-space. Let $v_1, v_2, \ldots, v_m$ be an E-basis for V and let $e_1, e_2, \ldots, e_n$ be an F-basis for E. It is routine to check that $\{v_i e_j\}$ is an F-basis for V and so V is finite dimensional over F, $\mathfrak{Z}$ is defined, and (a) follows.

Let $\mathfrak{Y}$ be a (unique up to similarity) irreducible F-representation of G such that $\mathfrak{X}$ is a constituent of $\mathfrak{Y}^E$. By Theorem 9.6, choose $b \in F[G] \subseteq E[G]$, such that $\mathfrak{Y}(b) = \mathfrak{Y}(1)$ and $\mathfrak{Y}_0(b) = 0$ for every irreducible F-representation $\mathfrak{Y}_0$ not similar to $\mathfrak{Y}$. Since $\mathfrak{X}$ is a constituent of $\mathfrak{Y}^E$ and $\mathfrak{Y}(b)$ is an identity matrix, then so is $\mathfrak{X}(b)$ an identity matrix. Thus b acts as an identity on V and hence $\mathfrak{Z}(b)$ is an identity matrix and $\mathfrak{Y}_i(b) \neq 0$ for every irreducible constituent $\mathfrak{Y}_i$ of $\mathfrak{Z}$. Thus every $\mathfrak{Y}_i$ is similar to $\mathfrak{Y}$ and (b) is proved.

For (c), let $g \in G$ and write $v_i g = \sum_j v_j \alpha_{ij}$, where $\alpha_{ij} \in E$ and $\chi(g) = \sum_i \alpha_{ii}$. Now write

$$\alpha_{ij} e_\mu = \sum_\nu e_\nu \beta_{ij\mu\nu}$$

for $1 \leq \mu, \nu \leq n$ and $\beta_{ij\mu\nu} \in F$. Let $\mathfrak{Z}$ afford the character ψ. Then

$$(v_i e_\mu) g = \sum_{j,\nu} v_j e_\nu \beta_{ij\mu\nu} \qquad \text{and} \qquad \psi(g) = \sum_{i,\mu} \beta_{ii\mu\mu}.$$

We have

$$\left(\sum_i \alpha_{ii}\right) e_\mu = \sum_{i,\nu} e_\nu \beta_{ii\mu\nu}$$

and since we are assuming that $\sum_i \alpha_{ii} = \chi(g) \in F$, we conclude that

$$\sum_i \alpha_{ii} = \sum_i \beta_{ii\mu\mu}$$

for each μ, $1 \leq \mu \leq n$. The result now follows. ∎

(9.19) COROLLARY Let $F \subseteq E$ with $|E:F| = n < \infty$. Let $\mathfrak{X}$ be an irreducible E-representation of G and let $\mathfrak{Y}$ be an irreducible F-representation such that $\mathfrak{X}$ is a constituent of $\mathfrak{Y}^E$. Then $\deg \mathfrak{Y}$ divides $n(\deg \mathfrak{X})$.

Proof Let $\mathfrak{Z}$ be an F-representation obtained by viewing an $E[G]$-module corresponding to $\mathfrak{X}$ as an $F[G]$-module. By Corollary 9.7 and Lemma 9.18(b), we conclude that $\mathfrak{Y}$ is the unique irreducible constituent of $\mathfrak{Z}$ and so $\deg \mathfrak{Y}$ divides $\deg \mathfrak{Z}$. The result now follows from 9.18(a). ∎

The following corollary is what survives of Theorem 9.14 when the hypothesis of prime characteristic is dropped.

(9.20) COROLLARY Let $\mathfrak{X}$ be an absolutely irreducible E-representation of G which affords the character χ. Let $F \subseteq E$ be such that $F(\chi) = F$. Then there exists an irreducible F-representation $\mathfrak{Y}$, such that $\mathfrak{X}$ is the unique (up to similarity) irreducible constituent of $\mathfrak{Y}^E$. In particular, $\mathfrak{Y}$ affords $m\chi$ for some integer m.

Proof If F has prime charactersitic, then by Theorem 9.14, there exists $\mathfrak{Y}$ such that $\mathfrak{Y}^E$ is similar to $\mathfrak{X}$ and there is nothing to prove. Assume then that $\text{char}(F) = 0$.

It will suffice to show for some positive integer n that the character $n\chi$ is afforded by some F-representation $\mathfrak{Z}$. This is sufficient since by the linear independence of $\text{Irr}_L(G)$ for a splitting field $L \supseteq E$, it follows that every irreducible constituent of $\mathfrak{Z}^L$ affords χ and so is similar to $\mathfrak{X}^L$. We may thus take $\mathfrak{Y}$ to be any irreducible constituent of $\mathfrak{Z}$ and quote Corollary 9.7 to complete the proof.

To produce $\mathfrak{Z}$, let $K \supseteq F$ be a splitting field with $|K:F| = n < \infty$. By Lemma 9.13, $\chi \in \text{Irr}_K(G)$. Let $\mathfrak{Z}$ be the F-representation which results by taking a $K[G]$-module which affords χ and viewing it as an $F[G]$-module. Then $\mathfrak{Z}$ affords $n\chi$ by Lemma 9.18(c) and the result follows. ∎

Of course it follows in the above situation that if $\mathfrak{Y}_0$ is any irreducible F-representation such that $\mathfrak{X}$ is a constituent of $(\mathfrak{Y}_0)^E$, then $\mathfrak{Y}_0$ is similar to $\mathfrak{Y}$ and hence $\mathfrak{X}$ is the unique irreducible constituent of $(\mathfrak{Y}_0)^E$.

(9.21) THEOREM Let $F \subseteq E$, where E is a splitting field for G. Let $\mathfrak{Y}$ be an irreducible F-representation of G. Then

(a) The irreducible constituents of $\mathfrak{Y}^E$ all occur with equal multiplicity m.

(b) If $\text{char}(E) \neq 0$, then $m = 1$.

(c) The characters $\chi_i \in \text{Irr}_E(G)$ afforded by the irreducible constituents of $\mathfrak{Y}^E$ constitute a Galois conjugacy class over F and so the fields $F(\chi_i)$ are all equal.

(d) Let $L = F(\chi_i)$. The irreducible constituents of $\mathfrak{Y}^L$ occur with multiplicity 1.

(e) If $\mathfrak{Z}$ is any irreducible constituent of $\mathfrak{Y}^L$ then $\mathfrak{Z}^E$ has a unique irreducible constituent. Its multiplicity is m.

(f) $\mathfrak{Y}^L$ and $\mathfrak{Y}^E$ are completely reducible.

Proof Let $\mathfrak{X}$ be an irreducible constituent of $\mathfrak{Y}^E$ and suppose $\mathfrak{X}$ affords $\chi \in \text{Irr}_E(G)$. Let $L = F(\chi)$ and let $\mathfrak{Z}$ be an irreducible constituent of $\mathfrak{Y}^L$ such that $\mathfrak{X}$ is a constituent of $\mathfrak{Z}^E$. By Corollary 9.20, $\mathfrak{X}$ is the unique irreducible constituent of $\mathfrak{Z}^E$. Let m be its multiplicity, so that $\mathfrak{Z}$ affords the character $m\chi$. If $\text{char}(E) \neq 0$, then Theorem 9.14 yields $m = 1$ and $\mathfrak{Z}$ is absolutely irreducible.

Let $\chi = \chi_1, \chi_2, \ldots, \chi_n$ be the distinct Galois conjugates of χ over F so that $n = |L:F|$ by Lemma 9.17(c). For $\sigma \in \mathcal{G}(L/F)$, form the L-representation $\mathfrak{Z}^\sigma$ defined by $\mathfrak{Z}^\sigma(g) = \mathfrak{Z}(g)^\sigma$. As σ runs over $\mathcal{G}(L/F)$ we obtain n representations $\mathfrak{Z}^\sigma$ affording the characters $m\chi_i$, $1 \leq i \leq n$. Since $m = 1$ when $\text{char}(E) \neq 0$, the $m\chi_i$ are distinct in all cases and thus the $\mathfrak{Z}^\sigma$ are pairwise nonsimilar. Also, each $(\mathfrak{Z}^\sigma)^E$ has a unique irreducible constituent and it occurs with multiplicity m. This follows from the characters if $\text{char}(E) = 0$ and it holds in prime characteristic since then all of the $\mathfrak{Z}^\sigma$ are absolutely irreducible.

We claim that the $n = |L:F|$ representations $\mathfrak{Z}^\sigma$ are exactly the irreducible constituents of $\mathfrak{Y}^L$ and that each occurs with multiplicity 1. Since the irreducible constituents of $(\mathfrak{Z}^\sigma)^E$ and $(\mathfrak{Z}^\tau)^E$ are nonsimilar if $\sigma \neq \tau$, statements (a)–(e) will follow when the claim is established.

Since $\mathfrak{Z}$ is a constituent of $\mathfrak{Y}^L$ and $(\mathfrak{Y}^L)^\sigma = \mathfrak{Y}^L$ for $\sigma \in \mathscr{G}(L/F)$, it follows that $\mathfrak{Z}^\sigma$ is a constituent of $\mathfrak{Y}^L$ for every σ. Therefore $n(\deg \mathfrak{Z}) \leq \deg \mathfrak{Y}$ since $n = |\mathscr{G}(L/F)|$. By Corollary 9.19, $\deg \mathfrak{Y}$ divides $n(\deg \mathfrak{Z})$ and we thus have equality. Therefore the $\mathfrak{Z}^\sigma$ are the only irreducible constituents of $\mathfrak{Y}^L$ and each has multiplicity 1 as claimed.

All that remains is to show that $\mathfrak{Y}^L$ and $\mathfrak{Y}^E$ are completely reducible. This follows from Maschke's Theorem 1.9 if $\text{char}(E) = 0$ so we assume $\text{char}(E) \neq 0$. We may assume without loss that $\mathfrak{Z}$ is a bottom constituent of $\mathfrak{Y}^L$ (since some constituent certainly is). Since $(\mathfrak{Y}^L)^\sigma = \mathfrak{Y}^L$, it follows that $\mathfrak{Z}^\sigma$ is a bottom constituent for every $\sigma \in \mathscr{G}(L/F)$. Now let V be an $L[G]$-module corresponding to $\mathfrak{Y}^L$ and let W be the sum of all of the irreducible submodules of V. By Theorem 1.10, W is completely reducible and it suffices to show $W = V$.

Since every irreducible constituent of $\mathfrak{Y}^L$ is a bottom constituent, every composition factor of V occurs as a composition factor of W. Since the composition factors of V all have multiplicity 1, V/W is necessarily trivial. The proof for $\mathfrak{Y}^E$ is similar since $\mathfrak{Z}^E$ is irreducible in this case. ∎

We obtain some consequences now. The first generalizes Theorem 9.14.

(9.22) COROLLARY Let F be any field. Then the characters of nonsimilar irreducible F-representations of G are nonzero, distinct and linearly independent over F.

Proof Let $E \supseteq F$ be a splitting field for G. By Theorem 9.21, the characters of nonsimilar irreducible F-representations of G are nonzero multiples of sums of disjoint subsets of $\text{Irr}_E(G)$. The result follows from 9.12 ∎

(9.23) COROLLARY Let $F \subseteq E$ be fields of prime characteristic. Let $\mathfrak{X}$ be an irreducible E-representation of G which affords the character χ. Let $\mathfrak{Y}$ be an irreducible F-representation such that $\mathfrak{X}$ is a constituent of $\mathfrak{Y}^E$. Then $\deg \mathfrak{Y} = |F(\chi):F| \deg \mathfrak{X}$. In particular, if $F(\chi) = F$, then $\mathfrak{X}$ is similar to $\mathfrak{Y}^E$.

Proof Let $L \supseteq E$ be a splitting field for G and let $\zeta \in \text{Irr}_L(G)$ be the character of an irreducible constituent of $\mathfrak{X}^L$. Let $\mathscr{S}$ and $\mathscr{T}$ be the Galois conjugacy classes of ζ over E and F respectively. Since E has prime characteristic, it suffices by Theorem 9.21 to show that $|\mathscr{T}| = |F(\chi):F||\mathscr{S}|$.

By Lemma 9.17(b), we have $\mathscr{S} \subseteq \mathscr{T}$ and thus $\chi = \sum \mathscr{S}$ takes on values in $F(\zeta)$ and so $F(\chi) \subseteq F(\zeta)$. Since $|\mathscr{T}| = |F(\zeta):F|$ by 9.17(c), we must show that $|\mathscr{S}| = |F(\zeta):F(\chi)|$. Let $\mathscr{G} = \mathscr{G}(F(\zeta)/F(\chi))$. If $\sigma \in \mathscr{G}$, then

$$\sum_{\eta \in \mathscr{S}} \eta = \chi = \chi^\sigma = \sum_{\eta \in \mathscr{S}} \eta^\sigma$$

and it follows from the linear independence of $\mathrm{Irr}_L(G)$ and Lemma 9.16 that $\mathscr{G}$ permutes $\mathscr{S}$. Since only the identity of $\mathscr{G}$ can fix ζ, this yields $|\mathscr{S}| \geq |\mathscr{G}| = |F(\zeta):F(\chi)|$. However, $F(\chi) \subseteq E \cap F(\zeta)$ and thus

$$|\mathscr{S}| = |E(\zeta):E| = |F(\zeta):E \cap F(\zeta)| \leq |F(\zeta):F(\chi)|$$

and the proof is complete. ∎

Problems

(9.1) Let $F \subseteq E$ be fields and let V be a finite dimensional E-space. Let $W \subseteq V$ be an F-subspace. We write $V = W *_F E$ provided $V = WE$ and $\dim_E(V) = \dim_F(W)$.

(a) If $V = W *_F E$ and $U \subseteq W$ is an F-subspace, show that $UE = U *_F E$.

(b) If $\mathfrak{X}$ is an F-representation of G, show that $\dim_F(\mathfrak{X}(F[G])) = \dim_E(\mathfrak{X}^E(E[G]))$.

Note The symbol $*_F$ denotes an internal tensor product.

(9.2) Let $\mathfrak{X}$ be an F-representation of G and let $E \supseteq F$. Let V be an $E[G]$-module corresponding to $\mathfrak{X}^E$.

(a) Show that $V = W *_F E$ for some G-invariant, F-subspace $W \subseteq V$ such that the $F[G]$-module W corresponds to $\mathfrak{X}$.

(b) Let $U \subseteq W$ be an $F[G]$-submodule. Let U and W/U correspond to the F-representations $\mathfrak{Y}$ and $\mathfrak{Z}$, respectively. Show that UE is an $E[G]$-submodule of V which corresponds to $\mathfrak{Y}^E$ and that V/UE corresponds to $\mathfrak{Z}^E$.

(9.3) Let W be an $F[G]$-module and let $F \subseteq E$ with $|E:F| = n < \infty$. Let W^E be an $E[G]$-module corresponding to $\mathfrak{X}^E$ where $\mathfrak{X}$ is an F-representation corresponding to W. Now view W^E as an F-space. Show that the resulting $F[G]$-module is isomorphic to the direct sum of n copies of W.

Note It follows from the Krull–Schmidt theorem that if V and W are $F[G]$-modules and $V \oplus V \oplus \cdots \oplus V \cong W \oplus W \oplus \cdots \oplus W$, where each direct sum has n terms, then $V \cong W$. We conclude via Problem 9.3 that if $\mathfrak{X}$ and $\mathfrak{Y}$ are F-representations of G and $\mathfrak{X}^E$ is similar to $\mathfrak{Y}^E$, then $\mathfrak{X}$ is similar to $\mathfrak{Y}$, provided $|E:F| < \infty$.

(9.4) Let $F \subseteq E$ be a field extension of possibly infinite degree. Let $\mathfrak{X}$ and $\mathfrak{Y}$ be F-representations of G and assume $\mathfrak{X}^E$ and $\mathfrak{Y}^E$ are similar.

(a) Show that there exist matrices $M_1, M_2, \ldots, M_k$ over F and elements $e_1, e_2, \ldots, e_k \in E$ such that $\mathfrak{X}(g)M_i = M_i\mathfrak{Y}(g)$ for all $g \in G$ and $1 \leq i \leq k$ and such that $e_1M_1 + \cdots + e_kM_k$ is nonsingular over E.
(b) If $|F| > \deg \mathfrak{X}$, show that $\mathfrak{X}$ is similar to $\mathfrak{Y}$.

Hint [For (b)] Let $f(x_1, x_2, \ldots, x_k)$ be a nonzero polynomial over F and assume that the degree of f in x_i is $< |F|$ for each i. Then there exists $a_1, a_2, \ldots, a_k \in F$ such that $f(a_1, a_2, \ldots, a_k) \neq 0$.

(9.5) Under the hypotheses of Problem 9.4, show that $\mathfrak{X}$ and $\mathfrak{Y}$ are similar without assuming anything about $|F|$ or $|E:F|$.

Hint Combine the results of Problem 9.4(b) and the note following Problem 9.3.

(9.6) Let $F \subseteq E$ and let V be an irreducible $E[G]$-module which affords the character χ. Assume that $E = F(\chi)$. Show that V is irreducible when viewed as an $F[G]$-module.

(9.7) Let F have prime characteristic and let $\mathfrak{X}$ be an irreducible F-representation of G. Let D be the centralizer of $\mathfrak{X}(G)$ in the matrix algebra $M_n(F)$ where $n = \deg \mathfrak{X}$. Show that D is a field.

Hint Let $E \supseteq F$ be a splitting field and consider the centralizer of $\mathfrak{X}^E$ in $M_n(E)$. Use Theorems 9.21 and 9.2.

(9.8) In the situation of Problem 9.7, let χ be the character of an irreducible constituent of $\mathfrak{X}^E$ where $E \supseteq F$ is a splitting field for G. Show that the natural isomorphism between F and $F \cdot 1$, the scalar matrices in $M_n(F)$, extends to an isomorphism $F(\chi) \cong D$.

Hint Let $L = F(\chi)$ and let V be an $L[G]$-module affording χ. View V as an $F[G]$-module. Then $D \cong \mathbf{E}_{F[G]}(V)$. Use Problem 9.6.

(9.9) Let $\mathfrak{X}$ be an F-representation of G and assume $\mathfrak{X}^E$ is completely reducible for some $E \supseteq F$. Show that $\mathfrak{X}$ is completely reducible.

Hint Consider $\mathfrak{X}(J(F[G]))$. Use Problem 1.10.

(9.10) Let $\mathfrak{X}$ be a completely reducible F-representation of G and let $E \supseteq F$. Show that $\mathfrak{X}^E$ is completely reducible.

Note The analogous statement for algebras other than group algebras does not hold, in general.

(9.11) Let $F \subseteq E$ and view $F[G] \subseteq E[G]$. Show that $J(E[G]) = J(F[G]) *_F E$ (in the notation of Problem 9.1).

Hint To obtain $J(E[G]) \subseteq (J(F[G])E$, use Problems 9.2(b), 9.10, and 1.10 to argue that $E[G]^\circ/(J(F[G]))E$ is a completely reducible $E[G]$-module.

(9.12) If $\mathfrak{X}$ is an F-representation of G let $s(\mathfrak{X})$ denote $\dim_F(\mathfrak{X}(F[G]))$. Let $\mathfrak{Y}_1, \ldots, \mathfrak{Y}_k$ be the distinct irreducible constituents of $\mathfrak{X}$. Show that

$$\sum_i s(\mathfrak{Y}_i) \leq s(\mathfrak{X})$$

with equality if $\mathfrak{X}$ is completely reducible.

(9.13) Let $\mathfrak{X}$ be an irreducible F-representation of G. Let $\mathfrak{Y}$ be an irreducible constituent of $\mathfrak{X}^E$, where $E \supseteq F$ is a splitting field. Show that $ms(\mathfrak{X}) = \deg(\mathfrak{X}) \deg(\mathfrak{Y})$, where $s(\mathfrak{X})$ is as in Problem 9.12 and m is the multiplicity of $\mathfrak{Y}$ as a constituent of $\mathfrak{X}^E$.

(9.14) Let $\mathfrak{X}$ be an irreducible F-representation of G and let D be the centralizer of $\mathfrak{X}(F[G])$ in the matrix ring $M_n(F)$, where $n = \deg(\mathfrak{X})$. Let $d = \dim_F(D)$. Show that $s(\mathfrak{X}) = n^2/d$.

Hint Let V be an $F[G]$-module corresponding to $\mathfrak{X}$ so that V becomes a vector space over the division ring D. Express $s(\mathfrak{X}) = \dim_F(\mathfrak{X}(F[G]))$ in terms of $\dim_D(V)$ and d. Use Theorem 1.16.

Note Applying Problem 9.14 in the situation of Problem 9.13 we obtain $mn^2/d = ms(\mathfrak{X}) = n \deg(\mathfrak{Y})$ and $mn = d \deg(\mathfrak{Y})$. Since $n = m \deg(\mathfrak{Y})|F(\chi):F|$, where $\chi \in \mathrm{Irr}_E(G)$ is afforded by $\mathfrak{Y}$, we obtain $d = m^2|F(\chi):F|$. Compare this with Problem 9.8. This formula can be used to generalize Problem 9.8 to the characteristic zero case.

(9.15) Let $\mathfrak{X}$ be an irreducible F-representation of G and let χ be the character afforded by an irreducible constituent $\mathfrak{Y}$ of $\mathfrak{X}^E$, where $E \supseteq F$ is a splitting field for G. Let D be the centralizer of $\mathfrak{X}(F[G])$ as in Problem 9.14. Show that the isomorphism $F \cong F \cdot 1 \subseteq D$ extends to an isomorphism $F(\chi) \cong \mathbf{Z}(D)$.

Hints Let $L = F(\chi)$ and let $\mathfrak{Z}$ be the irreducible constituent of $\mathfrak{X}^L$ such that $\mathfrak{Y}$ is a constituent of $\mathfrak{Z}^E$. Let V be an $L[G]$-module corresponding to $\mathfrak{Z}$. Let $D_0 = \mathbf{E}_{L[G]}(V)$ and $D_1 = \mathbf{E}_{F[G]}(V)$. Check that $D_0 \subseteq D_1 \cong D$. Use the note following Problem 9.14 to show that $D_0 = D_1$.

Let $Z = \mathbf{Z}(D_0) \supseteq L \cdot 1$. Observe that $D_0 = \mathbf{E}_{Z[G]}(V)$. Let $\mathfrak{W}$ be the Z-representation corresponding to V viewed as a $Z[G]$-module, and compute the multiplicity of $\mathfrak{Y}$ as a constituent of $\mathfrak{W}^E$. Now use the note following Problem 9.14 again to show $Z = L \cdot 1$.

(9.16) Let $\mathfrak{X}$ be an irreducible F-representation of G and let $E \supseteq F$. Show that $g \in G$ lies in $\ker(\mathfrak{X})$ iff $g \in \ker(\mathfrak{Y})$ for every irreducible constituent $\mathfrak{Y}$ of $\mathfrak{X}^E$.

(9.17) Let F have prime characteristic p. Show that $\mathbf{O}_p(G)$ (the largest normal p-subgroup of G) is the intersection of the kernels of the irreducible F-representations of G.

(9.18) Let $k = \text{char}(F)$.

(a) If $k \neq 2$, show that Q_8 has a unique (up to similarity) nonlinear irreducible F-representation.
(b) If $k \neq 0$, show that F is a splitting field for Q_8.
(c) If $k = 0$, show that F is a splitting field for Q_8 iff -1 is a sum of two squares in F.

(9.19) Let $F \subseteq E$, where E is a splitting field for G. Does there necessarily exist a splitting field K such that $F \subseteq K \subseteq E$ and $|K:F| < \infty$? Consider zero and nonzero characteristic separately.

Hint Consider $F = \mathbb{Q}$ and $E = \mathbb{Q}(\alpha, \beta) \subseteq \mathbb{C}$, where α, β are suitable transcendentals. Take $G = Q_8$.

(9.20) Let G be cyclic of order n and let F have characteristic not dividing n. Let $E = F(\varepsilon)$, where ε is a primitive nth root of unity. Let $m = |E:F|$. Show that every faithful irreducible F-representation of G has degree m and that there are exactly $\varphi(n)/m$ similarity classes of such representations.

Note If $|F| = q < \infty$, then m can be characterized as the least positive integer such that $n \mid (q^m - 1)$.

(9.21) Let $G = \text{GL}(n, p^e)$ be the full group of nonsingular $n \times n$ matrices over $\text{GF}(p^e)$. Show that G has an irreducible representation of degree ne over $\text{GF}(p)$.

(9.22) Let $F \subseteq E$ be fields of prime characteristic and let $\mathfrak{X}$ be an E-representation of G. Assume that $\mathfrak{X}$ affords a character χ such that $F(\chi) = F$. Do not assume $\mathfrak{X}$ is irreducible.

(a) Show that χ is afforded by some F-representation.
(b) Find an example where $\mathfrak{X}$ is not similar to $\mathfrak{Y}^E$ for any F-representation $\mathfrak{Y}$.

10 The Schur index

The main question considered in this chapter is the following. If $\chi \in \mathrm{Irr}(G)$, then for which fields $F \subseteq \mathbb{C}$ is χ afforded by an F-representation? If $F \subseteq \mathbb{C}$ is not one of these fields, we wish to measure the extent to which χ fails to be afforded over F. This and the results of Chapter 9 (and, in particular, Theorem 9.21) suggest the following definition.

(10.1) DEFINITION Let $F \subseteq E$, where E is any splitting field for G. Let $\chi \in \mathrm{Irr}_E(G)$. Choose an irreducible E-representation $\mathfrak{X}$ which affords χ and an irreducible F-representation $\mathfrak{Y}$ such that $\mathfrak{X}$ is a constituent of $\mathfrak{Y}^E$. Then the multiplicity of $\mathfrak{X}$ as a constituent of $\mathfrak{Y}^E$ is the *Schur index* of χ over F. It is denoted by $m_F(\chi)$.

Note that given χ as above, representations $\mathfrak{X}$ and $\mathfrak{Y}$ do exist and are unique up to similarity and so $m_F(\chi)$ is well defined. Actually, $m_F(\chi)$ does not really depend on E. If $L \supseteq F(\chi)$ is another splitting field, then $\chi \in \mathrm{Irr}_L(G)$ by Lemma 9.13. A routine argument shows that $m_F(\chi)$ is the same when computed in E or in L.

By Theorem 9.21(b), Schur indices are always trivial (that is, equal to 1) in prime characteristic. For this reason we now restrict our attention to the characteristic zero case. In fact, it is really no loss to confine ourselves to subfields of the complex numbers, $\mathbb{C}$.

Many important results about Schur indices appear to depend on deep facts about division algebras and number theory. Nevertheless, as this chapter will demonstrate, much can be done by elementary means. In particular, some of the results of Chapter 8 will prove invaluable.

We collect some facts.

(10.2) COROLLARY Let $\chi \in \mathrm{Irr}(G)$ and $F \subseteq \mathbb{C}$. We have

(a) $m_F(\chi) = m_{F(\chi)}(\chi)$.

(b) Let $\mathscr{S}$ be the Galois conjugacy class of χ over F. Then $m_F(\chi)(\sum \mathscr{S})$ is the character of an irreducible F-representation of G.

(c) If Ξ is the character of any F-representation, then $m_F(\chi)$ divides $[\Xi, \chi]$.

(d) $m_F(\chi)$ is the smallest integer m such that $m\chi$ is afforded by an $F(\chi)$-representation.

(e) $m_F(\chi)$ is the unique integer m such that $m\chi$ is afforded by an irreducible $F(\chi)$-representation.

(f) If $F \subseteq E \subseteq \mathbb{C}$, then $m_E(\chi)$ divides $m_F(\chi)$.

(g) If $F \subseteq E \subseteq \mathbb{C}$ and $|E:F| = n < \infty$, then $m_F(\chi)$ divides $nm_E(\chi)$.

(h) $m_F(\chi)$ divides $\chi(1)$.

Proof Part (a) follows from Theorem 9.21(e) and part (b) is a consequence of 9.21(a, c). Let $\mathfrak{Y}$ be the F-representation in (b). Then $\mathfrak{Y}$ is the unique (up to similarity) irreducible F-representation whose character ψ satisfies $[\psi, \chi] \neq 0$. Since $m_F(\chi) = [\psi, \chi]$, part (c) follows.

To prove (d) and (e), we may assume (by (a)) that $F = F(\chi)$. Now $\mathscr{S} = \{\chi\}$ and the representation $\mathfrak{Y}$ of the preceding paragraph affords $m\chi$. Parts (d) and (e) are now immediate.

Now let $F \subseteq E \subseteq \mathbb{C}$. Any character afforded by an F-representation is also afforded by an E-representation. Now (f) follows from (b) and (c). Assume $|E:F| = n < \infty$. Then $n_0 = |E(\chi):F(\chi)| = |E:E \cap F(\chi)|$ which divides n. Since $m_E(\chi)\chi$ is afforded by an $E(\chi)$-representation, we conclude from Lemma 9.18(c) that $n_0 m_E(\chi)\chi$ is afforded by an $F(\chi)$-representation. Now (c) and (a) yield (g).

Finally, let ρ be the character of the regular F-representation of G. Then $[\rho, \chi] = \chi(1)$ and (h) follows from (c). The proof is complete. ∎

A useful method for obtaining information about Schur indices is to use induced representations. Let $F \subseteq \mathbb{C}$ and $H \subseteq G$. If ϑ is a character of H which is afforded by an F-representation of H, then ϑ^G is afforded by an F-representation of G. (See Theorems 5.8 and 5.9.) This idea underlies much of the remainder of the chapter. We use it to prove the following celebrated result.

(10.3) THEOREM (*Brauer*) Let G have exponent n and let $F = \mathbb{Q}(e^{2\pi i/n})$. Then F is a splitting field for G and every $\chi \in \mathrm{Irr}(G)$ is afforded by an F-representation.

Proof The two conclusions are equivalent by Corollary 9.11. Let $\chi \in \mathrm{Irr}(G)$. Since $F(\chi) = F$, it suffices by 10.2(d) to show that $m_F(\chi) = 1$. By Brauer's Theorem 8.4 we may write

$$\chi = \sum a_\lambda \lambda^G,$$

where λ runs over linear characters of subgroups of G and $a_\lambda \in \mathbb{Z}$. Now a linear $\lambda \in \mathrm{Irr}(H)$ is obviously afforded by an F-representation of H and hence λ^G is afforded by an F-representation of G. By 10.2(c), we have $m_F(\chi) \mid [\lambda^G, \chi]$ and since

$$1 = [\chi, \chi] = \sum_\lambda a_\lambda[\lambda^G, \chi],$$

it follows that $m_F(\chi) = 1$ and the proof is complete. ∎

We wish to exploit the ideas in the above proof in order to obtain some more delicate information about the Schur index.

(10.4) LEMMA Let $H \subseteq G$ and $F \subseteq \mathbb{C}$. Suppose $\psi \in \mathrm{Irr}(H)$ and $\chi \in \mathrm{Irr}(G)$. Then $m_F(\chi)$ divides $m_F(\psi)|F(\chi, \psi) : F(\chi)|[\psi^G, \chi]$.

Proof We may replace F by $E = F(\chi)$. This is so since $m_E(\chi) = m_F(\chi)$ by 10.2(a) and $m_E(\psi) | m_F(\psi)$ by 10.2(f) (and of course $F(\chi, \psi) = E(\psi)$). Thus we assume $F = F(\chi)$.

Let $\mathscr{S}$ be the Galois conjugacy class of ψ over F so that $m_F(\psi) \sum \mathscr{S}$ is afforded by an F-representation of G. Therefore $m_F(\chi)$ divides

$$m_F(\psi) \sum_{\eta \in \mathscr{S}} [\eta^G, \chi]$$

by 10.2(c).

Now if $\eta \in \mathscr{S}$, then $\psi = \eta^\sigma$ for some $\sigma \in \mathscr{G}(F(\psi)/F)$. We have $\chi = \chi^\sigma$ and

$$[\eta^G, \chi] = [(\eta^G)^\sigma, \chi^\sigma] = [\psi^G, \chi]$$

and thus $m_F(\chi)$ divides $m_F(\psi)|\mathscr{S}|[\psi^G, \chi]$. Since $|\mathscr{S}| = |F(\psi) : F|$, the proof is complete. ∎

The next theorem is a variation on a result of Brauer and Witt. It allows us to analyze $m_F(\chi)$ one prime at a time in terms of certain sections of a group with sharply defined properties. (A *section* of G is a factor group, H/K, where $K \lhd H \subseteq G$.)

(10.5) DEFINITION Let $F \subseteq \mathbb{C}$. Then (H, X, ϑ) is an *F-triple* provided

(a) H is a group, $X \lhd H$, $X = \mathbf{C}_H(X)$;
(b) $\vartheta \in \mathrm{Irr}(H)$ is faithful;
(c) the irreducible (linear) constituents of ϑ_X are Galois conjugate over $F(\vartheta)$.

(10.6) COROLLARY Let (H, X, ϑ) be an F-triple and let λ be a linear constituent of ϑ_X. Then

(a) λ is faithful and X is cyclic;
(b) $I_H(\lambda) = X$, $\lambda^H = \vartheta$ and $F(\vartheta) \subseteq F(\lambda)$;
(c) ϑ_X is afforded by an irreducible $F(\vartheta)$-representation;
(d) $H/X \cong \mathcal{G}(F(\lambda)/F(\vartheta))$.

Proof By part (c) of the definition, all linear constituents of ϑ_X have the same kernel. Thus $\ker \lambda \subseteq \ker \vartheta = 1$ and (a) follows. Since λ takes on distinct values at distinct elements of X, it follows that if $\lambda^h = \lambda$, then $h \in \mathbf{C}(X) = X$. Thus $X = I_H(\lambda)$ and λ^G is irreducible by Theorem 6.11(a). Now (b) follows. Also ϑ_X is the sum of $\vartheta(1) = |H : X|$ distinct linear characters. These must constitute the full Galois conjugacy class of λ over $F(\vartheta)$ since if $\sigma \in \mathcal{G}(F(\lambda)/F(\vartheta))$, then $[\lambda^\sigma, \vartheta_X] = [\lambda, \vartheta_X] \neq 0$. Since $m_{F(\vartheta)}(\lambda) = 1$, Corollary 10.2(b) asserts that ϑ_X is afforded by an irreducible $F(\vartheta)$-representation and (c) is proved.

Finally, for each $h \in H$, there exists a unique $\sigma_h \in \mathcal{G}(F(\lambda)/F(\vartheta))$ such that $\lambda^{h^{-1}} = \lambda^{\sigma_h}$. Now $\lambda^{h^{-1}k^{-1}} = (\lambda^{\sigma_h})^{k^{-1}} = (\lambda^{k^{-1}})^{\sigma_h} = \lambda^{\sigma_k\sigma_h}$. Therefore, $\sigma_{kh} = \sigma_k\sigma_h$ and we have a homomorphism, $\alpha\colon H \to \mathcal{G}(F(\lambda)/F(\vartheta))$, defined by $\alpha(h) = \sigma_h$. Now $\ker \alpha = I_H(\lambda) = X$. Since it was shown above that the H-conjugates of λ constitute a full Galois conjugacy class over $F(\vartheta)$, it follows that α maps onto $\mathcal{G}(F(\lambda)/F(\vartheta))$ and the proof is complete. ∎

(10.7) THEOREM Let $\chi \in \operatorname{Irr}(G)$ and $F \subseteq \mathbb{C}$. Assume p^a divides $m_F(\chi)$ for some prime p. Then there exists an F-triple (H, X, ϑ) such that

(a) H is a section of G;
(b) $p^a \mid m_F(\vartheta)$;
(c) H/X is a p-group;
(d) $p \nmid |F(\chi, \vartheta) : F(\chi)|$.

Proof By Lemma 9.17(b), it follows that if (H, X, ϑ) is an E-triple for some $E \supseteq F$, then it is an F-triple. By 10.2(f, a) we may therefore replace F by $F(\chi)$ and so we assume $F = F(\chi)$.

Use induction on $|G|$. If there exists $K < G$ and $\psi \in \operatorname{Irr}(K)$ such that $p \nmid [\psi^G, \chi]$ and $p \nmid |F(\psi) : F|$, then by Lemma 10.4 it follows that $p^a \mid m_F(\psi)$ and we may apply the inductive hypothesis to K with respect to ψ and obtain an F-triple (H, X, ϑ) which satisfies (a), (b), and (c) and such that $p \nmid |F(\psi, \vartheta) : F(\psi)|$. Since also $p \nmid |F(\psi) : F|$, we conclude that $p \nmid |F(\vartheta) : F|$ and the proof is complete in this case. We therefore assume that no such pair (K, ψ) exists.

By Solomon's Theorem 8.7 we can write $1_G = \sum a_Q(1_Q)^G$, where $a_Q \in \mathbb{Z}$ and Q runs over $\mathcal{H}$, the set of quasi-elementary subgroups of G. It follows that

$$\chi = \chi 1_G = \sum a_Q \chi(1_Q)^G = \sum a_Q(\chi_Q)^G$$

and thus

$$1 = [\chi, \chi] = \sum a_Q[(\chi_Q)^G, \chi].$$

Choose $Q \in \mathscr{H}$ such that p does not divide $[(\chi_Q)^G, \chi] = [\chi_Q, \chi_Q]$. If ψ is any irreducible constituent of χ_Q, then since $F(\chi) = F$, every element of the Galois conjugacy class of ψ over F has equal multiplicity as a constituent of χ_Q and we may write $\chi_Q = \sum b_\Delta \Delta$, where $b_\Delta \in \mathbb{Z}$ and Δ runs over sums of Galois conjugacy classes over F in $\mathrm{Irr}(Q)$. Thus p does not divide $\sum b_\Delta{}^2[\Delta, \Delta]$ and we choose Δ such that $p \nmid b_\Delta{}^2[\Delta, \Delta]$. Write $\Delta = \sum \mathscr{S}$, where $\mathscr{S}$ is a Galois conjugacy class over F and let $\psi \in \mathscr{S}$. Then $b_\Delta = [\chi_Q, \psi] = [\chi, \psi^G]$ is not divisible by p and $[\Delta, \Delta] = |\mathscr{S}| = |F(\psi) : F|$ is not divisible by p. By the result of the second paragraph of the proof, we conclude that $G = Q$ is quasi-elementary.

Thus G has a cyclic normal q-complement C for some prime q. By Theorem 6.15 we have $\chi(1)$ is a power of q. Now if $a = 0$, the result is trivial, taking $H = X = 1$, and so we assume $a > 0$ and $p \mid m_F(\chi)$. Since $m_F(\chi) \mid \chi(1)$ by 10.2(h), it follows that $q = p$ and G/C is a p-group. Let $X \lhd G$, with $X \supseteq C$ be chosen maximal such that X is abelian. By the inductive hypothesis applied to $G/\ker \chi$, we may assume that χ is faithful. We shall show that (G, X, χ) is an F-triple to complete the proof.

We have $X \subseteq \mathbf{C}_G(X) \lhd G$. If $X < \mathbf{C}_G(X)$, then since G/X is a p-group, there exists $U \lhd G$ with $X \subseteq U \subseteq \mathbf{C}_G(X)$ and $|U : X| = p$. Thus U is abelian and this contradicts the choice of X. Thus part (a) of Definition 10.5 is established.

Now let λ be a linear constituent of χ_X and let $S = \{g \in G \mid \lambda^g$ is Galois conjugate to λ over $F\}$. If $s_1, s_2 \in S$, there exist $\sigma_1, \sigma_2 \in \mathscr{G}(F(\lambda)/F)$ with $\lambda^{s_i} = \lambda^{\sigma_i}$. It follows that $\lambda^{s_1 s_2} = \lambda^{\sigma_2 \sigma_1}$ and thus $s_1 s_2 \in S$ and S is a subgroup of G. Let $T = I_G(\lambda) \subseteq S$. By Theorem 6.11(b), there exists a unique $\eta \in \mathrm{Irr}(T)$ such that $\eta^G = \chi$ and $[\eta_X, \lambda] \neq 0$. Let $\psi = \eta^S$ so that $\psi^G = \chi$ and $\psi \in \mathrm{Irr}(S)$. We claim that $F(\psi) = F$.

Let $\sigma \in \mathscr{G}(F(\psi, \lambda)/F)$. Then $\chi^\sigma = \chi$ and so $[\lambda^\sigma, \chi_X] = [\lambda, \chi_X] \neq 0$. Thus $\lambda^\sigma = \lambda^g$ for some $g \in G$ and since λ^σ is Galois conjugate to λ over F, we have $g \in S$. Since Galois conjugate characters have the same inertia groups, we have $T \lhd S$ and so $(\eta^\sigma)^{g^{-1}} \in \mathrm{Irr}(T)$. Now

$$((\eta^\sigma)^{g^{-1}})^G = (\eta^\sigma)^G = \chi^\sigma = \chi$$

and

$$[((\eta^\sigma)^{g^{-1}})_X, \lambda] = [(\eta^\sigma)_X, \lambda^g] = [(\eta^\sigma)_X, \lambda^\sigma] \neq 0.$$

By the uniqueness of η we conclude that $(\eta^\sigma)^{g^{-1}} = \eta$ and so

$$\psi^\sigma = (\eta^\sigma)^S = (\eta^g)^S = \eta^S = \psi$$

since $g \in S$. Since $F(\psi, \lambda)$ is a Galois extension of F, it follows that ψ takes values in F and thus $F(\psi) = F$ as claimed. By the result of the second paragraph of the proof applied to (S, ψ), it follows that $S = G$. Thus condition (c) of Definition 10.5 is satisfied and the proof is complete. ∎

Before proceeding to our major applications of Theorem 10.7 we derive two easier consequences which demonstrate its power. These are included in Theorem 10.9.

(10.8) LEMMA Let $H \subseteq G$ and suppose H has a complement in G. Let $\varphi \in \mathrm{Irr}(H)$ and suppose $\varphi^G = \chi \in \mathrm{Irr}(G)$. Then $m_F(\chi)$ divides $\varphi(1)$ for every $F \subseteq \mathbb{C}$.

Proof Let $U \subseteq G$ with $UH = G$ and $U \cap H = 1$. Since $(1_U)^G$ is afforded by an F-representation, we have $m_F(\chi) \mid [(1_U)^G, \chi]$. Now $[(1_U)^G, \chi] = [(1_U)^G, \varphi^G] = [((1_U)^G)_H, \varphi]$. However, $((1_U)^G)_H = (1_{U \cap H})^H = \rho_H$, the regular character of H. Thus $[(1_U)^G, \chi] = [\rho_H, \varphi] = \varphi(1)$ and the result follows. ∎

(10.9) THEOREM Let $\chi \in \mathrm{Irr}(G)$ and assume $p \mid m_F(\chi)$ for some $F \subseteq \mathbb{C}$ and prime p. Then the Sylow p-subgroups of G are not elementary abelian and $pm_F(\chi)$ divides $|G|$.

Proof Let p^a be the p-part of $m_F(\chi)$ so that $a > 0$. Since $m_F(\chi) \mid \chi(1)$, the second statement will follow if p^{a+1} divides $|G|$. Let (H, X, ϑ) be the F-triple whose existence is guaranteed by Theorem 10.7. Thus $\vartheta = \lambda^H$ for some linear $\lambda \in \mathrm{Irr}(X)$ by Corollary 10.6(b). Since $m_F(\vartheta) > 1$, we conclude from Lemma 10.8 that X is not complemented in H.

Since H/X is a p-group, it follows that a Sylow p-subgroup of H is not elementary abelian. Also, if $P \in \mathrm{Syl}_p(H)$, then $P \cap X \neq 1$ and so $\vartheta(1) = |H : X| < |P|$. Since $p^a \leq m_F(\vartheta) \leq \vartheta(1)$, we have $p^a < |P|$ and so $p^{a+1} \mid |H|$. The result now follows. ∎

There is an important case when we can decide whether or not $m_F(\vartheta) = 1$ for an F-triple (H, X, ϑ). If H/X is cyclic, the problem "reduces" to a question in field theory. If $E \supseteq F$ is a Galois field extension and $\alpha \in E$, we define $N_{E/F}(\alpha) = \prod_{\sigma \in \mathscr{G}(E/F)} \alpha^\sigma$. The image of this *norm map* is clearly contained in F.

(10.10) THEOREM Let (H, X, ϑ) be an F-triple for some $F \subseteq \mathbb{C}$ and assume $H = XC$, where C is a cyclic group. Let λ be a linear constituent of ϑ_X and write $E = F(\lambda)$ and $K = F(\vartheta)$. Let $m = |X \cap C|$ and let ε be a primitive mth root of unity in $\mathbb{C}$. Then $\varepsilon \in K$. Also, $m_F(\vartheta) = 1$ iff ε lies in the image of $N_{E/K}$.

Proof Since $X \cap C \subseteq \mathbf{Z}(H)$ and $\vartheta \in \mathrm{Irr}(H)$ is faithful, we can choose a generator y of $X \cap C$, such that $\vartheta(y) = \varepsilon\vartheta(1)$ and thus $\varepsilon \in F(\vartheta) = K$. Write $|X : X \cap C| = s$ and $|H : X| = |C : X \cap C| = t$ and choose generators x and c for X and C such that $x^s = y = c^t$. Note that $\lambda(x)^s = \lambda(y) = \varepsilon$. By Corollary

10.6 we have

$$t = \vartheta(1) = |\mathcal{G}(E/K)| = |E:K|.$$

Write $x^c = x^r$ for some integer r.

By Definition 10.5 there exists $\sigma \in \mathcal{G}(E/K)$ such that $\lambda^{c^{-1}} = \lambda^\sigma$. It follows that the orbits of λ under C and under $\langle\sigma\rangle$ are identical. Since C is transitive on the t distinct linear constituents of ϑ_X, we conclude that $|\langle\sigma\rangle| \geq t = |\mathcal{G}(E/K)|$ and thus $\mathcal{G}(E/K) = \langle\sigma\rangle$. Therefore,

$$N_{E/K}(\alpha) = \alpha\alpha^\sigma\alpha^{\sigma^2}\cdots\alpha^{\sigma^{t-1}}$$

for $\alpha \in E$. Also, $\lambda(x)^r = \lambda(x^r) = \lambda^{c^{-1}}(x) = \lambda(x)^\sigma$.

Let V be the (unique up to isomorphism) irreducible $K[X]$-module which affords ϑ_X. (See Corollary 10.6(c).) Now $m_F(\vartheta) = 1$ iff ϑ is afforded by a $K[H]$-module (by Corollary 10.2(d)) and this happens iff there exists a K-linear action of C on V such that for $v \in V$ we have

(a) $v \cdot c^t = v\varepsilon$;
(b) $((v \cdot c^{-1}) \cdot x) \cdot c = v \cdot x^r$.

(The sufficiency of this condition involves an easy generators and relations argument on H.)

View E as an $E[X]$ module affording λ so that $\beta \cdot u = \beta\lambda(u)$ for $u \in X$. With this same action of X on E, we now view E as a $K[X]$-module of dimension $|E:K| = t$. By Lemma 9.18(b) it follows that V is an irreducible constituent of the $K[X]$-module E. (This is because λ is a constituent of $\mathfrak{Y}^E$ where $\mathfrak{Y}$ is a K-representation corresponding to V.) Since $\dim_K(V) = \vartheta(1) = t = \dim_K(E)$, it follows that $V \cong E$ as $K[X]$ modules. We conclude that $m_F(\vartheta) = 1$ iff there exists a K-linear transformation $\hat{c}$ of E such that for all $\beta \in E$

(a′) $\beta \cdot \hat{c}^t = \beta\varepsilon$;
(b′) $((\beta \cdot \hat{c}^{-1})\lambda(x)) \cdot \hat{c} = \beta(\lambda(x))^r$.

Now suppose that $m_F(\vartheta) = 1$ and let $\hat{c}$ be as above. Write $\delta = \lambda(x)$, $\alpha = 1 \cdot \hat{c}$ and let $n \geq 0$ be an integer. We show that $(\delta^n) \cdot \hat{c} = \alpha\delta^{nr}$ by induction on n. This is clear for $n = 0$. For $n > 0$, the inductive hypothesis and (b′) yield

$$(\delta^n) \cdot \hat{c} = (((\alpha\delta^{(n-1)r}) \cdot \hat{c}^{-1})\delta) \cdot \hat{c} = \alpha\delta^{(n-1)r}\delta^r = \alpha\delta^{nr}$$

as claimed. Since $\delta^\sigma = \delta^r$ we have $(\delta^n) \cdot \hat{c} = \alpha(\delta^n)^\sigma$. Since $E = K(\lambda)$ and $\delta = \lambda(x)$, the δ^n span E over K. The map $\beta \to \alpha\beta^\sigma$ for $\beta \in E$ is K-linear and agrees with $\hat{c}$ on the δ^n. We conclude that $\beta \cdot \hat{c} = \alpha\beta^\sigma$ for all $\beta \in E$. Now (a′) yields

$$\varepsilon = 1 \cdot (\hat{c})^t = \alpha\alpha^\sigma\alpha^{\sigma^2}\cdots\alpha^{\sigma^{t-1}} = N_{E/K}(\alpha)$$

as desired.

Conversely, suppose $N_{E/K}(\alpha) = \varepsilon$ for some $\alpha \in E$. Define $\hat{c}$ on E by $\beta \cdot \hat{c} = \alpha\beta^\sigma$ and observe that $\hat{c}$ is K-linear. Check that (a′) and (b′) are satisfied. Thus $m_F(\vartheta) = 1$ and the proof is complete. ∎

(10.11) LEMMA Let p be a prime and let $1 \neq k \in \mathbb{Z}$. Assume that $p \mid (k - 1)$ and if $p = 2$, assume $4 \mid (k - 1)$. Write

$$f(e) = (k^{p^e} - 1)/(k - 1)$$

for $0 \leq e \in \mathbb{Z}$. Then p^e is the exact power of p dividing $f(e)$.

Proof Since $f(0) = 1$, we assume $e \geq 1$ and use induction on e. Since $k^{p^{e-1}} = 1 + (k - 1)f(e - 1)$, we have

$$\begin{aligned} k^{p^e} &= 1 + p(k - 1)f(e - 1) \\ &\quad + \frac{p(p-1)}{2}(k - 1)^2 f(e - 1)m + (k - 1)^3 f(e - 1)n \end{aligned}$$

for suitable integers m, n. Thus

$$f(e) = f(e - 1)\left[p + \frac{p(p-1)}{2}(k - 1)m + (k - 1)^2 n\right]$$

and it suffices to show that the second and third terms in the brackets are divisible by p^2. Since $p \mid (k - 1)$, this is clear if $p \neq 2$. If $p = 2$, the hypothesis that $4 \mid (k - 1)$ yields the result. ∎

We next give our principal result about the Schur index. Note that it generalizes Brauer's Theorem 10.3.

(10.12) THEOREM (*Goldschmidt–Isaacs*) Let $\chi \in \operatorname{Irr}(G)$, where G has exponent n and let ε be a primitive nth root of unity in $\mathbb{C}$. Let $F \subseteq \mathbb{C}$ and assume that $\mathscr{G}(F(\varepsilon)/F(\chi))$ has a cyclic Sylow p-subgroup P. Then $p \nmid m_F(\chi)$ except possibly when $p = 2$, $P > 1$ and $\sqrt{-1} \notin F(\chi)$.

Proof We may replace F by $F(\chi)$ and assume $F(\chi) = F$. Suppose $p \mid m_F(\chi)$ and let (H, X, ϑ) be the F-triple whose existence is given by Theorem 10.7 so that $p \mid m_F(\vartheta)$. Let λ be a linear constituent of ϑ_X so that $F(\chi) = F \subseteq F(\vartheta) \subseteq F(\lambda) \subseteq F(\varepsilon)$. By 10.6(d), $H/X \cong \mathscr{G}(F(\lambda)/F(\vartheta))$ which is a section of $\mathscr{G}(F(\varepsilon)/F(\chi))$. Since H/X is a p-group, it must be cyclic. Also, since $\vartheta(1) > 1$, we have $H > X$ and we conclude that $P > 1$.

There exists a cyclic p-group C with $XC = H$. Let $|C| = p^c$ and $|C \cap X| = p^b$ and write $E = F(\lambda)$ and $K = F(\vartheta)$. Since $m_F(\vartheta) \neq 1$, Theorem 10.10 yields that no primitive p^bth root of unity lies in the image of $N_{E/K}$, but that K does contain such a root. Since $N_{E/K}(1) = 1$, we have $b > 0$ and K contains a primitive pth root of unity.

Let γ be a primitive p^cth root of unity in $\mathbb{C}$. Then $|K(\gamma):K|$ divides p^{c-b} and $K(\gamma) \subseteq F(\varepsilon)$. Also $|E:K| = |H:X| = p^{c-b}$ and $E \subseteq F(\varepsilon)$. Since $\mathscr{G}(F(\varepsilon)/K)$ is abelian and has a cyclic Sylow p-subgroup, its subgroups of p-power index are linearly ordered by inclusion. The fundamental theorem of Galois theory now yields that $K(\gamma) \subseteq E$ (since $|K(\gamma):K| \leq |E:K|$) and so $\gamma \in E$. We compute $N_{E/K}(\gamma)$.

Write $t = p^{c-b}$. If $\gamma \in K$, then $N_{E/K}(\gamma) = \gamma^t$, which is a primitive p^bth root of unity, a contradiction. Thus $\gamma \notin K$. Write $\mathscr{G}(E/K) = \langle \sigma \rangle$, where $o(\sigma) = t$ and $\gamma^\sigma = \gamma^k$ for some integer $k \neq 1$. Then $\gamma^{\sigma^i} = \gamma^{k^i}$ and

$$N_{E/K}(\gamma) = \gamma\gamma^k\gamma^{k^2}\cdots\gamma^{k^{t-1}} = \gamma^q,$$

where $q = (k^t - 1)/(k - 1)$. If p^{c-b} is the exact p-power divisor of q, then γ^q is a primitive p^bth root of unity which is not the case. By Lemma 10.11 then, $p \nmid (k-1)$ if $p \neq 2$ and $4 \nmid (k-1)$ if $p = 2$.

Let $\delta \in K$ be a primitive p^ath root of unity with a chosen as large as possible. Then $0 < b \leq a$. Since $\gamma \notin K$, we have $a < c$ and $\delta \in \langle \gamma \rangle$ so that $\delta = \delta^\sigma = \delta^k$ and $p^a | (k-1)$. Since $a > 0$, we have $p = 2$ and $a = 1$ so that $\sqrt{-1} \notin K$. The proof is complete. ∎

(10.13) COROLLARY (*Fong*) Let G have exponent $n = mp^a$, where p is prime and $p \nmid m$. Suppose $F \subseteq \mathbb{C}$ and F contains a primitive mth root of unity. Let $\chi \in \operatorname{Irr}(G)$. Then $m_F(\chi) = 1$ unless $p = 2$ and $\sqrt{-1} \notin F$, in which case $m_F(\chi) \leq 2$.

Proof If $p = 2$ and $\sqrt{-1} \notin F$, let $E = F[\sqrt{-1}]$ so that $|E:F| = 2$ and $m_F(\chi) | (2m_E(\chi))$ by Corollary 10.2(g). Thus it suffices to assume that $\sqrt{-1} \in F$ if $p = 2$ and to prove $m_F(\chi) = 1$.

Let ε be a primitive nth root of unity in $\mathbb{C}$. The hypotheses on F now guarantee that $\mathscr{G}(F(\varepsilon)/F)$ is cyclic. If q is a prime divisor of $m_F(\chi)$, then Theorem 10.12 yields $q = 2$ and $\sqrt{-1} \notin F$. Thus $p \neq 2$, $4 \nmid m$, and $m_F(\chi)$ is a power of 2. We have $4 \nmid n$ and thus a Sylow 2-subgroup of G is elementary abelian and $2 \nmid m_F(\chi)$ by Theorem 10.9. The result follows. ∎

(10.14) COROLLARY (*Roquette*) Let G be a p-group and $\chi \in \operatorname{Irr}(G)$. Let $F \subseteq \mathbb{C}$. Then $m_F(\chi) = 1$ unless $p = 2$ and $\sqrt{-1} \notin F$ in which case $m_F(\chi) \leq 2$.

Proof Immediate from 10.13. ∎

Note that taking G to be the quaternion group of order 8 shows that the exceptional cases in the three preceding results can, in fact, occur.

(10.15) COROLLARY (*L. Solomon*) Let k be the product of the distinct prime divisors of $|G|$ and let $F \subseteq \mathbb{C}$. Assume that F contains a primitive $(2k)$th root of unity. Then $m_F(\chi) = 1$ for all $\chi \in \operatorname{Irr}(G)$.

Proof Note that if $2 \,|\, |G|$, then $\sqrt{-1} \in F$. Let ε be a primitive nth root of unity, where n is the exponent of G. Since every prime divisor of n divides k, the hypothesis on F guarantees that $\mathcal{G}(F(\varepsilon)/F)$ is cyclic. If $\chi \in \mathrm{Irr}(G)$ and p is a prime divisor of $m_F(\chi)$, then Theorem 10.12 yields $p = 2$ and $\sqrt{-1} \notin F$. Since $m_F(\chi) \,|\, \chi(1)$ we conclude that $2 \,|\, |G|$ and we have a contradiction. ∎

Most of the above results are directed to showing that Schur indices are small. We have not yet seen an example to show that $m_F(\chi)$ can be greater than 2. In fact every positive integer can occur as a Schur index. Here, we shall settle for an indication of how to prove that every prime can occur. We need two facts from number theory

(a) Let $F = \mathbb{Q}(\varepsilon)$, where ε is a root of unity. Let $R = \mathbb{Z}[\varepsilon]$ and let I be a proper ideal of the ring R. Let $\alpha \in F$. Then $\alpha = u/v$ for some $u, v \in R$ with not both $u, v \in I$.

(b) (*Dirichlet*) Let $a, b \in \mathbb{Z}$ with $(a, b) = 1$. Then there exists a prime of the form $ak + b$ for some $k \in \mathbb{Z}$.

Let p be a prime. By (b) above, choose a prime $q = p^2k + (p + 1)$ for some k. Then $p \,|\, (q - 1)$ but $p^2 \nmid (q - 1)$. We now construct the semidirect product $G = QP$, where Q is cyclic of order q and P is cyclic of order p^2 and acts nontrivially (but not faithfully) on Q. There exists faithful $\chi \in \mathrm{Irr}(G)$ with $\chi(1) = p$. We claim that $m_{\mathbb{Q}}(\chi) = p$.

(10.16) THEOREM Let $G = XP$, where $X \lhd G$ is cyclic of order pq, $P \not\!\lhd\, G$ is cyclic of order p^2 and $|G| = p^2q$ for primes p and q such that $p^2 \nmid (q - 1)$. Then there exists faithful $\chi \in \mathrm{Irr}(G)$ such that $m_{\mathbb{Q}}(\chi) = p$.

Proof Note that $X = \mathbf{C}_G(X)$. Let λ be a faithful linear character of X and let $\chi = \lambda^G$. Then $\chi \in \mathrm{Irr}(G)$, $\chi(1) = p$, and (G, X, χ) is a $\mathbb{Q}$-triple. Since $m_{\mathbb{Q}}(\chi) \,|\, p$ by 10.2(h), it suffices to show that $m_{\mathbb{Q}}(\chi) \neq 1$. Let $K = \mathbb{Q}(\chi)$, $E = \mathbb{Q}(\lambda)$, and $\omega \in K$ be a primitive pth root of unity. By Theorem 10.10, it suffices to show that ω is not in the image of $N_{E/K}$.

Let ν be a primitive qth root of unity in E and write $\varepsilon = \omega\nu$ so that ε is a primitive (pq)th root and $E = \mathbb{Q}(\varepsilon)$. Let $R = \mathbb{Z}[\varepsilon]$ and let $I \supseteq qR$ be a maximal ideal of R. Since R/I is a field of characteristic q, the only qth root of unity in R/I is 1. Thus if $\sigma \in \mathcal{G}(E/K)$, then $\nu \equiv 1 \equiv \nu^\sigma \bmod I$ and since $\omega \in K$, we have $\varepsilon^\sigma = \omega\nu^\sigma \equiv \omega\nu = \varepsilon \bmod I$. Since $R = \mathbb{Z}[\varepsilon]$, it follows that $r^\sigma \equiv r \bmod I$ for all $r \in R$. Since $|E : K| = p$, we conclude that $N_{E/K}(r) \equiv r^p \bmod I$ for all $r \in R$.

Suppose, by way of contradiction, that $\omega = N_{E/K}(\alpha)$ for some $\alpha \in E$. By Fact (a), write $\alpha = b/c$, where $b, c \in R$ but not both $b, c \in I$. We conclude that

$b^p \equiv N_{E/K}(b) = N_{E/K}(c)N_{E/K}(\alpha) \equiv c^p\omega \bmod I$. Since not both $b, c \in I$, we conclude that $\omega^* = (b^*/c^*)^p$, where $*: R \to R/I$ is the natural homomorphism.

Now $1 + x + x^2 + \cdots + x^{p-1} = \prod_{i=1}^{p-1} (x - \omega^i)$ and setting $x = 1$, we see that $1 - \omega$ divides p in R. Since $p^* \neq 0$ in R/I, we conclude that $\omega^* \neq 1$. Thus b^*/c^* is a primitive p^2 root of unity in R/I.

Since $p \mid (q - 1)$, there exists $a \in \mathbb{Z}$ such that a^* is a primitive pth root of unity in R/I. Since $(\varepsilon^*)^p = 1$, we have $\varepsilon^* = (a^*)^k$ for some k and hence $\varepsilon \in \mathbb{Z} + I$. Thus $R = \mathbb{Z}[\varepsilon] = \mathbb{Z} + I$ and $R/I \cong \mathbb{Z}/(\mathbb{Z} \cap I) = \mathbb{Z}/q\mathbb{Z}$. Since $p^2 \nmid (q - 1)$, $\mathbb{Z}/q\mathbb{Z}$ does not contain a primitive p^2 root of unity and this contradiction completes the proof. ∎

Suppose $F \subseteq E \subseteq \mathbb{C}$, with $|E:F| = n < \infty$ and let $\chi \in \mathrm{Irr}(G)$ with $F(\chi) = F$. By 10.2(g), $m_F(\chi)$ divides $nm_E(\chi)$. In particular, if χ is afforded by an E-representation, then $m_E(\chi) = 1$ and so $m_F(\chi)$ divides $|E:F|$. This suggests the question of finding minimal fields $E \supseteq F$ over which χ is afforded.

(10.17) THEOREM Let $\chi \in \mathrm{Irr}(G)$ and $F \subseteq \mathbb{C}$ with $F(\chi) = F$. Let $m = m_F(\chi)$. Then there exists $E \supseteq F$ such that χ is afforded by an E-representation and $|E:F| = m$.

Proof Let V be an irreducible $F[G]$-module which affords $m\chi$ (by Corollary 10.2(b)). Let $D = \mathbf{E}_{F[G]}(V)$, the centralizer ring of V, so that D is a division ring by Schur's lemma. Let E be any maximal subfield of D. Then E acts on V and V may be viewed as an E-space. Since E commutes with the action of G, we may view V as an $E[G]$-module. Now $\mathbf{E}_{E[G]}(V)$ is the centralizer of E in D. The maximality of E thus yields $\mathbf{E}_{E[G]}(V) = E$ and thus the $E[G]$-module V corresponds to an absolutely irreducible E-representation $\mathfrak{X}$ by Theorem 9.2(a, c). Let $\mathfrak{Z}$ be the F-representation corresponding to the $F[G]$-module V. By Lemma 9.18, $\mathfrak{X}$ is a constituent of $\mathfrak{Z}^E$. Since $\mathfrak{Z}$ affords $m\chi$ and $\mathfrak{X}$ is absolutely irreducible, we conclude that $\mathfrak{X}$ affords χ. Finally, by 9.18(a) we conclude that $m = |E:F|$ and the proof is complete. ∎

We can, of course, find an F-isomorphism of E into $\mathbb{C}$ in the above situation, and so the result remains true if we add the condition that $E \subseteq \mathbb{C}$. However, it is not always true that if $F \subseteq L \subseteq \mathbb{C}$, where L is a splitting field for G and $\chi \in \mathrm{Irr}(G)$ with $F(\chi) = F$, then there exists a field E with $F \subseteq E \subseteq L$, $|E:F| = m_F(\chi)$ and such that χ is afforded by an E-representation.

We close this chapter by stating a few more facts about the Schur index. The proofs of some of these seem to require a fairly deep knowledge of number theory and division algebras.

(a) (*Fein*) If $\chi \in \mathrm{Irr}(G)$, then $m_{\mathbb{Q}}(\chi)$ divides $n[\chi^n, 1_G]$ for every positive integer n.

(b) (*Brauer–Speiser*) (Corollary of (a)) If $\chi \in \text{Irr}(G)$ is real valued, then $m_{\mathbb{Q}}(\chi) \leq 2$.

(c) (*Fein–Yamada*) If $\chi \in \text{Irr}(G)$, then $m_{\mathbb{Q}}(\chi)$ divides the exponent of G and $m_{\mathbb{Q}}(\chi)^2$ divides $|G|$.

Problems

(10.1) Let $H \subseteq G$, $\vartheta \in \text{Irr}(H)$ and $\chi \in \text{Irr}(G)$. Suppose $F \subseteq \mathbb{C}$.

(a) If $\chi_H = \vartheta$, show that $m_F(\chi) | m_F(\vartheta)$ and $m_F(\vartheta) \leq |G : H| m_F(\chi)$.
(b) If $\vartheta^G = \chi$, show that $m_F(\vartheta) | m_F(\chi)$ and $m_F(\chi) \leq |G : H| m_F(\vartheta)$.

(10.2) Let $N \lhd G$ and $\varphi \in \text{Irr}(N)$. Let $F \subseteq \mathbb{C}$ and put $H = I_G(\varphi)$. Let ϑ be an irreducible constituent of φ^H and let $\chi = \vartheta^G$. Show that $m_F(\chi)$ divides $|\mathbf{N}_G(H) : H| m_F(\vartheta)$.

Hint Consider $\{g \in G \mid \varphi^g$ and φ are Galois conjugate over $F(\chi)\}$.

(10.3) Let $\chi \in \text{Irr}(G)$ and $F \subseteq \mathbb{C}$. Define $l_F(\chi)$ to be the greatest common divisor of the integers $n_\lambda = |F(\chi, \lambda) : F(\chi)| [\lambda^G, \chi]$ as λ runs over linear characters of subgroups of G. Show that $m_F(\chi) | l_F(\chi)$ but that examples exist where $m_F(\chi) < l_F(\chi)$.

Hint For the example, take $G = Q_8$.

(10.4) Let $\chi \in \text{Irr}(G)$ and $F \subseteq \mathbb{C}$. We say that χ is *F-semiprimitive* if there does not exist $H < G$ and $\psi \in \text{Irr}(H)$ such that $\psi^G = \chi$ and $F(\psi) = F(\chi)$.

(a) If χ is F-semiprimitive and $N \lhd G$, show that the irreducible constituents of χ_N are Galois conjugate over $F(\chi)$.

(b) Let G be nilpotent and assume $\sqrt{-1} \in F$ if $|G|$ is even. Show that $\chi \in \text{Irr}(G)$ is F-semiprimitive iff $\chi(1) = 1$.

(c) In the situation of (b) and the notation of Problem 10.3, show that $l_F(\chi) = 1$ for all $\chi \in \text{Irr}(G)$.

Hint A noncyclic p-group in which every normal abelian subgroup is cyclic is necessarily a 2-group and contains a cyclic maximal subgroup.

Note Since $m_F(\chi) | l_F(\chi)$, this problem provides an alternate (and more elementary) proof of Corollary 10.14. Since $l_F(\chi)$ can, in general, exceed $m_F(\chi)$, this result strengthens Corollary 10.14 slightly.

(10.5) Let $F \subseteq \mathbb{C}$ be a field in which -1 is a sum of two squares. Let G be a 2-group and $\chi \in \text{Irr}(G)$. Show $m_F(\chi) = 1$.

Hint Reduce to the case where $\chi(1) = 2$ and $Q_8 \subseteq G$.

Note This strengthens Corollary 10.14 in a different direction and suggests a strengthening of Theorem 10.12. In fact, B. Fein has proved that the exceptional case in Theorem 10.12 can only occur if -1 is not a sum of two squares in F.

(10.6) Let $\chi \in \mathrm{Irr}(G)$ and $F \subseteq \mathbb{C}$. Let $H = G \times G \times \cdots \times G$ be a direct product of copies of G and let $\psi = \chi \times \chi \times \cdots \times \chi \in \mathrm{Irr}(H)$. Show that $m_F(\psi)$ divides $m_F(\chi)$.

Note In fact if H is the product of $m_F(\chi)$ copies of G, then $m_F(\psi) = 1$. Conversely, if $m_F(\psi) = 1$ and H is the product of n copies of G, then $m_F(\chi) \mid n$ provided $|F : \mathbb{Q}| < \infty$.

(10.7) (*Fein*) Let H be the product of n copies of G and let χ and ψ be as in Problem 10.6. Show that $m_{\mathbb{Q}}(\psi)$ divides $[\chi^n, 1_G]$.

Note Problem 10.7 and the note preceding it prove that $m_{\mathbb{Q}}(\chi)$ divides $n[\chi^n, 1_G]$ for every $n > 0$ and $\chi \in \mathrm{Irr}(G)$.

(10.8) Let $H = G \times G$ and let $\chi \in \mathrm{Irr}(G)$. Let $\psi = \chi \times \bar{\chi} \in \mathrm{Irr}(H)$. Show that $m_{\mathbb{Q}}(\psi) = 1$.

(10.9) Let $G = Q_8 \times \mathbb{Z}_3$ and let $\chi \in \mathrm{Irr}(G)$ be faithful. Show that $m_{\mathbb{Q}}(\chi) = 1$.

(10.10) Let $G = H \times K$ and $\psi \in \mathrm{Irr}(H)$ and $\vartheta \in \mathrm{Irr}(K)$. Let $\chi = \psi \times \vartheta$. Let $F \subseteq \mathbb{C}$.

(a) Show that $m_F(\chi)$ divides $m_F(\psi)m_F(\vartheta)$.

(b) Show that equality occurs in (a) provided $(m_F(\psi), \vartheta(1)|F(\vartheta) : F|) = 1$ and $(m_F(\vartheta), \psi(1)|F(\psi) : F|) = 1$.

(c) Let p, q be primes such that $p \nmid (q - 1)$ and $q \nmid (p - 1)$. Show that pq occurs as a Schur index.

(10.11) Let $\chi \in \mathrm{Irr}(G)$ and $p \mid m_{\mathbb{Q}}(\chi)$ for some odd prime p. Show that G contains an element of order pq, where q is some prime such that $p \mid (q - 1)$.

(10.12) Let p be an odd prime and let P be a nonabelian p-group of order p^3 and exponent p. It is possible to find $H \subseteq \mathrm{Aut}(P)$ such that $H \cong Q_8$ and $\mathbf{C}_P(H) = \mathbf{Z}(P) = \mathbf{C}_P(t)$, where t is the involution in H. Let $G = PH$, the semidirect product. Let $\vartheta \in \mathrm{Irr}(P)$ with $\vartheta(1) = p$.

(a) Show that ϑ can be extended to $\hat{\vartheta} \in \mathrm{Irr}(G)$ such that $\hat{\vartheta}(t) = \pm 1$ and $\hat{\vartheta}(t) \equiv p \bmod 4$.

(b) Show that $[\hat{\vartheta}_H, \chi] = (p - \hat{\vartheta}(t))/4$, where $\chi \in \mathrm{Irr}(H)$, $\chi(1) = 2$.

(c) Conclude that -1 is a sum of two squares in $\mathbb{Q}(e^{2\pi i/p})$ if $p \equiv 3$ or $5 \bmod 8$.

Hints Obtain $\hat{\vartheta}$ via Corollary 6.28. Compute $\hat{\vartheta}(t)$ by working in $N = \langle P, t\rangle$. Note that $|\mathbf{C}_N(t)| = 2p$ and that the $p - 1$ Galois conjugates of $\hat{\vartheta}_N$ are all equal at t. If $p \equiv 3$ or $5 \bmod 8$, show that $m_F(\chi) = 1$, where $F = \mathbb{Q}(e^{2\pi i/p})$.

Note Conversely, if -1 is a sum of two squares in $\mathbb{Q}(e^{2\pi i/p})$, then $p \not\equiv 7 \bmod 8$. Some primes $\equiv 1 \bmod 8$ can occur, however.

(10.13) Let $P \in \mathrm{Syl}_p(G)$ be abelian of exponent p^a. Show that $p^a \nmid m_{\mathbb{Q}}(\chi)$ for $\chi \in \mathrm{Irr}(G)$.

Hint If $X \lhd G$ is cyclic and G/X is a p-group, then there exists $V \subseteq G$ such that $V \cap X = 1$ and $|G : VX| < p^a$.

(10.14) Show that the following two statements can be added to the conclusions of Theorem 10.7 provided $p \neq 2$.

(e) If Y is the p-complement in X, then $\mathbf{C}_H(Y) = X$.

(f) If $U \in \mathrm{Syl}_p(X)$ and $P \in \mathrm{Syl}_p(H)$, then $U \subseteq \Phi(P)$, the Frattini subgroup.

Hints If $\mathbf{C}_P(Y) > U$, show that H has a noncyclic normal abelian p-subgroup. If $M \subseteq P$ and $MU = P$, then ϑ_{MY} is irreducible.

(10.15) Suppose $m_F(\chi) = \chi(1)$ for some $\chi \in \mathrm{Irr}(G)$ and $F \subseteq \mathbb{C}$.

(a) If $H \subseteq G$, show that all irreducible constituents of χ_H are Galois conjugate over $F(\chi)$ and that $m_F(\vartheta) = \vartheta(1)$ for each such constituent ϑ.

(b) If $\chi \neq 1_G$, show that G is not simple.

(c) If $2 \nmid \chi(1)$ or $\sqrt{-1} \in F$, show that $G/\ker\chi$ is solvable.

Hint Use Lemma 10.4.

(10.16) Suppose that $m_F(\chi) = \chi(1)$ for all $\chi \in \mathrm{Irr}(G)$ with $F \subseteq \mathbb{C}$. Show that every subgroup of G is normal.

11 Projective representations

Let $N \lhd G$ and suppose $\vartheta \in \mathrm{Irr}(N)$ is invariant in G. For each irreducible constituent χ of ϑ^G we have that $\chi_N = e(\chi)\vartheta$, where $e(\chi)$, the *ramification*, is a positive integer. In Chapter 6 we obtained some information about these mysterious integers. If ϑ is extendible to G [which is equivalent to saying that some $e(\chi) = 1$], then the $e(\chi)$'s are exactly the degrees of the irreducible characters of G/N. (See Corollary 6.17.) We shall see that, in general, the $e(\chi)$'s are the degrees of irreducible "projective representations" of G/N.

(11.1) DEFINITION Let G be a group and F a field. Let $\mathfrak{X}: G \to \mathrm{GL}(n, F)$ be such that for every $g, h \in G$, there exists a scalar $\alpha(g, h) \in F$ such that

$$\mathfrak{X}(g)\mathfrak{X}(h) = \mathfrak{X}(gh)\alpha(g, h).$$

Then $\mathfrak{X}$ is a *projective F-representation* of G. Its *degree* is n and the function $\alpha: G \times G \to F$ is the associated *factor set* of $\mathfrak{X}$.

Note that the "factor set" α has nonzero values and is uniquely determined by $\mathfrak{X}$. Both of these observations follow from the fact that the matrices $\mathfrak{X}(g)$ are nonsingular.

Let $Z(n, F) \subseteq \mathrm{GL}(n, F)$ be the group of nonzero scalar matrices. [Note that $Z(n, F) = \mathbf{Z}(\mathrm{GL}(n, F))$.] By definition, $\mathrm{PGL}(n, F) = \mathrm{GL}(n, F)/Z(n, F)$ is the *projective general linear* group. If $\mathfrak{X}$ is a projective F-representation of G of degree n, then the composition of $\mathfrak{X}$ with the canonical homomorphism $\mathrm{GL}(n, F) \to \mathrm{PGL}(n, F)$ is a homomorphism $G \to \mathrm{PGL}(n, F)$. Conversely, if $\pi: G \to \mathrm{PGL}(n, F)$ is any homomorphism, we can define a projective representation $\mathfrak{X}$ of G by setting $\mathfrak{X}(g)$ equal to any element of the coset $\pi(g)$ of $Z(n, F)$ in $\mathrm{GL}(n, F)$.

Before proceeding with the study of projective representations, we digress to give an instance of how they can arise in the study of ordinary representations.

(11.2) THEOREM Let $N \lhd G$ and suppose $\mathfrak{Y}$ is an irreducible $\mathbb{C}$-representation of N whose character is invariant in G. Then there exists a projective $\mathbb{C}$-representation $\mathfrak{X}$ of G such that for all $n \in N$ and $g \in G$ we have

(a) $\mathfrak{X}(n) = \mathfrak{Y}(n)$;
(b) $\mathfrak{X}(ng) = \mathfrak{X}(n)\mathfrak{X}(g)$;
(c) $\mathfrak{X}(gn) = \mathfrak{X}(g)\mathfrak{X}(n)$.

Furthermore, if $\mathfrak{X}_0$ is another projective representation satisfying (a), (b), and (c), then $\mathfrak{X}_0(g) = \mathfrak{X}(g)\mu(g)$ for some function $\mu: G \to \mathbb{C}^\times$, which is constant on cosets of N.

Proof For $g \in G$ and $n \in N$, write $\mathfrak{Y}(gng^{-1}) = \mathfrak{Y}^g(n)$. Since $\mathfrak{Y}$ affords a G-invariant character, we conclude that $\mathfrak{Y}$ and $\mathfrak{Y}^g$ are similar representations of N.

Now choose a transversal T for N in G (that is, a set of coset representatives). Take $1 \in T$. For each $t \in T$, choose a nonsingular matrix P_t such that $P_t \mathfrak{Y} P_t^{-1} = \mathfrak{Y}^t$. Take $P_1 = I$. Since every element of G is uniquely of the form nt for $n \in N$ and $t \in T$ we can define $\mathfrak{X}$ on G by $\mathfrak{X}(nt) = \mathfrak{Y}(n)P_t$. Properties (a) and (b) are immediate and (c) follows since

$$\begin{aligned}\mathfrak{X}(nt)\mathfrak{X}(m) = \mathfrak{Y}(n)P_t\mathfrak{Y}(m) = \mathfrak{Y}(n)\mathfrak{Y}^t(m)P_t &= \mathfrak{Y}(ntmt^{-1})P_t \\ &= \mathfrak{X}(ntmt^{-1} \cdot t) = \mathfrak{X}(nt \cdot m).\end{aligned}$$

Properties (a), (b), and (c) yield

$$\mathfrak{X}(g)\mathfrak{Y}(n) = \mathfrak{X}(gn) = \mathfrak{X}(gng^{-1} \cdot g) = \mathfrak{Y}(gng^{-1})\mathfrak{X}(g)$$

and

$$\mathfrak{X}(g)\mathfrak{Y}(n)\mathfrak{X}(g)^{-1} = \mathfrak{Y}(gng^{-1})$$

for all $g \in G$ and $n \in N$. If A is any nonsingular matrix such that

$$A\mathfrak{Y}(n)A^{-1} = \mathfrak{Y}(gng^{-1})$$

for all $n \in N$, then $A^{-1}\mathfrak{X}(g)$ commutes with all $\mathfrak{Y}(n)$ for $n \in N$ and thus $A^{-1}\mathfrak{X}(g)$ is a scalar matrix by Corollary 1.6. If $\mathfrak{X}_0$ also satisfies (a), (b), and (c), we may take $A = \mathfrak{X}_0(g)$ and conclude that $\mathfrak{X}_0(g) = \mathfrak{X}(g)\mu(g)$ for some $\mu(g) \in \mathbb{C}^\times$. Also

$$\mathfrak{X}(g)\mathfrak{X}(h)\mathfrak{Y}(n)\mathfrak{X}(h)^{-1}\mathfrak{X}(g)^{-1} = \mathfrak{X}(g)\mathfrak{Y}(hnh^{-1})\mathfrak{X}(g)^{-1} = \mathfrak{Y}(ghnh^{-1}g^{-1}).$$

Comparing this with

$$\mathfrak{X}(gh)\mathfrak{Y}(n)\mathfrak{X}(gh)^{-1} = \mathfrak{Y}(ghnh^{-1}g^{-1})$$

yields $\mathfrak{X}(g)\mathfrak{X}(h) = \mathfrak{X}(gh)\alpha(g, h)$ for some $\alpha(g, h) \in \mathbb{C}^\times$ and thus $\mathfrak{X}$ is a projective representation.

All that remains now is to check that μ is constant on cosets of N. We have

$$\mathfrak{X}(n)\mathfrak{X}(g)\mu(g) = \mathfrak{X}_0(n)\mathfrak{X}_0(g) = \mathfrak{X}_0(ng) = \mathfrak{X}(ng)\mu(ng).$$

Since $\mathfrak{X}(n)\mathfrak{X}(g) = \mathfrak{X}(ng)$ is nonsingular, the result follows. ∎

Of course, the above argument does not really require that $\mathfrak{Y}$ be a complex representation. Using Theorem 9.2 and Lemma 9.12, any absolutely irreducible representation will do.

We begin our study of projective representations by considering the associated factor sets.

(11.3) LEMMA Let $\alpha: G \times G \to F^\times$ be the associated factor set of a projective F-representation of G. Then

$$\alpha(xy, z)\alpha(x, y) = \alpha(x, yz)\alpha(y, z)$$

for all $x, y, z \in G$.

Proof Let $\mathfrak{X}$ be the projective representation. We have

$$\mathfrak{X}(x)\mathfrak{X}(y)\mathfrak{X}(z) = \mathfrak{X}(xy)\mathfrak{X}(z)\alpha(x, y) = \mathfrak{X}(xyz)\alpha(xy, z)\alpha(x, y).$$

Also

$$\mathfrak{X}(x)\mathfrak{X}(y)\mathfrak{X}(z) = \mathfrak{X}(x)\mathfrak{X}(yz)\alpha(y, z) = \mathfrak{X}(xyz)\alpha(x, yz)\alpha(y, z).$$

The result now follows because all of the matrices are nonsingular. ∎

(11.4) DEFINITION Let A be a possibly infinite abelian group and let G be any group. Then an *A-factor set* of G is a function $\alpha: G \times G \to A$ such that

$$\alpha(xy, z)\alpha(x, y) = \alpha(x, yz)\alpha(y, z)$$

for all $x, y, z \in G$.

Thus the factor set of a projective F-representation is an $F^\times$-factor set where $F^\times$ denotes the multiplicative group of F. Conversely, we shall show that every $F^\times$-factor set is associated with a projective F-representation. To do this we introduce the "twisted group algebra."

Let G be a finite group and F a field. Let α be an $F^\times$-factor set of G. Let $F^\alpha[G]$ be the F-vectorspace with basis $\{\bar{g} \mid g \in G\}$. (That is, there is a specified basis of $F^\alpha[G]$ which is in one-to-one correspondence with G.) Define multiplication in $F^\alpha[G]$ by $\bar{g} \cdot \bar{h} = \overline{gh}\alpha(g, h)$ and extend via the distributive law. To establish that the multiplication thus defined is associative, it suffices to check it on the basis elements $\bar{g}$ for $g \in G$. That it holds there is immediate from the definition of a factor set. The finite dimensional algebra $F^\alpha[G]$ is the *twisted group algebra* with respect to α. Note that if α is the trivial $F^\times$-factor set, that is, $\alpha(g, h) = 1$ for all g, h, we can identify $F^\alpha[G]$ with $F[G]$.

In order to see that $F^\alpha[G]$ has a 1, we need the following.

(11.5) LEMMA Let α be an A-factor set of G. Then $\alpha(1, x) = \alpha(1, 1) = \alpha(x, 1)$ for all $x \in G$.

Proof We have

$$\alpha(1 \cdot 1, x)\alpha(1, 1) = \alpha(1, 1 \cdot x)\alpha(1, x).$$

Canceling $\alpha(1, x)$ yields $\alpha(1, 1) = \alpha(1, x)$ for $x \in G$. That $\alpha(1, 1) = \alpha(x, 1)$ follows symmetrically. ∎

Let α be an $F^\times$-factor set of G and let $v = \alpha(1, 1)^{-1} \in F$. Now $(v\bar{1})\bar{g} = v\bar{g}\alpha(1, g) = \bar{g}$ and similarly $\bar{g}(v\bar{1}) = \bar{g}$. Thus $v\bar{1}$ is the unit element in $F^\alpha[G]$. It is now immediate that the elements $\bar{g} \in F^\alpha[G]$ for $g \in G$ all have inverses.

Now let α be an $F^\times$-factor set of G and let $\mathfrak{Y}$ be any representation of the algebra $F^\alpha[G]$. Define $\mathfrak{X}(g) = \mathfrak{Y}(\bar{g})$. Then $\mathfrak{X}(g)$ is nonsingular and

$$\mathfrak{X}(g)\mathfrak{X}(h) = \mathfrak{Y}(\bar{g})\mathfrak{Y}(\bar{h}) = \mathfrak{Y}(\bar{g} \cdot \bar{h}) = \mathfrak{Y}(\overline{gh}\alpha(g, h)) = \mathfrak{X}(gh)\alpha(g, h)$$

so that $\mathfrak{X}$ is a projective representation of G with factor set α.

Conversely, if $\mathfrak{X}$ is a projective F-representation of G with factor set α, we can define a representation $\mathfrak{Y}$ of $F^\alpha[G]$ by setting $\mathfrak{Y}(\bar{g}) = \mathfrak{X}(g)$ and extending by linearity. In other words, the projective F-representations of G having factor set α are in a natural one-to-one correspondence with the representations of the twisted group algebra $F^\alpha[G]$. The situation is analogous to the connection between ordinary representations and the ordinary group algebra.

Exactly as in the case with ordinary representations, we define two projective representations $\mathfrak{X}$ and $\mathfrak{Y}$ to be *similar* if $\mathfrak{Y} = P^{-1}\mathfrak{X}P$ for some nonsingular matrix P. Also $\mathfrak{X}$ is *irreducible* if it is not similar to a projective representation in the form

$$\begin{pmatrix} * & * \\ 0 & * \end{pmatrix}.$$

Note that similar projective representations have equal factor sets and correspond to similar representations of the appropriate twisted group algebra. Also, irreducible projective representations correspond to irreducible representations of the algebra. Since every finite dimensional algebra has irreducible representations, we have proved the following.

(11.6) COROLLARY Let α be an $F^\times$-factor set of G. Then G has irreducible projective F-representations with factor set α.

The set of A-factor sets of G forms a group under pointwise multiplication. In the language of group cohomology, this group is denoted $Z^2(G, A)$, the group of "2-cocycles." If $\mu: G \to A$ is an arbitrary function, we can define $\delta(\mu): G \times G \to A$ by

$$\delta(\mu)(g, h) = \mu(g)\mu(h)\mu(gh)^{-1}.$$

It is routine to check that $\delta(\mu)$ is a factor set.

Note that δ is a homomorphism from the group of A-valued functions on G (with pointwise multiplication) into $Z^2(G, A)$. The image of δ is the subgroup $B^2(G, A) \subseteq Z^2(G, A)$ which is called the group of "2-coboundaries." The factor group $Z^2(G, A)/B^2(G, A)$ can be identified with the second cohomology group $H^2(G, A)$. (More generally in cohomology theory, one assumes that G acts on A. Here we are considering only the "trivial action" case.)

The superscript 2 above refers to the fact that we are discussing functions of two variables on G. Since that will be the only situation considered here, we write $Z(G, A)$ for the group of A-factor sets $B(G, A)$ for the image of δ and we define $H(G, A) = Z(G, A)/B(G, A)$.

We say that two A-factor sets of G are *equivalent* if they are congruent mod $B(G, A)$. Thus $H(G, A)$ is the set of equivalence classes of A-factor sets on G.

If $\mathfrak{X}$ is a projective F-representation on G with factor set α, and $\mu: G \to F^\times$ is any function, define $\mathfrak{Y} = \mathfrak{X}\mu$ by $\mathfrak{Y}(g) = \mathfrak{X}(g)\mu(g)$. It is trivial to check that $\mathfrak{Y}$ is a projective representation of G with factor set $\beta = \alpha\delta(\mu)$. Thus $\mathfrak{X}$ and $\mathfrak{Y}$ have equivalent factor sets.

If $\mathfrak{X}$ and $\mathfrak{Y}$ are projective F-representations of G, we say that $\mathfrak{X}$ and $\mathfrak{Y}$ are *equivalent* if $\mathfrak{Y}$ is similar to $\mathfrak{X}\mu$ for some function $\mu: G \to F^\times$. This is easily seen to define an equivalence relation which preserves irreduciblility.

We can now give a necessary and sufficient condition for an invariant irreducible character of a normal subgroup to be extendible.

(11.7) THEOREM Let $N \lhd G$ and let $\vartheta \in \text{Irr}(N)$ be invariant in G. Let $\mathfrak{Y}$ be a representation affording ϑ and let $\mathfrak{X}$ be a projective representation of G satisfying conditions (a), (b), and (c) of Theorem 11.2. Let α be the factor set of $\mathfrak{X}$. Define $\beta \in Z(G/N, \mathbb{C}^\times)$ by $\beta(gN, hN) = \alpha(g, h)$. Then β is well-defined and its image $\bar{\beta} \in H(G/N, \mathbb{C}^\times)$ depends only on ϑ. Also, ϑ is extendible to G iff $\bar{\beta} = 1$.

Proof For $m, n \in N$ and $g, h \in G$, we have

$$\alpha(gn, hm)\mathfrak{X}(gnhm) = \mathfrak{X}(gn)\mathfrak{X}(hm) = \mathfrak{X}(g)\mathfrak{X}(nhm)$$

using 11.2(b) and (c). Furthermore,

$$\begin{aligned}\mathfrak{X}(g)\mathfrak{X}(nhm) &= \mathfrak{X}(g)\mathfrak{X}(hn^hm) = \mathfrak{X}(g)\mathfrak{X}(h)\mathfrak{X}(n^hm)\\ &= \alpha(g, h)\mathfrak{X}(gh)\mathfrak{X}(n^hm) = \alpha(g, h)\mathfrak{X}(ghn^hm)\end{aligned}$$

by 11.2(c). Since $\mathfrak{X}(gnhm) = \mathfrak{X}(ghn^hm)$ is nonsingular, we have $\alpha(gn, hm) = \alpha(g, h)$ and β is well defined. That β is a factor set is clear.

If $\mathfrak{X}_0$ were chosen in place of $\mathfrak{X}$, we have $\mathfrak{X}_0 = \mathfrak{X}\mu$ where $\mu: G \to \mathbb{C}^\times$ is constant on cosets of N so that we can define $\nu(gN) = \mu(g)$. If $\mathfrak{X}_0$ has factor set α_0, then

$$\begin{aligned}\alpha_0(g, h) &= \alpha(g, h)\mu(g)\mu(h)\mu(gh)^{-1}\\ &= \beta(gN, hN)\nu(gN)\nu(hN)\nu(ghN)^{-1}\end{aligned}$$

and hence $\bar{\beta}$ is independent of the choice of $\mathfrak{X}$. If $P\mathfrak{Y}P^{-1}$ were chosen in place of $\mathfrak{Y}$, we can replace $\mathfrak{X}$ by $P\mathfrak{X}P^{-1}$ which leaves α, β and $\bar{\beta}$ unchanged. Thus $\bar{\beta}$ is uniquely determined once ϑ is given.

If ϑ is extendible to G, we can choose $\mathfrak{X}$ and $\mathfrak{Y}$ so that $\mathfrak{X}$ is a representation. Thus $\alpha = 1$ and hence $\bar{\beta} = 1$.

Conversely, if $\bar{\beta} = 1$, there exists $\nu: G/N \to \mathbb{C}^\times$ such that

$$\beta(gN, hN) = \nu(gN)\nu(hN)\nu(ghN)^{-1}.$$

Define $\mu: G \to \mathbb{C}^\times$ by $\mu(g) = \nu(gN)$. Now define $\mathfrak{X}_0(g) = \mathfrak{X}(g)\mu(g)^{-1}$ so that $\mathfrak{X}_0$ is a projective representation of G with factor set

$$\alpha_0(g, h) = \alpha(g, h)\mu(g)^{-1}\mu(h)^{-1}\mu(gh) = 1$$

and thus $\mathfrak{X}_0$ is a representation of G. Furthermore, since $\mathfrak{X}(1) = \mathfrak{Y}(1) = I$, we have

$$1 = \alpha(1, 1) = \mu(1)\mu(1)\mu(1)^{-1} = \mu(1).$$

Thus $\mu(n) = \nu(1) = \mu(1) = 1$ for all $n \in N$ and $\mathfrak{X}_0(n) = \mathfrak{X}(n)\mu(n)^{-1} = \mathfrak{Y}(n)$. Thus $\mathfrak{X}_0$ is an extension of $\mathfrak{Y}$ and the proof is complete. ▮

A word of caution about Theorem 11.7 is appropriate. In the notation of the theorem, if $\alpha \in B(G, \mathbb{C}^\times)$ so that $\bar{\alpha} = 1$ in $H(G, \mathbb{C}^\times)$, it does *not* follow that ϑ is extendible to G. For instance, take $G = Q_8$, $N = \mathbf{Z}(G)$, and ϑ the non-principal linear character of N. Then ϑ is not extendible to G and yet $H(G, \mathbb{C}^\times) = 1$. (See Problem 11.18.)

The theory of projective representations is closely related to that of central extensions.

(11.8) DEFINITION A *central extension* of a group G is a (possibly infinite) group Γ together with a homomorphism π of Γ onto G such that $\ker \pi \subseteq \mathbf{Z}(\Gamma)$.

(11.9) LEMMA Let (Γ, π) be a central extension of G with $A = \ker \pi$. Let X be a set of coset representatives for A in Γ and write $X = \{x_g \mid g \in G\}$, where $\pi(x_g) = g$. Define $\alpha: G \times G \to A$ by $x_g x_h = \alpha(g, h)x_{gh}$. Then $\alpha \in Z(G, A)$. Furthermore, the equivalence class of α is independent of the choice of X.

Proof That α is an A-factor set follows by computing $x_g x_h x_k$ two ways, using the associative law in Γ. If $Y = \{y_g\}$ is another set of coset representatives, then $y_g = \mu(g)x_g$ for some $\mu(g) \in A$. Now

$$
\begin{aligned}
y_g y_h = \mu(g)\mu(h)x_g x_h &= \mu(g)\mu(h)\alpha(g, h)x_{gh} \\
&= \mu(g)\mu(h)\mu(gh)^{-1}\alpha(g, h)y_{gh}
\end{aligned}
$$

and the result follows. ∎

If A and U are abelian groups and $\lambda \in \mathrm{Hom}(A, U)$, then for each $\alpha \in Z(G, A)$, we define $\lambda(\alpha)$ by $\lambda(\alpha)(g, h) = \lambda(\alpha(g, h))$. It is routine to check that $\lambda(\alpha) \in Z(G, U)$.

(11.10) COROLLARY Let (Γ, π) be a finite central extension of G and let A, $X = \{x_g\}$ and α be as in Lemma 11.9. Let $\mathfrak{Y}$ be an (ordinary) F-representation of Γ such that the restriction $\mathfrak{Y}_A$ is the scalar representation λI for some $\lambda \in \mathrm{Hom}(A, F^\times)$. Define $\mathfrak{X}(g) = \mathfrak{Y}(x_g)$ for $g \in G$. Then $\mathfrak{X}$ is a projective F-representation of G with factor set $\lambda(\alpha)$. Furthermore,

$$\mathfrak{Y}(y) = \mathfrak{X}(\pi(y))\mu(y)$$

for all $y \in \Gamma$, where $\mu: \Gamma \to F^\times$ is the function defined by $\mu(y) = \lambda(yx_{\pi(y)}^{-1})$. Also $\mathfrak{X}$ is irreducible iff $\mathfrak{Y}$ is and the equivalence class of $\mathfrak{X}$ is independent of the choice of coset representatives X.

Proof We have

$$\mathfrak{X}(g)\mathfrak{X}(h) = \mathfrak{Y}(x_g)\mathfrak{Y}(x_h) = \mathfrak{Y}(\alpha(g, h)x_{gh}) = \lambda(\alpha(g, h))\mathfrak{X}(gh)$$

and $\mathfrak{X}$ is a projective representation with factor set $\lambda(\alpha)$. If $y \in \Gamma$, we have $y = ax_g$, where $g = \pi(y)$ and $a \in A$. Thus

$$\mathfrak{Y}(y) = \mathfrak{X}(g)\lambda(a) = \mathfrak{X}(\pi(y))\lambda(yx_{\pi(y)}^{-1}).$$

In particular, $\mathfrak{X}(G)$ and $\mathfrak{Y}(\Gamma)$ span the same vectorspace of matrices over F and the assertion about irreducibility follows.

Finally, if $\mathfrak{X}_1$ is the projective representation determined by an alternate choice of coset representatives, we have

$$\mathfrak{X}_1(\pi(y)) = \mathfrak{Y}(y)\mu_1(y)^{-1} = \mathfrak{X}(\pi(y))\mu(y)\mu_1(y)^{-1}$$

and $\mathfrak{X}$ and $\mathfrak{X}_1$ are equivalent. ∎

Note that if $\mathfrak{Y}$ is an irreducible $\mathbb{C}$-representation (or any absolutely irreducible F-representation), then the condition that $\mathfrak{Y}_A$ be a scalar representation in Corollary 11.10 is automatically satisfied.

Henceforth we shall consider only $\mathbb{C}$-representations.

(11.11) DEFINITION Let (Γ, π) be a finite central extension of G. Let $\mathfrak{X}$ be a projective $\mathbb{C}$-representation of G. We say that $\mathfrak{X}$ can be *lifted* to Γ if there exists an ordinary representation $\mathfrak{Y}$ of Γ and a function $\mu: \Gamma \to \mathbb{C}^\times$ such that

$$\mathfrak{Y}(x) = \mathfrak{X}(\pi(x))\mu(x)$$

for all $x \in \Gamma$. Furthermore, (Γ, π) has the *projective lifting property* for G if every projective $\mathbb{C}$-representation of G can be lifted to Γ.

Note that if $\mathfrak{X}$ is lifted to the representation $\mathfrak{Y}$ on Γ, then $\mathfrak{Y}_A$ is necessarily a scalar representation and by Corollary 11.10 we can construct a projective representation $\mathfrak{X}_1$ of G by choosing coset representatives for $\ker \pi$ in Γ. Then

$$\mathfrak{X}_1(\pi(x))\mu_1(x) = \mathfrak{Y}(x) = \mathfrak{X}(\pi(x))\mu(x)$$

and hence $\mathfrak{X}_1$ is equivalent to $\mathfrak{X}$. Thus if (Γ, π) has the projective lifting property for G, then all projective $\mathbb{C}$-representations of G are equivalent to ones obtained via the construction in Corollary 11.10.

We shall prove a theorem of Schur which asserts that every finite group has a finite central extension with the projective lifting property. The point of Schur's theorem is that it allows us to apply what we know about ordinary representations to the study of projective representations. For instance, in the situation of Corollary 11.10, if $\mathfrak{Y}$ is irreducible and $F = \mathbb{C}$, then we have $\deg \mathfrak{X} = \deg \mathfrak{Y}$ divides $|\Gamma : A| = |G|$ by Theorem 3.12. Thus a consequence of Schur's theorem is that the degree of every irreducible projective $\mathbb{C}$-representation of G divides $|G|$.

(11.12) DEFINITION The *Schur multiplier* of G is the group $H(G, \mathbb{C}^\times)$. It is denoted $M(G)$.

For a finite abelian group A we use the notation $\hat{A}$ to denote the group $\mathrm{Irr}(A)$. If (Γ, π) is a central extension of G with $A = \ker \pi$ finite, we construct a homomorphism $\eta: \hat{A} \to M(G)$ as follows. Choose a set X of coset representatives for A in Γ and write $X = \{x_g \mid g \in G\}$, where $\pi(x_g) = g$. Let $\alpha \in Z(G, A)$ be defined by $x_g x_h = \alpha(g, h)x_{gh}$ as in Lemma 11.9. For $\lambda \in \hat{A}$ define $\eta(\lambda) = \overline{\lambda(\alpha)}$, where $\lambda(\alpha) \in Z(G, \mathbb{C}^\times)$ is defined by $\lambda(\alpha)(g, h) = \lambda(\alpha(g, h))$ and the bar denotes the canonical map $Z(G, \mathbb{C}^\times) \to H(G, \mathbb{C}^\times) = M(G)$. Note that η is a homomorphism.

The map $\eta: \hat{A} \to M(G)$ is independent of the choice of coset representatives X. This follows since another choice would yield a factor set $\beta \in Z(G, A)$, which is equivalent to α by Lemma 11.9. Since $\alpha\beta^{-1} \in B(G, A)$, it follows that $\lambda(\alpha)\lambda(\beta)^{-1} = \lambda(\alpha\beta^{-1}) \in B(G, \mathbb{C}^{\times})$. Thus $\lambda(\alpha)$ and $\lambda(\beta)$ are equivalent in $Z(G, \mathbb{C}^{\times})$ and $\overline{\lambda(\alpha)} = \overline{\lambda(\beta)}$.

We shall call the unique homomorphism η: $\hat{A} \to M(G)$ constructed above the *standard map*.

(11.13) THEOREM Let (Γ, π) be a finite central extension of G and let η be the associated standard map. Let $\mathfrak{X}$ be a projective $\mathbb{C}$-representation of G with factor set γ. Then $\mathfrak{X}$ can be lifted to Γ iff $\bar{\gamma}$ lies in the image of η. In particular, (Γ, π) has the projective lifting property iff η maps onto $M(G)$.

Proof. Let $A = \ker \pi$ and let $X = \{x_g \mid g \in G\}$ be a set of coset representatives for A in Γ with $\pi(x_g) = g$. Write $x_g x_h = \alpha(g, h)x_{gh}$ so that $\alpha \in Z(G, A)$. Now suppose $\bar{\gamma} = \eta(\lambda)$ for some $\lambda \in \hat{A}$. Then $\lambda(\alpha)$ is equivalent to γ and we have

$$\lambda(\alpha(g, h)) = \gamma(g, h)\mu(g)\mu(h)\mu(gh)^{-1}$$

for some function μ: $G \to \mathbb{C}^{\times}$.

Define $\mathfrak{Y}$ on Γ by $\mathfrak{Y}(ax_g) = \lambda(a)\mathfrak{X}(g)\mu(g)$ for $a \in A$ and $g \in G$. We have

$$\begin{aligned}\mathfrak{Y}(x_g)\mathfrak{Y}(x_h) &= \mathfrak{X}(g)\mathfrak{X}(h)\mu(g)\mu(h) = \mathfrak{X}(gh)\gamma(g, h)\mu(g)\mu(h)\\ &= \lambda(\alpha(g, h))\mathfrak{X}(gh)\mu(gh) = \mathfrak{Y}(\alpha(g, h)x_{gh}) = \mathfrak{Y}(x_g x_h).\end{aligned}$$

Since $\mathfrak{Y}(ax_g) = \lambda(a)\mathfrak{Y}(x_g)$, it follows that $\mathfrak{Y}$ is a representation. Thus $\mathfrak{Y}$ lifts $\mathfrak{X}$ to Γ.

Conversely, if $\mathfrak{X}$ can be lifted, we have $\mathfrak{Y}(x) = \mathfrak{X}(\pi(x))\mu(x)$ for some representation $\mathfrak{Y}$ of Γ and function μ: $\Gamma \to \mathbb{C}$. Thus $\mathfrak{X}(1) = \mu(1)^{-1}\mathfrak{Y}(1)$ and for every $a \in A$ we have $\mathfrak{Y}(a) = \mathfrak{X}(1)\mu(a) = \mu(a)\mu(1)^{-1}\mathfrak{Y}(1)$ and so $\lambda(a) = \mu(a)\mu(1)^{-1}$ is a linear character of A. Now write $\nu(g) = \mu(x_g)$. We have

$$\lambda(\alpha(g, h))\mathfrak{Y}(x_{gh}) = \mathfrak{Y}(x_g)\mathfrak{Y}(x_h) = \mathfrak{X}(g)\mathfrak{X}(h)\nu(g)\nu(h).$$

Thus

$$\lambda(\alpha(g, h))\nu(gh)\mathfrak{X}(gh) = \gamma(g, h)\nu(g)\nu(h)\mathfrak{X}(gh)$$

and $\lambda(\alpha)$ is equivalent to γ. Therefore

$$\eta(\lambda) = \overline{\lambda(\alpha)} = \bar{\gamma}$$

and the proof is complete. ∎

Given an arbitrary finite group G we shall show that $M(G)$ is finite and construct a central extension (Γ, π) of G with $A = \ker \pi$ such that the standard map η: $\hat{A} \to M(G)$ is an isomorphism. This will thus be a group with the projective lifting property for G with smallest possible order, namely $|G||M(G)|$. Such a group Γ is called a *Schur representation group* for G.

An abelian group Q is *divisible* if for every $x \in Q$ and positive integer n there exists $y \in Q$, with $y^n = x$. For instance, $F^{\times}$ is divisible if F is algebraically closed.

(11.14) LEMMA Let A be an (infinite) abelian group and let $Q \subseteq A$ with Q divisible. Assume $|A:Q| < \infty$. Then Q is complemented in A.

Proof Use induction on $|A:Q|$. We may assume $A > Q$ and choose $a \in A - Q$. Let $n = o(aQ)$ in A/Q and let $u = a^n \in Q$. Let $v \in Q$ with $v^n = u$ by divisibility and let $b = av^{-1}$ so that $b^n = 1$. Since $aQ = bQ$, it follows that $n = o(bQ)$ in A/Q and thus $\langle b \rangle \cap Q = 1$.

Now let $\bar{A} = A/\langle b \rangle$ so $\bar{Q} = Q\langle b \rangle/\langle b \rangle$ satisfies $\bar{Q} \cong Q$ and $|\bar{A}:\bar{Q}| = |A:Q\langle b \rangle| < |A:Q|$. By the inductive hypothesis, $\bar{Q}$ is complemented in $\bar{A}$ and thus there exists $B \subseteq A$ with $B \cap Q\langle b \rangle = \langle b \rangle$ and $QB = A$. Now $Q \cap B = Q \cap Q\langle b \rangle \cap B = Q \cap \langle b \rangle = 1$ and the proof is complete. ∎

The above result remains true without the assumption that $|A:Q|$ is finite. We will not need that more general fact, however.

(11.15) THEOREM Let F be an algebraically closed field and G a finite group. Then $H(G, F^\times)$ is finite and each of its elements has order dividing $|G|$. Furthermore, $B(G, F^\times)$ is complemented in $Z(G, F^\times)$.

Proof First, we argue that $B(G, F^\times)$ is divisible. If $\beta \in B(G, F^\times)$ and n is a positive integer, write $\beta = \delta(\mu)$ for some function $\mu: G \to F^\times$. For each $g \in G$, choose $v(g) \in F^\times$ such that $v(g)^n = \mu(g)$. Then $\delta(v)^n = \delta(\mu) = \beta$. We can thus apply Lemma 11.14 when we show that $|H(G, F^\times)| < \infty$.

Let $\alpha \in Z(G, F^\times)$ and define

$$\mu(g) = \prod_{x \in G} \alpha(g, x).$$

For fixed $g, h \in G$, we have $\alpha(g, hx)\alpha(h, x) = \alpha(gh, x)\alpha(g, h)$. Taking the product over all $x \in G$ yields

$$\mu(g)\mu(h) = \mu(gh)\alpha(g, h)^{|G|}$$

and thus $\alpha^{|G|} \in B(G, F^\times)$. This shows that $H(G, F^\times)$ has exponent dividing $|G|$.

Now let $U = \{\alpha \in Z(G, F^\times) \mid \alpha^{|G|} = 1\}$. For $\alpha \in Z(G, F^\times)$, let $A = \langle B(G, F^\times), \alpha \rangle$. By the result of the previous paragraph $|A : B(G, F^\times)|$ divides $|G|$. Thus by Lemma 11.14, $B(G, F^\times)$ is complemented in A and the complement is a subgroup of U. Thus $\alpha \in B(G, F^\times)U$ and hence $B(G, F^\times)U = Z(G, F^\times)$.

Now every element of U is a function from $G \times G$ into $\{y \in F \mid y^{|G|} = 1\}$. Since this is a finite set, it follows that $|U| < \infty$ and thus

$$|H(G, F^\times)| = |B(G, F^\times)U : B(G, F^\times)| \leq |U| < \infty$$

and the proof is complete. ∎

Next we need a general method for constructing central extensions.

(11.16) LEMMA Let A be an arbitrary abelian group and let $\alpha \in Z(G, A)$. Then there exists a central extension (Γ, π) of G with $\ker \pi = A$ and such that a set of coset representatives $\{x_g \mid g \in G\}$ exists with $\pi(x_g) = g$ and $x_g x_h = \alpha(g, h)x_{gh}$.

Proof Let $\Gamma = G \times A$ as a set and define $(g, a)(h, b) = (gh, \alpha(g, h)ab)$. That this multiplication is associative follows from the fact that α is an A-factor set. Let $\alpha(1, 1)^{-1} = z \in A$. Then $(1, z)(g, a) = (g, \alpha(1, g)za) = (g, a)$ by Lemma 11.5. Also

$$(g^{-1}, a^{-1}\alpha(g^{-1}, g)^{-1}z)(g, a) = (1, z)$$

and hence Γ is a group with $1 = (1, z)$.

Clearly $\pi: \Gamma \to G$ defined by $\pi(g, a) = g$ is a homomorphism with $\ker \pi = \tilde{A} = \{(1, a) \mid a \in A\}$. That $\tilde{A} \subseteq \mathbf{Z}(G)$ follows since $\alpha(1, g) = \alpha(g, 1)$ by Lemma 11.5.

Since $(1, za)(1, zb) = (1, zab)$, it follows that $a \leftrightarrow (1, za)$ defines an isomorphism $A \cong \tilde{A}$. We identify A and $\tilde{A}$ via this isomorphism.

Let $X = \{(g, 1) \mid g \in G\}$. Then π maps X one-to-one and onto G and so X is a set of coset representatives for A in Γ. Now

$$(g, 1)(h, 1) = (gh, \alpha(g, h)) = (1, z\alpha(g, h))(gh, 1)$$

by Lemma 11.5. Since $(1, z\alpha(g, h)) = \alpha(g, h)$ under our identification, the proof is complete. ∎

(11.17) THEOREM (*Schur*) Given G, there exists a finite central extension (Γ, π) which has the projective lifting property for G. Furthermore, (Γ, π) can be chosen such that $\ker \pi = A \cong M(G)$ and the standard map $\hat{A} \to M(G)$ is an isomorphism.

Proof Let M be a complement for $B(G, \mathbb{C}^\times)$ in $Z(G, \mathbb{C}^\times)$ by Theorem 11.15. Let $A = \hat{M}$. Define $\alpha(g, h) \in A$ by $\alpha(g, h)(\gamma) = \gamma(g, h)$ for $\gamma \in M$. It is clear that in fact $\alpha(g, h) \in \hat{M} = A$. Next

$$(\alpha(gh, k)\alpha(g, h))(\gamma) = \gamma(gh, k)\gamma(g, h)$$

using the definition of multiplication in $\hat{M}$. Similarly,

$$(\alpha(g, hk)\alpha(h, k))(\gamma) = \gamma(g, hk)\gamma(h, k)$$

and since γ runs over a set of factor sets, it follows that $\alpha \in Z(G, A)$.

Now let (Γ, π) and $X = \{x_g \mid g \in G\}$ be as in Lemma 11.16 so that $x_g x_h = \alpha(g, h)x_{gh}$. Let $\eta: \hat{A} \to M(G)$ be the standard map. We show that η maps onto.

For $\bar{\gamma} \in M(G) = Z(G, \mathbb{C}^\times)/B(G, \mathbb{C}^\times)$, there exists $\gamma \in M \cap \bar{\gamma}$ since M complements $B(G, \mathbb{C}^\times)$ in $Z(G, \mathbb{C}^\times)$. Now define λ on $A = \hat{M}$ to be the evaluation map at γ. Note that $\lambda \in \hat{A}$. Now

$$\lambda(\alpha(g, h)) = \alpha(g, h)(\gamma) = \gamma(g, h)$$

so that $\lambda(\alpha) = \gamma$. Therefore

$$\eta(\lambda) = \overline{\lambda(\alpha)} = \bar{\gamma}$$

as desired.

By Theorem 11.13, it follows that Γ has the projective lifting property for G. Also,

$$|A| = |\hat{A}| \geq |\eta(\hat{A})| = |M(G)| = |M| = |A|$$

and so η must be one-to-one and $\hat{A} \cong M(G)$. However $A \cong \hat{A}$ by Problem 2.7(c) and the proof is complete. ∎

(11.18) COROLLARY Let $\mathfrak{X}$ be an irreducible projective $\mathbb{C}$-representation of G. Then $\deg(\mathfrak{X})$ divides $|G|$.

Proof See the discussion preceding Definition 11.12. ∎

Before going on to exploit Schur's theorem, we give another result which is useful in the computation of $M(G)$. We have defined a Schur representation group of G to be a minimal central extension with the projective lifting property. The next theorem will yield another characterization.

(11.19) THEOREM Let (Γ, π) be a finite central extension of G with $A = \ker \pi$ and let $\eta: \hat{A} \to M(G)$ be the standard map. Let $A_0 = A \cap \Gamma'$. Then

$$\ker \eta = \{\lambda \in \hat{A} \mid A_0 \subseteq \ker \lambda\}.$$

In particular, η is one-to-one iff $A \subseteq \Gamma'$.

Proof Let $X = \{x_g \mid g \in G\}$ be as usual and write $x_g x_h = \alpha(g, h)x_{gh}$ with $\alpha \in Z(G, A)$. Suppose $\lambda \in \ker \eta$. Then $1 = \eta(\lambda) = \overline{\lambda(\alpha)}$ and so $\lambda(\alpha) \in B(G, \mathbb{C}^\times)$. Thus

$$\lambda(\alpha(g, h)) = \mu(g)\mu(h)\mu(gh)^{-1}$$

for some function $\mu: G \to \mathbb{C}^\times$. Now define $\hat{\lambda}$ on Γ by $\hat{\lambda}(ax_g) = \lambda(a)\mu(g)$ and check that

$$\hat{\lambda}(x_g)\hat{\lambda}(x_h) = \hat{\lambda}(x_g x_h).$$

(This is essentially the same calculation as in the proof of Theorem 11.13.) It follows that $\hat{\lambda}$ is a linear character of Γ which extends λ. Now $\Gamma' \subseteq \ker \hat{\lambda}$ and so $A_0 \subseteq \ker \lambda$.

Conversely, let $A_0 \subseteq \ker \lambda$. Then λ is extendible to $\lambda_0 \in \mathrm{Irr}(A\Gamma'/\Gamma')$ by $\lambda_0(ax) = \lambda(a)$ for $x \in \Gamma'$. Since Γ/Γ' is abelian, λ_0 is extendible to $\lambda_1 \in \mathrm{Irr}(\Gamma/\Gamma')$. Now

$$\lambda_1(x_g)\lambda_1(x_h) = \lambda_1(x_g x_h) = \lambda(\alpha(g, h))\lambda_1(x_{gh}).$$

Define $\mu(g) = \lambda_1(x_g)$. Then

$$\lambda(\alpha(g, h)) = \mu(g)\mu(h)\mu(gh)^{-1}$$

and the result follows. ∎

(11.20) COROLLARY Let (Γ, π) be a finite central extension of G with $A = \ker \pi$.

(a) If $A \subseteq \Gamma'$, then A is isomorphic to a subgroup of $M(G)$.
(b) Assume $|A| = |M(G)|$. Then $A \subseteq \Gamma'$ iff Γ has the projective lifting property for G. In this case $M(G) \cong A$.

Proof (a) follows since $\hat{A} \cong A$ and (b) is immediate from Theorems 11.19 and 11.13. ∎

Thus we see that Γ is a Schur representation group for G iff $\Gamma/A \cong G$ for some $A \subseteq \mathbf{Z}(\Gamma)$ such that $|A| = |M(G)|$ and $A \subseteq \Gamma'$.

(11.21) COROLLARY Let p be a prime divisor of $|M(G)|$. Then a Sylow p-subgroup of G is noncyclic.

Proof Let Γ be a Schur representation group for G (which exists by Theorem 11.17). We may assume that $\Gamma/A = G$ with $A \subseteq \mathbf{Z}(\Gamma)$. Let $P/A \in \mathrm{Syl}_p(G)$ and assume P/A is cyclic. Since A is central, it follows that P is abelian and thus Γ has an abelian Sylow p-subgroup. By Theorem 5.6, $p \nmid |\Gamma' \cap \mathbf{Z}(\Gamma)|$. Since $A \subseteq \Gamma'$ by Corollary 11.20, we have $p \nmid |A|$. Thus $p \nmid |M(G)|$. ∎

The next corollary is quite important. It can, of course, be proved without all of the complex machinery which we have developed.

(11.22) COROLLARY Let $N \lhd G$ with G/N cyclic and let $\vartheta \in \mathrm{Irr}(N)$ be invariant in G. Then ϑ is extendible to G.

Proof By Corollary 11.21, $M(G/N) = H(G/N, \mathbb{C}^\times)$ is trivial. The result now follows from Theorem 11.7. ∎

The above theory, together with some rather technical computations yields a tool which is very useful when studying the characters of groups with normal subgroups. Let $N \lhd G$ and let $\vartheta \in \mathrm{Irr}(N)$ be invariant in G. Under these hypotheses we say that (G, N, ϑ) is a *character triple*. The analysis of this situation is much easier if ϑ is linear. We shall use a Schur representation group for G/N to replace (G, N, ϑ) by another character triple (Γ, A, λ) in which $\Gamma/A \cong G/N$ and λ is linear. Of course, this would not be of much value unless we knew that the character theory of Γ was somehow closely tied to the character theory of G. For instance, we would want λ^Γ and ϑ^G to have the same numbers of irreducible constituents with the same ramifications.

We define below the notion of "isomorphism" of character triples. This rather complicated definition makes precise some of the ways in which the character theory of two character triples can be related. We first introduce some notation. If (G, N, ϑ) is a character triple, let $\mathrm{Ch}(G\,|\,\vartheta)$ denote the set of (possibly reducible) characters χ of G such that χ_N is a multiple of ϑ. Let $\mathrm{Irr}(G\,|\,\vartheta)$ be the irreducible characters among these, so that $\mathrm{Irr}(G\,|\,\vartheta)$ is the set of irreducible constituents of ϑ^G. Note that if $N \subseteq H \subseteq G$, then (H, N, ϑ) is a character triple and $\chi_H \in \mathrm{Ch}(H\,|\,\vartheta)$ whenever $\chi \in \mathrm{Ch}(G\,|\,\vartheta)$. If $\tau\colon U \to V$ is an isomorphism of groups and $\varphi \in \mathrm{Irr}(U)$, let $\varphi^\tau \in \mathrm{Irr}(V)$ denote the corresponding character, so that $\varphi^\tau(u^\tau) = \varphi(u)$.

(11.23) DEFINITION Let (G, N, ϑ) and (Γ, M, φ) be character triples and let $\tau\colon G/N \to \Gamma/M$ be an isomorphism. For $N \subseteq H \subseteq G$, let H^τ denote the inverse image in Γ of $\tau(H/N)$. For every such H, suppose there exists a map $\sigma_H\colon \mathrm{Ch}(H\,|\,\vartheta) \to \mathrm{Ch}(H^\tau\,|\,\varphi)$ such that the following conditions hold for H, K with $N \subseteq K \subseteq H \subseteq G$ and $\chi, \psi \in \mathrm{Ch}(H\,|\,\vartheta)$.

(a) $\sigma_H(\chi + \psi) = \sigma_H(\chi) + \sigma_H(\psi)$;
(b) $[\chi, \psi] = [\sigma_H(\chi), \sigma_H(\psi)]$;
(c) $\sigma_K(\chi_K) = (\sigma_H(\chi))_{K^\tau}$;
(d) $\sigma_H(\chi\beta) = \sigma_H(\chi)\beta^\tau$ for $\beta \in \mathrm{Irr}(H/N)$.

Let σ denote the union of the maps σ_H. Then (τ, σ) is an *isomorphism* from (G, N, ϑ) to (Γ, M, φ).

Note that if (τ, σ) is an isomorphism from (G, N, ϑ) to (Γ, M, φ), then σ_H is determined by its restriction to $\mathrm{Irr}(H\,|\,\vartheta)$. (This follows from (a).) By (b) it follows that σ_H maps $\mathrm{Irr}(H\,|\,\vartheta)$ one-to-one into $\mathrm{Irr}(H^\tau\,|\,\varphi)$. Therefore, to construct an isomorphism (τ, σ) it suffices to define σ_H on $\mathrm{Irr}(H\,|\,\vartheta)$ to be one-to-one, then extend the definition by (a) and check that (c) and (d) hold for $\chi \in \mathrm{Irr}(H\,|\,\vartheta)$.

(11.24) LEMMA Let $(\tau, \sigma)\colon (G, N, \vartheta) \to (\Gamma, M, \varphi)$ be an isomorphism of character triples. Then σ_H is a bijection of $\mathrm{Ch}(H\,|\,\vartheta)$ onto $\mathrm{Ch}(H^\tau\,|\,\varphi)$ for all H with $N \subseteq H \subseteq G$. Furthermore, $\chi(1)/\vartheta(1) = \sigma_H(\chi)(1)/\varphi(1)$ for all $\chi \in \mathrm{Ch}(H\,|\,\vartheta)$.

Proof If $\sigma_H(\chi_1) = \sigma_H(\chi_2)$ for $\chi_i \in \mathrm{Ch}(H\,|\,\vartheta)$, we have $[\chi_i, \psi] = [\sigma_H(\chi_i), \sigma_H(\psi)]$ is independent of i for all $\psi \in \mathrm{Irr}(H\,|\,\vartheta)$. It follows that $\chi_1 = \chi_2$ and hence σ_H is one-to-one.

For $\chi \in \mathrm{Ch}(H\,|\,\vartheta)$ write $e(\chi) = \chi(1)/\vartheta(1)$ and similarly set $e(\eta) = \eta(1)/\varphi(1)$ for $\eta \in \mathrm{Ch}(H^\tau\,|\,\varphi)$. Note that $\sigma_N(\vartheta) \in \mathrm{Irr}(M\,|\,\varphi)$ and so $\sigma_N(\vartheta) = \varphi$. We have

$\chi_N = e(\chi)\vartheta$ and $\eta_M = e(\eta)\varphi$ and thus

$$e(\sigma_H(\chi))\varphi = (\sigma_H(\chi))_M = \sigma_N(\chi_N) = \sigma_N(e(\chi)\vartheta) = e(\chi)\varphi$$

and thus $e(\sigma_H(\chi)) = e(\chi)$ as claimed.

By Frobenius reciprocity, we have

$$\vartheta^H = \sum_{\chi \in \mathrm{Irr}(H|\vartheta)} e(\chi)\chi$$

and comparing degrees yields $\sum e(\chi)^2\vartheta(1) = |H:N|\vartheta(1)$ so that $\sum e(\chi)^2 = |H:N|$ where χ runs over $\mathrm{Irr}(H|\vartheta)$. Similarly $\sum e(\eta)^2 = |H^\tau:M| = |H:N|$ for $\eta \in \mathrm{Irr}(H^\tau|\varphi)$. Since σ_H maps $\mathrm{Irr}(H|\vartheta)$ one-to-one into $\mathrm{Irr}(H^\tau|\varphi)$, we have

$$|H:N| = \sum e(\chi)^2 = \sum e(\sigma_H(\chi))^2 \le \sum e(\eta)^2 = |H:N|.$$

It follows that every $\eta \in \mathrm{Irr}(H^\tau|\varphi)$ is of the form $\sigma_H(\chi)$ for some $\chi \in \mathrm{Irr}(H|\vartheta)$. The result now follows. ∎

(11.25) COROLLARY Isomorphism is an equivalence relation on character triples.

Proof. The reflexive and transitive properties are obvious. If

$$(\tau, \sigma): (G, N, \vartheta) \to (\Gamma, M, \varphi)$$

is an isomorphism, then σ_H: $\mathrm{Ch}(H|\vartheta) \to \mathrm{Ch}(K|\varphi)$ is one-to-one and onto where $K = H^\tau$. We can, therefore, define σ^{-1} by $(\sigma^{-1})_K = (\sigma_H)^{-1}$ for $M \subseteq K \subseteq \Gamma$ where $H = K^{\tau^{-1}}$. It is routine to check that

$$(\tau^{-1}, \sigma^{-1}): (\Gamma, M, \varphi) \to (G, N, \vartheta)$$

is an isomorphism. ∎

A nearly trivial example of an isomorphism of character triples is given by the following result.

(11.26) LEMMA Let (G, N, ϑ) be a character triple and let μ: $G \to \Gamma$ be an onto homomorphism with $\ker \mu \subseteq \ker \vartheta$. Let $M = \mu(N)$ and let $\varphi \in \mathrm{Irr}(M)$ be the character corresponding to $\vartheta \in \mathrm{Irr}(N/\ker \mu)$. Then (G, N, ϑ) and (Γ, M, φ) are isomorphic character triples.

Proof We have τ: $G/N \to \Gamma/M$ defined naturally from μ. For $N \subseteq H \subseteq G$ and $\chi \in \mathrm{Ch}(H|\vartheta)$ we obtain $\ker \mu \subseteq \ker \chi$ and we may view χ as a character of $H/ker\ \mu$. Let $\sigma_H(\chi)$ be the corresponding character of $\mu(H) \cong H/\ker \mu$. Check that (τ, σ) is the desired isomorphism. ∎

Another example of an isomorphism of character triples is provided by the following.

(11.27) LEMMA Let (G, N, φ) be a character triple and let $\eta \in \mathrm{Irr}(G)$ be such that $\eta_N \varphi = \vartheta \in \mathrm{Irr}(N)$. For $N \subseteq H \subseteq G$, define σ_H: $\mathrm{Ch}(H \mid \varphi) \to \mathrm{Ch}(H \mid \vartheta)$ by $\sigma_H(\psi) = \psi\eta_H$. Let i: $G/N \to G/N$ be the identity map. Then

$$(i, \sigma): (G, N, \varphi) \to (G, N, \vartheta)$$

is an isomorphism of character triples.

Proof It is clear that σ_H does map $\mathrm{Ch}(H \mid \varphi)$ to $\mathrm{Ch}(H \mid \vartheta)$ and that properties (a), (c), and (d) of Definition 11.23 hold. It thus suffices to show that σ_H maps $\mathrm{Irr}(H \mid \varphi)$ one-to-one into $\mathrm{Irr}(H \mid \vartheta)$. This follows from Theorem 6.16. ∎

Given invariant λ, $\vartheta \in \mathrm{Irr}(N)$ for $N \lhd G$, with $\lambda(1) = 1$, it follows that (G, N, λ) and (G, N, ϑ) are isomorphic if $\vartheta\lambda^{-1}$ is extendible to G.

(11.28) THEOREM Let (G, N, ϑ) be a character triple and let (Γ, π) be a finite central extension of G/N having the projective lifting property. Let $A = \ker \pi$. Then (G, N, ϑ) is isomorphic to (Γ, A, λ) for some $\lambda \in \hat{A}$.

Proof By Theorem 11.7, the triple (G, N, ϑ) determines an element $\bar{\beta} \in H(G/N, \mathbb{C}^\times) = M(G/N)$. Since Γ has the projective lifting property for G/N, we can find $\lambda \in \hat{A}$ such that $\eta(\lambda) = \bar{\beta}^{-1}$ where η is the standard map (Theorem 11.13).

Now let $G^* \subseteq G \times \Gamma$ be defined by $G^* = \{(g, x) \mid \bar{g} = \pi(x)\}$, where $\bar{g} = gN$, the image of g in G/N. Note that G^* is a subgroup of $G \times \Gamma$. Let $L = N \times A$ and observe that $L \lhd G^*$. Define ϑ^* and λ^* on L by $\vartheta^*(n, a) = \vartheta(n)$ and $\lambda^*(n, a) = \lambda(a)$. Note that $\vartheta^*, \lambda^* \in \mathrm{Irr}(L)$ are invariant in G^*.

We have projection homomorphisms μ_G: $G^* \to G$ and μ_Γ: $G^* \to \Gamma$. These maps are onto and $\ker \mu_G = 1 \times A \subseteq \ker \vartheta^*$ and $\ker \mu_\Gamma = N \times 1 \subseteq \ker \lambda^*$. It follows by Lemma 11.26 that (G^*, L, ϑ^*) is isomorphic to (G, N, ϑ) and (G^*, L, λ^*) is isomorphic to (Γ, A, λ). By Corollary 11.25, it suffices to show that (G^*, L, λ^*) and (G^*, L, ϑ^*) are isomorphic and by Theorem 11.27 we will be done when we show that $\vartheta^*(\lambda^*)^{-1}$ is extendible to G^*.

Let $\mathfrak{Y}$ be a representation of N affording ϑ and let $\mathfrak{X}$ be a projective representation of G as in Theorem 11.2. Let α be the factor set of G belonging to $\mathfrak{X}$ and let β be the corresponding factor set of G/N as in Theorem 11.7. Let $\{x_{\bar{g}} \mid \bar{g} \in G/N\}$ be a set of coset representatives for A in Γ with $\pi(x_{\bar{g}}) = \bar{g}$. Take $x_{\bar{1}} = 1$. Write $x_{\bar{g}} x_{\bar{h}} = \gamma(\bar{g}, \bar{h}) x_{\bar{g}\bar{h}}$ so that $\gamma \in Z(G/N, A)$. Since $\eta(\lambda) = \bar{\beta}^{-1}$, we have

$$\lambda(\gamma)\beta \in B(G/N, \mathbb{C}^\times)$$

and hence

$$\lambda(\gamma(\bar{g}, \bar{h}))\alpha(g, h) = \nu(\bar{g})^{-1}\nu(\bar{h})^{-1}\nu(\bar{g}\bar{h})$$

for some function ν: $G/N \to \mathbb{C}^\times$.

Every element of G^* is uniquely of the form $(g, ax_{\bar{g}})$ for $g \in G$ and $a \in A$. Define $\mathfrak{Z}$ on G^* by

$$\mathfrak{Z}(g, ax_{\bar{g}}) = \mathfrak{X}(g)\lambda(a)^{-1}\nu(\bar{g}).$$

We compute

$$\mathfrak{Z}(g, ax_{\bar{g}})\mathfrak{Z}(h, bx_{\bar{h}}) = \mathfrak{X}(gh)\lambda(ab)^{-1}\alpha(g, h)\nu(\bar{g})\nu(\bar{h})$$

and

$$\mathfrak{Z}(gh, ab\gamma(\bar{g}, \bar{h})x_{\bar{g}\bar{h}}) = \mathfrak{X}(gh)\lambda(ab)^{-1}\lambda(\gamma(\bar{g}, \bar{h}))^{-1}\nu(\bar{g}\bar{h}).$$

Since these two expressions are equal, we conclude that $\mathfrak{Z}$ is a representation of G^*.

Now $\gamma(\bar{1}, \bar{1}) = 1$ and $\alpha(1, 1) = 1$ and it follows that $\nu(\bar{n}) = \nu(1) = 1$ for $n \in N$. Therefore $\mathfrak{Z}(n, a) = \mathfrak{Y}(n)\lambda(a)^{-1}$ and $\mathfrak{Z}_L$ affords $\vartheta^*(\lambda^*)^{-1}$. The proof is now complete. ∎

We now consider some applications.

(11.29) COROLLARY Let $N \lhd G$ and $\chi \in \mathrm{Irr}(G)$. Let $\vartheta \in \mathrm{Irr}(N)$ be a constituent of χ_N. Then $\chi(1)/\vartheta(1)$ divides $|G : N|$.

Proof Let $T = I_G(\vartheta)$, the inertia group, and let $\psi \in \mathrm{Irr}(T)$ such that $\psi^G = \chi$ and $\psi_N = e\vartheta$ (Theorem 6.11). Since $\chi(1) = |G : T|\psi(1)$, it suffices to show that $\psi(1)/\vartheta(1)$ divides $|T : N|$. Let (Γ, A, λ) be a character triple isomorphic to (T, N, ϑ) with λ linear. Let $\zeta \in \mathrm{Irr}(\Gamma | \lambda)$ correspond to $\psi \in \mathrm{Irr}(T | \vartheta)$. Then $\psi(1)/\vartheta(1) = \zeta(1)/\lambda(1) = \zeta(1)$ by Lemma 11.24. Since $A \subseteq \mathbf{Z}(\zeta)$ we have $\zeta(1)$ divides $|\Gamma : A| = |T : N|$ by Theorem 3.12. The result follows. ∎

Note that by repeated application of this result we can weaken the hypothesis that N is normal in G and assume only that N is subnormal, that is, that there exist subgroups N_i such that

$$N \lhd N_1 \lhd \cdots \lhd N_k \lhd G.$$

In particular, if N is subnormal and abelian then since $\vartheta(1) = 1$ we obtain the following generalization of Ito's Theorem 6.15.

(11.30) COROLLARY Let $N \subseteq G$ be subnormal and abelian. Then $\chi(1)$ divides $|G : N|$ for every $\chi \in \mathrm{Irr}(G)$.

(11.31) COROLLARY Let $N \lhd G$ and let $\vartheta \in \mathrm{Irr}(N)$ be invariant in G. Suppose for every Sylow subgroup P/N of G/N that ϑ is extendible to P. Then ϑ is extendible to G.

Proof Let (Γ, A, λ) be a character triple isomorphic to (G, N, ϑ) and with λ linear. If $N \subseteq H \subseteq G$ and $A \subseteq K \subseteq \Gamma$ are such that H and K correspond, then ϑ is extendible to H iff λ is extendible to K. This follows from Lemma 11.24 since extendibility of ϑ is equivalent to existence of $\chi \in \mathrm{Irr}(H \mid \vartheta)$ with $\chi(1)/\vartheta(1) = 1$ and similarly for extendibility of λ. Thus λ is extendible to the inverse image in Γ of every Sylow subgroup of Γ/A. By Theorem 6.26, we conclude that λ is extendible to Γ and hence ϑ is extendible to G, as desired. ∎

We can use Corollary 11.31 to obtain an alternate proof of Gallagher's Theorem 8.15, a proof independent of the results of Chapter 8. We are assuming that (G, N, ϑ) is a character triple in which $(\vartheta(1), |G:N|) = 1$ and that $\det(\vartheta)$ is extendible to G. By Theorem 6.25, ϑ is extendible to H for every H with $N \subseteq H \subseteq G$ and H/N solvable. The result now follows from Corollary 11.31.

The following combines several of our extendibility criteria.

(11.32) THEOREM Let $N \lhd G$ and $\vartheta \in \mathrm{Irr}(N)$, with ϑ invariant in G. Assume for every prime divisor p of $(\vartheta(1)o(\vartheta), |G:N|)$ that a Sylow p-subgroup of G is abelian or if $p \neq 2$, that ϑ is p-rational. Then ϑ is extendible to G.

Proof By Corollary 11.31, it suffices to assume that G/N is a p-group for some prime p. If a Sylow p-subgroup of G is abelian, then ϑ is extendible to G by Theorem 8.26. If $p \neq 2$ and ϑ is p-rational, then ϑ is extendible by Theorem 6.30. In the remaining case, $(|G:N|, o(\vartheta)\vartheta(1)) = 1$ and Corollary 6.28 yields the result. ∎

As another application of the theory of projective representations, we prove a theorem of T. Berger about the characters of solvable groups. Recall that $\chi \in \mathrm{Irr}(G)$ is said to be quasi-primitive if χ_N is homogeneous (that is, a multiple of an irreducible) for every $N \lhd G$. Primitive characters are necessarily quasi-primitive and, in general, the converse is false.

(11.33) THEOREM (*Berger*) Let G be solvable and suppose $\chi \in \mathrm{Irr}(G)$ is quasi-primitive. Then χ is primitive.

In order to prove Berger's theorem, we shall exploit the fact that there exists a central extension of G with the projective lifting property. We break the proof into several intermediate results. The first of these is independent of projective representations.

(11.34) THEOREM Let $L \subseteq H < G$ with $L \lhd G$, H maximal and G/L solvable. Let $\vartheta \in \mathrm{Irr}(H)$ such that $\vartheta^G = \chi$ and ϑ_L are irreducible. Then there exists $M \lhd G$ such that χ_M is not homogeneous and $M \supseteq L$.

Proof We may assume without loss that $L = \mathrm{core}_G(H)$, the largest normal subgroup of G contained in H. Let K/L be a chief factor of G so that

K/L is an abelian p-group and $K \not\subseteq H$. Thus $G = KH$ by the maximality of H. Since K/L is abelian, we have $H \cap K \lhd K$ and since $H \cap K \lhd H$ we conclude that $H \cap K \lhd G$ and thus $H \cap K = L$.

Let $C = \mathbf{C}_G(K/L)$. Then $C \lhd G$ and so $C \cap H \lhd H$. However, $K \subseteq \mathbf{N}(C \cap H)$ and hence $C \cap H \lhd G$. The maximality of L forces $L = C \cap H$. We have $C = K(C \cap H) = K$.

If $K = G$ then $H = L \lhd G$ and since $\vartheta^G \in \mathrm{Irr}(G)$ we have $I_G(\vartheta) = L$ by Problem 6.1 and thus χ_L is not homogeneous. We assume, therefore, that $K < G$ and let M/K be a chief factor of G so that M/K is an abelian q-group for some prime q. If $q = p$, then $(K/L) \cap \mathbf{Z}(M/L) \neq 1$ and the minimality of K/L yields $K/L \subseteq \mathbf{Z}(M/L)$. This contradicts $\mathbf{C}_G(K/L) = K$ and we conclude $p \neq q$.

We suppose that χ_L and χ_M are homogeneous and write $\chi_M = e\eta$ for some $\eta \in \mathrm{Irr}(M)$. Now $\chi(1)/\vartheta(1) = |G : H| = |K : L|$ is a power of p. Since $\vartheta_L \in \mathrm{Irr}(L)$, it is a constituent of η_L and hence $\eta(1)/\vartheta(1)$ is an integer which divides $\chi(1)/\vartheta(1)$. Therefore $\eta(1)/\vartheta(1)$ is a power of p. Since M/K is a q-group for $q \neq p$, we conclude that $\eta_K \in \mathrm{Irr}(K)$ by Theorem 6.18 or Corollary 11.29.

Since χ_L is homogeneous, it is a multiple of ϑ_L and thus η_L is also. Let $R = M \cap H$ and let τ be any irreducible constituent of η_R. Then $[\tau_L, \vartheta_L] \neq 0$ and hence $\tau = \vartheta_R\beta$ for some $\beta \in \mathrm{Irr}(R/L)$ by Corollary 6.17. Since $M = KR$ and $K \cap R = L$, we can find $\gamma \in \mathrm{Irr}(M/K)$ such that $\gamma_R = \beta$. Now

$$\tau^M = (\vartheta_R\beta)^M = (\vartheta_R\gamma_R)^M = (\vartheta_R)^M\gamma.$$

However, since $MH = G$ and $M \cap H = R$, we have

$$(\vartheta_R)^M = (\vartheta^G)_M = \chi_M = e\eta$$

and thus $\tau^M = e\eta\gamma$. Since η is a constituent of τ^M by Frobenius reciprocity, we have $[\eta, \eta\gamma] \neq 0$. Since $\eta_K \in \mathrm{Irr}(K)$, Corollary 6.17 yields $\gamma = 1_M$ and hence $\beta = 1_R$ and $\tau = \vartheta_R$. Thus η_R is homogeneous and since $(\vartheta_R)^M = e\eta$ we have $\eta_R = e\vartheta_R$ and $\eta(1) = e\vartheta(1)$. This yields $\vartheta(1)|M:R| = ((\vartheta_R)^M)(1) = e^2\vartheta(1)$ and $[\eta_R, \eta_R] = e^2 = |M:R|$. It follows that η vanishes on $M - R$ by Lemma 2.29. As η is G-invariant, it vanishes on $M - R^g$ for all $g \in G$ and hence vanishes on $M - \bigcap R^g = M - L$. In particular, η vanishes on $M - K$ and so $|M:K| = [\eta_K, \eta_K]$ by Lemma 2.29. This contradicts $\eta_K \in \mathrm{Irr}(K)$ and completes the proof. ∎

(11.35) LEMMA Let $(\tau, \sigma): (G, N, \vartheta) \to (\Gamma, M, \varphi)$ be an isomorphism of character triples and let $N \subseteq H \subseteq G$. Suppose ψ is a character of H. Then $\psi \in \mathrm{Ch}(H|\vartheta)$ iff $\psi^G \in \mathrm{Ch}(G|\vartheta)$. Also, if $\psi \in \mathrm{Ch}(H|\vartheta)$, we have $\sigma_G(\psi^G) = (\sigma_H(\psi))^\Gamma$.

Proof The first assertion is immediate from Frobenius reciprocity. To prove that $\sigma_G(\psi^G) = (\sigma_H(\psi))^\Gamma$ when $\psi \in \mathrm{Ch}(H \mid \vartheta)$, it suffices to check that

$$[\sigma_G(\psi^G), \chi] = [(\sigma_H(\psi))^\Gamma, \chi]$$

for all $\chi \in \mathrm{Irr}(\Gamma \mid \varphi)$. Since σ_G maps $\mathrm{Irr}(G \mid \vartheta)$ onto $\mathrm{Irr}(\Gamma \mid \varphi)$ by Lemma 11.24, we have $\chi = \sigma_G(\xi)$ for some $\xi \in \mathrm{Irr}(G \mid \vartheta)$. Now

$$\begin{aligned}[\sigma_G(\psi^G), \sigma_G(\xi)] &= [\psi^G, \xi] = [\psi, \xi_H] \\ &= [\sigma_H(\psi), \sigma_H(\xi_H)] = [\sigma_H(\psi), (\sigma_G(\xi))_H] \\ &= [(\sigma_H(\psi))^\Gamma, \sigma_G(\xi)]\end{aligned}$$

and the result follows. ∎

(11.36) LEMMA Let $Z \subseteq \mathbf{Z}(\Gamma)$ be such that Γ has the projective lifting property for $G = \Gamma/Z$ with respect to the natural homomorphism $\pi\colon \Gamma \to G$. Suppose $K \lhd \Gamma$, $H \subseteq \Gamma$, $HK = \Gamma$ and $H \cap K \supseteq Z$. Let $\vartheta \in \mathrm{Irr}(H/Z)$ and assume $H \cap K \subseteq \mathbf{Z}(\vartheta)$. Then there exists $\chi \in \mathrm{Irr}(\Gamma)$ such that $K \subseteq \mathbf{Z}(\chi)$ and $\chi_H = \lambda\vartheta$ for some linear $\lambda \in \mathrm{Irr}(H)$.

Proof Let $\mathfrak{Y}$ be a representation of H which affords ϑ. Let $L = H \cap K \lhd H$. Since $L \subseteq \mathbf{Z}(\vartheta)$, $\mathfrak{Y}_L$ is a scalar representation and we can apply Corollary 11.10 to $H/(L \cap \ker \vartheta)$ and construct a projective representation $\mathfrak{X}$ of H/L such that $\mathfrak{Y}(h) = \mathfrak{X}(hL)\mu(h)$ for some function $\mu\colon H \to \mathbb{C}^\times$.

Now let $\tau\colon G \to H/L$ be the composite of the natural maps

$$G = \Gamma/Z \to \Gamma/K = HK/K \cong H/L$$

so that $\tau(hkZ) = hL$ for $h \in H$ and $k \in K$. Let $\mathfrak{X}^*$ be the projective representation of G obtained by composing τ with $\mathfrak{X}$ so that $\mathfrak{X}^*(hkZ) = \mathfrak{X}(hL) = \mathfrak{Y}(h)\mu(h)^{-1}$.

Now lift $\mathfrak{X}^*$ to the representation $\mathfrak{Z}$ of Γ so that $\mathfrak{Z}(x) = \mathfrak{X}^*(xZ)\nu(x)$ for $x \in \Gamma$, where $\nu\colon \Gamma \to \mathbb{C}^\times$. This yields

$$\mathfrak{Z}(hk) = \mathfrak{Y}(h)\mu(h)^{-1}\nu(hk)$$

for $h \in H$ and $k \in K$. Let $\mathfrak{Z}$ afford χ.

Restricting the above to H, we have $\mathfrak{Z}(h) = \mathfrak{Y}(h)\mu(h)^{-1}\nu(h)$ and hence $\lambda(h) = \mu(h)^{-1}\nu(h)$ defines a linear character $\lambda \in \mathrm{Irr}(H)$ and $\chi_H = \vartheta\lambda$. Thus $\chi \in \mathrm{Irr}(\Gamma)$.

For $k \in K$ we have

$$\mathfrak{Z}(k) = \mathfrak{Y}(1)\mu(1)^{-1}\nu(k),$$

which is a scalar matrix. Thus $K \subseteq \mathbf{Z}(\chi)$ and the proof is complete. ∎

The next result is a generalization of Theorem 11.34 which includes Berger's Theorem 11.33.

(11.37) THEOREM Let $L \subseteq H < G$ with $L \lhd G$ and G/L solvable. Let $\vartheta \in \text{Irr}(H)$, with $\vartheta^G = \chi \in \text{Irr}(G)$. Then there exists $M \lhd G$ such that $M \supseteq L$ and χ_M is not homogeneous.

Proof We may assume that there do not exist subgroups $L_0 \subseteq H_0 < G$ and $\vartheta_0 \in \text{Irr}(H_0)$ such that $L < L_0 \lhd G$ and $(\vartheta_0)^G = \chi$. We may also assume that H is maximal in G. We have $L = \text{core}_G(H)$. We assume that χ_L is homogeneous and write $\chi_L = a\varphi$ with $\varphi \in \text{Irr}(L)$ so that (G, L, φ) is a character triple and $\vartheta \in \text{Irr}(H \mid \varphi)$.

It follows from Lemma 11.35 that the hypotheses of the theorem remain invariant if (G, L, φ) is replaced by an isomorphic triple, and similarly for the conclusion. Therefore, by Theorem 11.28, we may assume that $\varphi(1) = 1$. Thus $L \subseteq \mathbf{Z}(\chi)$ and in particular, $L \subseteq \mathbf{Z}(\vartheta)$.

Let (Γ, π) be a finite central extension of G having the projective lifting property. Let $Z = \ker \pi$ and identify G with Γ/Z via π. For subgroups $U \subseteq G$, let U^* denote the inverse image of U in Γ so that $U^*/Z = U$ and $G^* = \Gamma$.

Let K/L be a chief factor of G so that K/L is abelian and $K \nsubseteq H$. Thus $KH = G$ and $K \cap H \lhd KH$ so that $K \cap H = L$. We have $\vartheta \in \text{Irr}(H^*/Z)$ and $L^* \subseteq \mathbf{Z}(\vartheta)$. Also $K^*H^* = \Gamma$ and $K^* \cap H^* = L^* \supseteq Z$. Lemma 11.36 thus applies and yields $\eta \in \text{Irr}(\Gamma)$ such that $\lambda\eta_{H^*} = \vartheta$ for some linear $\lambda \in \text{Irr}(H^*)$ and $K^* \subseteq \mathbf{Z}(\eta)$.

We have

$$\chi = \vartheta^\Gamma = (\lambda\eta_{H^*})^\Gamma = \lambda^\Gamma\eta$$

and hence $\lambda^\Gamma \in \text{Irr}(\Gamma)$. Since $\lambda_{L^*} \in \text{Irr}(L^*)$, Theorem 11.34 applies and we can find $M \supseteq L^*$ with $M \lhd \Gamma$ and $(\lambda^\Gamma)_M$ not homogeneous.

If $M = L^*$, let ν_1 and ν_2 be distinct irreducible constituents of $(\lambda^\Gamma)_M$. Let μ be an irreducible constituent of η_{L^*}. Since $L^* \subseteq K^* \subseteq \mathbf{Z}(\eta)$ we have $\mu(1) = 1$ and thus $\mu\nu_1$ and $\mu\nu_2$ are distinct irreducible constituents of $(\eta\lambda^\Gamma)_{L^*} = \chi_{L^*}$. This is a contradiction since χ_{L^*} is homogeneous.

Therefore, $M > L^*$. Since $(\lambda^\Gamma)_M$ is not homogeneous, Theorem 6.11 yields a subgroup $T \supseteq M$ with $T < \Gamma$ and $\psi \in \text{Irr}(T)$ such that $\psi^\Gamma = \lambda^\Gamma$. Thus

$$\chi = \eta\lambda^\Gamma = \eta\psi^\Gamma = (\eta_T\psi)^\Gamma.$$

Since $Z \subseteq \ker \chi$ we have $Z \subseteq \ker(\eta_T\psi)$ and thus in G, χ is induced from the proper subgroup $T/Z < G$. Also, $T/Z \supseteq M/Z > L$ and this contradicts the first sentence of the proof. It follows that χ_L is not homogeneous and the proof is complete. ∎

Problems

(11.1) Let $\alpha, \beta \in Z(G, F^\times)$, where F is a field. If α and β are equivalent, show that $F^\alpha[G] \cong F^\beta[G]$.

(11.2) Let $\alpha \in Z(G, A)$ and let $g \in G$. Show that $\alpha(x, y) = \alpha(y, x)$ for $x, y \in \langle g \rangle$.

Note In general, $xy = yx$ does not imply that $\alpha(x, y) = \alpha(y, x)$.

(11.3) Let $\alpha \in Z(G, A)$ and let $x, y \in G$ commute. Assume that A is divisible. Show that $\alpha(x, y) = \alpha(y, x)$ iff the restriction of α to $\langle x, y \rangle$ is equivalent to the trivial factor set.

Hint Use Lemma 11.16.

(11.4) Let $\alpha \in Z(G, A)$. Say that $g \in G$ is *α-special* if $\alpha(g, c) = \alpha(c, g)$ for every $c \in \mathbf{C}_G(g)$. Show that if g is α-special, then so is every conjugate of g in G.

Hint Let (Γ, π) be as in Lemma 11.16 with respect to G, A, and α. Then $\pi(x)$ is α-special iff $\pi(\mathbf{C}_\Gamma(x)) = \mathbf{C}_G(\pi(x))$.

Note If α and β are equivalent, then $g \in G$ is α-special iff it is β-special. Thus one can speak of $\bar{\alpha}$-special elements for $\bar{\alpha} \in H(G, A)$.

(11.5) Let $\alpha \in Z(G, A)$ and let $\mathscr{K}$ be a conjugacy class of G. Show that $\mathscr{K}$ consists of α-special elements (see Problem 11.4) iff there exists a function $\mu\colon \mathscr{K} \to A$ such that

$$\mu(g)\alpha(g, h) = \mu(g^h)\alpha(h, g^h)$$

for all $g \in \mathscr{K}$ and $h \in G$.

Hint See the hint for Problem 11.4. Consider the conjugacy class of x_g in Γ, where x_g is the coset representative of A in Γ corresponding to g.

(11.6) Let (Γ, π) be a finite central extension of G with $\ker \pi = A$. Choose a set of coset representatives for A in Γ and let $\alpha \in Z(G, A)$ be as in Lemma 11.9. Let $\lambda \in \hat{A}$. For each $\chi \in \mathrm{Irr}(\Gamma \mid \lambda)$, choose a representation $\mathfrak{Y}$ affording χ and construct the projective $\mathbb{C}$-representation $\mathfrak{X}$ as in Lemma 11.10. Show that this defines a bijection of $\mathrm{Irr}(\Gamma \mid \lambda)$ onto the set of similarity classes of irreducible projective $\mathbb{C}$-representations of G with factor set $\lambda(\alpha)$.

(11.7) Let $\alpha \in Z(G, \mathbb{C}^\times)$. Show that $\mathbb{C}^\alpha[G]$ is semisimple.

Hint Show that $\sum (\dim M)^2 = |G|$, where M runs over a representative set of irreducible $\mathbb{C}^\alpha[G]$-modules. Use Problem 11.6.

(11.8) Let $\alpha \in Z(G, F^\times)$. Show that $\dim \mathbf{Z}(F^\alpha[G])$ is equal to the number of conjugacy classes of α-special elements in G.

Hint Use Problem 11.5.

(11.9) Let (G, N, ϑ) be a character triple. If $\bar{g} \in G/N$, say that $\bar{g}$ is *ϑ-special* if ϑ is extendible to $\langle N, g, c \rangle$ for all $c \in G$ with $[g, c] \in N$. Check that this is well defined. Let $\bar{\beta} \in H(G/N, \mathbb{C}^{\times})$ be as in Theorem 11.7. Show that $\bar{g} \in G/N$ is ϑ-special iff it is $\bar{\beta}$-special. (See the note following Problem 11.4.)

Note It is clear (without appeal to Problem 11.4) that conjugates of ϑ-special elements in G/N are ϑ-special.

(11.10) (*Gallagher*) Let (G, N, ϑ) be a character triple. Show that $|\operatorname{Irr}(G|\vartheta)|$ is the number of ϑ-special conjugacy classes of G/N.

Hint Several of the previous problems are relevant.

(11.11) The character triple (G, N, ϑ) is *fully ramified* if there exists $\chi \in \operatorname{Irr}(G|\vartheta)$ such that $(\chi(1)/\vartheta(1))^2 = |G:N|$. Suppose (G, N, ϑ) is fully ramified and $G > N$. Show that no cyclic subgroup of G/N can be its own centralizer in G/N.

Hint Use Problem 11.10.

(11.12) Let (G, N, ϑ) be a character triple. For $x, y \in G$ with $[x, y] \in N$, define the complex number $\langle\langle x, y\rangle\rangle$ as follows: Let ψ be an extension of ϑ to $H = \langle N, y\rangle$. Write $\psi^x = \psi\lambda$ for $\lambda \in \operatorname{Irr}(H/N)$ and put $\langle\langle x, y\rangle\rangle = \lambda(y)$. Show that $\langle\langle\, ,\rangle\rangle$ is uniquely defined and that

(a) $\langle\langle x, y\rangle\rangle = \langle\langle xn, ym\rangle\rangle$ for $n, m \in N$;
(b) $\langle\langle x_1x_2, y\rangle\rangle = \langle\langle x_1, y\rangle\rangle\langle\langle x_2 y\rangle\rangle$ whenever $[x_1, y], [x_2, y] \in N$;
(c) $\langle\langle x, y\rangle\rangle = \vartheta([x, y])$ if ϑ is linear.

Note By (a) we can view $\langle\langle\, ,\rangle\rangle$ as being defined on commuting pairs of elements of G/N.

(11.13) Let (G, N, ϑ) be a character triple. If $N \subseteq H \subseteq G$, $\psi \in \operatorname{Ch}(H|\vartheta)$ and $\bar{g} = gN \in G/N$ we define $\psi^{\bar{g}} \in \operatorname{Ch}(H^g|\vartheta)$ by $\psi^{\bar{g}}(h^g) = \psi(h)$. Check that this is well defined. We say that the isomorphism.

$$(\tau, \sigma): (G, N, \vartheta) \to (\Gamma, M, \varphi)$$

is *strong* provided $(\sigma_H(\psi))^{\tau(\bar{g})} = \sigma_{H^g}(\psi^{\bar{g}})$ for all $\bar{g} \in G/N$, all H with $N \subseteq H \subseteq G$ and all $\psi \in \operatorname{Ch}(H|\vartheta)$. Show that the isomorphism constructed in Theorem 11.28 is strong.

(11.14) In the notation of Problem 11.12, show that $\langle\langle x, y\rangle\rangle = \langle\langle y, x\rangle\rangle^{-1}$ and $\langle\langle x, y_1y_2\rangle\rangle = \langle\langle x, y_1\rangle\rangle\langle\langle x, y_2\rangle\rangle$ if $[x, y_1], [x, y_2] \in N$.

Hint Viewing $\langle\langle\, ,\rangle\rangle$ as defined on commuting pairs of elements of G/N, it is invariant under strong isomorphisms of character triples. (See Problem 11.13.)

(11.15) Let (G, N, ϑ) be a character triple.

(a) If $g \in G$ and $\langle\langle g, x \rangle\rangle \neq 1$ for some $x \in G$, show that $\chi(g) = 0$ for all $\chi \in \text{Irr}(G \mid \vartheta)$.

(b) If $\bar{g} \in G/N$, show that $\bar{g}$ is ϑ-special iff $\langle\langle g, x \rangle\rangle = 1$ for all $x \in G$ such that $\langle\langle g, x \rangle\rangle$ is defined.

See Problems 11.12 and 11.9 for definitions.

(11.16) Let E be elementary abelian of order p^n. Show that $M(E)$ is elementary abelian of order $p^{n(n-1)/2}$.

Hint Consider a Schur representation group for E to get $|M(E)| \leq p^{n(n-1)/2}$. For equality, let $\{x_i \mid 1 \leq i \leq n\}$ be a generating set for E. Define α_{ij}: $E \times E \to \mathbb{C}$ by $\alpha_{ij}(\prod x_\mu{}^{a_\mu}, \prod x_\mu{}^{b_\mu}) = \varepsilon^{a_i b_j}$, where ε is a primitive pth root of unity.

(11.17) Let $G = A_5$. Show that $|M(G)| = 2$.

Note In fact, $|M(A_n)| = 2$ for all $n \geq 4$ except for $n = 6, 7$, where the Schur multipliers have order 6.

(11.18) Let $G = CH$, where H and C are cyclic, $C \lhd G$, and $H \cap C = \mathbf{Z}(G)$. Show that $M(G) = 1$.

Hint If Γ is a Schur representation group for G, show that $|G'| = |\Gamma'|$.

(11.19) Let (Γ_1, π_1) and (Γ_2, π_2) be Schur representation groups for G. Define $\Gamma \subseteq \Gamma_1 \times \Gamma_2$ by $\Gamma = \{(x, y) \mid \pi_1(x) = \pi_2(y)\}$ and show that $\Gamma'_1 \cong \Gamma' \cong \Gamma'_2$.

Hint Let $A_i = \ker \pi_i$ and define π: $\Gamma \to G$ by $\pi(x, y) = \pi_1(x)$, so that $\ker \pi = A_1 \times A_2$ and (Γ, π) is a central extension of G. Use Theorem 11.19 to show that $|\Gamma'| \leq |\Gamma'_1|$. Note that $|\Gamma' : \Gamma' \cap (A_1 \times A_2)| = |G'|$.

Note One cannot conclude that $\Gamma_1 \cong \Gamma_2$ as the two nonabelian groups of order p^3 show. If $G = G'$ then $\Gamma_i = \Gamma'_i$ and $\Gamma_1 \cong \Gamma_2$ in this case.

12 Character degrees

Let c.d.(G) denote* the set $\{\chi(1) \mid \chi \in \operatorname{Irr}(G)\}$, where G is a group. In this chapter we consider what can be said about G when c.d.(G) is known. We have already seen in Theorem 6.9 that if every $f \in$ c.d.(G) is a power of the prime p then G has an abelian normal p-complement. The first results in this chapter consider a weaker hypothesis, namely that p divides f for every $f \in$ c.d.(G) with $f > 1$.

(12.1) THEOREM Fix a prime p. Write $\mathscr{S}(G) = \{\chi \in \operatorname{Irr}(G) \mid p \nmid \chi(1)$ and $p \nmid o(\chi)\}$. Let $s(G) = \sum_{\mathscr{S}(G)} \chi(1)^2$. Then

$$|\mathbf{O}^p(G)| \equiv s(G) \bmod p.$$

Proof Let $N = \mathbf{O}^p(G)$. Since $\mathbf{O}^p(N) = N$, N can have no irreducible character with determinantal order divisible by p. Thus $\operatorname{Irr}(N) = \mathscr{S}(N) \cup \{\psi \in \operatorname{Irr}(N) \mid p \mid \psi(1)\}$ and hence $|N| = \sum_{\psi \in \operatorname{Irr}(N)} \psi(1)^2 \equiv s(N) \bmod p$. It will therefore suffice to show that $s(N) \equiv s(G) \bmod p$.

Now G acts on $\mathscr{S}(N)$ and the resulting orbits are of p-power size. Let $\mathscr{S}_0 = \{\psi \in \mathscr{S}(N) \mid \psi$ is G-invariant$\}$. Since all characters in an orbit have equal degrees, it follows that $s(N) \equiv \sum_{\psi \in \mathscr{S}_0} \psi(1)^2 \bmod p$. We show that restriction defines a one-to-one map of $\mathscr{S}(G)$ onto $\mathscr{S}_0$ and this will complete the proof.

If $\chi \in \mathscr{S}(G)$ then $(\chi(1), |G:N|) = 1$ and so $\chi_N \in \operatorname{Irr}(N)$ by Problem 6.7. Thus $\chi_N \in \mathscr{S}_0$. Conversely, if $\psi \in \mathscr{S}_0$ then by Corollary 6.28, ψ is extendible to G and a unique extension of ψ lies in $\mathscr{S}(G)$. The result now follows. ∎

* The initials c.d. stand for "character degrees."

(12.2) COROLLARY (*Thompson*) Suppose $p \mid \chi(1)$ for every nonlinear $\chi \in \operatorname{Irr}(G)$, where p is a prime. Then G has a normal p-complement.

Proof In the notation of Theorem 12.1, all $\chi \in \mathscr{S}(G)$ are linear and have kernels which contain $\mathbf{O}^{p'}(G)$. It follows that $\mathscr{S}(G) = \operatorname{Irr}(G/G'\mathbf{O}^{p'}(G))$ and $s(G) = |G : G'\mathbf{O}^{p'}(G)|$. Thus $p \nmid s(G)$ and hence by 12.1, $p \nmid |\mathbf{O}^{p}(G)|$. Thus $\mathbf{O}^{p}(G)$ is a normal p-complement for G. ∎

The following lemma is very useful for inductive proofs of theorems giving information about G when c.d.(G) is known.

(12.3) LEMMA Let G be solvable and assume that G' is the unique minimal normal subgroup of G. Then all nonlinear irreducible characters of G have equal degree f and one of the following situations obtains:

(a) G is a p-group, $\mathbf{Z}(G)$ is cyclic and $G/\mathbf{Z}(G)$ is elementary abelian of order f^2.

(b) G is a Frobenius group with an abelian Frobenius complement of order f. Also, G' is the Frobenius kernel and is an elementary abelian p-group.

Proof If $\mathbf{Z}(G) \neq 1$, we must have that $\mathbf{Z}(G)$ is a cyclic p-group and $G' \subseteq \mathbf{Z}(G)$ with $|G'| = p$. Every $\chi \in \operatorname{Irr}(G)$ with $\chi(1) > 1$ is faithful and satisfies $\chi(1)^2 = |G : \mathbf{Z}(G)|$ by Theorem 2.31. If $x, y \in G$, then since $[x, y] \in \mathbf{Z}(G)$, we have $[x^p, y] = [x, y]^p = 1$ since $|G'| = p$. Thus $x^p \in \mathbf{Z}(G)$ for all $x \in G$. This completes the proof in situation (a).

Now suppose that $\mathbf{Z}(G) = 1$. Certainly, G' is an elementary abelian p-group for some prime p. Choose $q \mid |G|$, a prime different from p and let $Q \in \operatorname{Syl}_q(G)$. Then $G'Q \lhd G$ and it follows by the Frattini argument that $G = G'N$ where $N = \mathbf{N}_G(Q)$. Now $N \cap G' \lhd N$ since $G' \lhd G$ and $N \cap G' \lhd G'$ since G' is abelian. Thus $N \cap G' \lhd G$. Now $Q \ntriangleleft G$ and so $N < G$ and $N \not\supseteq G'$. By the minimality of G', it follows that $N \cap G' = 1$ and thus N is abelian.

If $1 \neq x \in N$, then N normalizes $\mathbf{C}_{G'}(x)$ since N centralizes x. Since $\mathbf{C}_{G'}(x) \lhd G'$ we have $\mathbf{C}_{G'}(x) \lhd NG' = G$. If $\mathbf{C}_{G'}(x) \neq 1$, then x centralizes G' and it follows that $x \in \mathbf{Z}(G)$, a contradiction. It follows that $\mathbf{C}_{G'}(x) = 1$ and thus G is a Frobenius group with kernel G' and complement N by Problem 7.1. We conclude that all nonlinear $\chi \in \operatorname{Irr}(G)$ are of the form λ^G for linear $\lambda \in \operatorname{Irr}(G')$. Thus $|G : G'| = |N|$ is the common degree of all nonlinear irreducible characters of G. The proof is complete. ∎

The way Lemma 12.3 is applied in practice is the following. Given nonabelian G, let $K \lhd G$ be maximal such that G/K is nonabelian. Then $(G/K)'$

is the unique minimal normal subgroup of G/K. Thus either G/K is nonsolvable or else satisfies the hypotheses of Lemma 12.3.

If χ is a character of G, we introduce the notation $\mathbf{V}(\chi) = \langle g \in G \mid \chi(g) \neq 0 \rangle$. Then $\mathbf{V}(\chi)$, the *vanishing-off* subgroup, is the smallest subgroup, $V \subseteq G$ such that χ vanishes on $G - V$. Of course, $\mathbf{V}(\chi) \lhd G$.

The relevance of $\mathbf{V}(\chi)$ to character degrees is this. Suppose $\chi \in \mathrm{Irr}(G)$ and $\mathbf{V}(\chi) \subseteq N \lhd G$. Then by Lemma 2.29, we have $[\chi_N, \chi_N] = |G:N|$. Now write $\chi_N = e\sum_{i=1}^{t} \vartheta_i$ where the $\vartheta_i \in \mathrm{Irr}(N)$ are distinct and of equal degree (Theorem 6.2). We have $|G:N| = [\chi_N, \chi_N] = e^2 t$ and $\chi(1) = et\vartheta(1)$, where ϑ is one of the ϑ_i. It follows that $|G:N|\vartheta(1)^2$ divides $\chi(1)^2$. In particular, $|G:N| \leq \chi(1)^2$ and $\chi(1)$ is divisible by every prime divisor of $|G:N|$.

The following result is useful when $K \lhd G$ and G/K satisfies the conditions of case (b) of Lemma 12.3.

(12.4) THEOREM Let $K \lhd G$ be such that G/K is a Frobenius group with kernel N/K, an elementary abelian p-group. Let $\psi \in \mathrm{Irr}(N)$. Then one of the following holds.

(a) $|G:N|\psi(1) \in \mathrm{c.d.}(G)$.
(b) $\mathbf{V}(\psi) \subseteq K$ and thus $|N:K|$ divides $\psi(1)^2$.

Proof For $\lambda \in \mathrm{Irr}(N/K)$, let $T(\lambda)$ denote $I_G(\psi\lambda)$, so that $T(\lambda) \supseteq N$. If $T(\lambda) = N$ for some λ, then $(\lambda\psi)^G \in \mathrm{Irr}(G)$ and hence $|G:N|\psi(1) \in \mathrm{c.d.}(G)$. We suppose then, that $T(\lambda) > N$ for all λ.

Let $W = \mathbf{V}(\psi)K$ so that $K \subseteq W \subseteq N$ and let $S = \mathbf{N}_G(W) \supseteq N$. Since $I_G(\psi)$ normalizes $\mathbf{V}(\psi)$, we have $I_G(\psi) \subseteq S$. However, $\mathbf{V}(\psi) = \mathbf{V}(\lambda\psi)$ for all $\lambda \in \mathrm{Irr}(N/K)$ since the λ are linear, and it follows that $T(\lambda) \subseteq S$ for all λ.

Now view N/K as an $F[S/N]$-module, where F is the field with p elements. Since G/K is a Frobenius group, we have $|G/N|$ divides $|N/K| - 1$ and thus $p \nmid |G/N|$. In particular, $p \nmid |S/N|$ and hence N/K is completely reducible as an $F[S/N]$-module by Maschke's Theorem 1.9. Since W/K is a submodule, we can write $N/K = (W/K) \times (U/K)$, where S normalizes U.

Now suppose $W > K$ so that $U < N$. Let $\lambda \in \mathrm{Irr}(N/U)$ with $\lambda \neq 1_N$. Since G/K is a Frobenius group, we have $I_G(\lambda) = N$ and thus $I_S(\lambda) = N$. Therefore, $|\mathrm{Irr}(N/U)| \geq 1 + |S/N|$. We claim that there exist distinct λ, $\mu \in \mathrm{Irr}(N/U)$ with $T(\lambda) \cap T(\mu) > N$. If not, then since $N < T(\lambda) \subseteq S$ for all $\lambda \in \mathrm{Irr}(N/U)$ we have

$$|S/N| - 1 \geq \sum (|T(\lambda)/N| - 1) \geq |\mathrm{Irr}(N/U)| > |S/N|,$$

where the sum runs over $\lambda \in \mathrm{Irr}(N/U)$. This contradiction shows that λ, $\mu \in \mathrm{Irr}(N/U)$ exist with $T(\lambda) \cap T(\mu) > N$ and $\lambda \neq \mu$, as claimed.

Now let $x \in (T(\lambda) \cap T(\mu)) - N$ and write $\nu = \lambda\bar{\mu}$. Then $\lambda^x = \mu^x\nu^x$ and

$$\psi\lambda = (\psi\lambda)^x = \psi^x\lambda^x = (\psi^x\mu^x)\nu^x = (\psi\mu)^x\nu^x = \psi\mu\nu^x$$

and thus $\psi\nu = \psi\nu^x$ and we have $\mathbf{V}(\psi) \subseteq \ker(\bar{\nu}\nu^x)$. Also $U \subseteq \ker(\bar{\nu}\nu^x)$ and thus $N = \mathbf{V}(\psi)U \subseteq \ker(\bar{\nu}\nu^x)$ so that $\bar{\nu}\nu^x = 1_N$ and $\nu^x = \nu$. Since $\lambda \neq \mu$, we have $\nu \neq 1_N$ and $I_G(\nu) = N$. This contradicts $x \notin N$ and thus proves that $W = K$ and $\mathbf{V}(\psi) \subseteq K$ as desired. That this implies $|N:K| \mid \psi(1)^2$ follows from the remarks preceding the statement of the theorem. ∎

In particular, in the situation of Theorem 12.4, we have for $\psi \in \text{Irr}(N)$ that either $|G:N|\psi(1) \in \text{c.d.}(G)$ or $p \mid \psi(1)$. This most important consequence of 12.4 could have been obtained with somewhat less work than was needed to prove the full strength of the theorem.

(12.5) THEOREM Let $\text{c.d.}(G) = \{1, m\}$. Then at least one of the following occurs.

(a) G has an abelian normal subgroup of index m.

(b) $m = p^e$ for a prime p and G is the direct product of a p-group and an abelian group.

Proof Suppose there exists $K \lhd G$ such that G/K satisfies conclusion (b) of Lemma 12.3. Let $N/K = (G/K)'$ so that $|G:N| = m$ and N/K is a p-group. Since G/K is a Frobenius group, $m \mid (|N/K| - 1)$ and so $p \nmid m$. We claim that N is abelian.

Let $\psi \in \text{Irr}(N)$ and let χ be an irreducible constituent of ψ^G. Then $\chi(1) = 1$ or m and $\psi(1) \mid \chi(1)$. In particular, $p \nmid \psi(1)$. By Theorem 12.4, $m\psi(1) \in \text{c.d.}(G)$ and thus $\psi(1) = 1$. This establishes the claim and situation (a) of the theorem holds in this case.

We now suppose that no $K \lhd G$ as above exists. Let π be the set of prime divisors of m. By Corollary 12.2, G has a normal p-complement for every $p \in \pi$. If $|\pi| > 1$, then G has no irreducible character of p-power degree and hence $G/\mathbf{O}^p(G)$ is abelian for all p. It follows that G' is a π'-group. Since the elements of $\text{c.d.}(G')$ divide elements of $\text{c.d.}(G)$, we conclude that $\text{c.d.}(G') = \{1\}$ and G' is abelian. In particular, G is solvable. Now let $K \lhd G$ be maximal such that G/K is nonabelian. Thus G/K satisfies the hypotheses of Lemma 12.3. By assumption, we are in case (a) of the lemma and this contradicts m not being a prime power.

We may now assume that $\pi = \{p\}$. Let A be the normal p-complement of G so that A is abelian. Let $\lambda \in \text{Irr}(A)$ and let $T = I_G(\lambda)$. If ψ is any irreducible constituent of λ^G, then by Clifford's theorem, $|G:T|$ divides $\psi(1)$, and thus $|G:T| \leq m$. Now $(1_T)^G$ is not irreducible and has degree $\leq m$. It follows that all of its irreducible constituents are linear and thus $G' \subseteq \ker((1_T)^G) \subseteq T$. (Note that we have just done Problem 5.14(c).)

Now if $D = \bigcap_{\lambda \in \text{Irr}(A)} I_G(\lambda)$, then $A \subseteq \mathbf{Z}(\xi)$ for all $\xi \in \text{Irr}(D)$ and thus $A \subseteq \mathbf{Z}(D)$. Thus $D = A \times P$ for $P \in \text{Syl}_p(D)$. Since $G' \subseteq D$, we have $D \lhd G$ and hence $P \lhd G$. If $G' \subseteq P$ then $[G, A] \subseteq A \cap P = 1$ so that $A \subseteq \mathbf{Z}(G)$

the result follows. Otherwise, let $K \supseteq P$, $K \lhd G$ with K maximal such that G/K is nonabelian. Since G is solvable, the hypotheses of Lemma 12.3 are satisfied. By assumption we are in case (a) of the lemma and G/K is a p-group. Thus $K \supseteq \mathbf{O}^p(G) = A$. We have $K \supseteq AP = D \supseteq G'$ and this contradicts G/K being nonabelian and completes the proof. ∎

(12.6) COROLLARY If $|\text{c.d.}(G)| = 2$, then G' is abelian.

Proof Let $\text{c.d.}(G) = \{1, m\}$. If $A \lhd G$ with $|G : A| = m$ then G/A can have no nonlinear irreducible characters and so is abelian. Thus if case (a) of Theorem 12.5 holds, we are done.

In case (b) of 12.5, G is nilpotent and hence is an M-group by Corollary 6.14. The result follows by Theorem 5.12. ∎

In the case that m is a prime we can sharpen Theorem 12.5 to give a necessary and sufficient condition on G that $\text{c.d.}(G) = \{1, m\}$. The following results prove more than is needed for this and will be used again. Theorem 12.7 is actually a generalization of Theorem 6.16.

(12.7) THEOREM Let $N \lhd G$ and suppose $\vartheta_1, \vartheta_2 \in \text{Irr}(N)$ are invariant in G and $\vartheta_1\vartheta_2 \in \text{Irr}(N)$. Let χ_i be an irreducible constituent of $(\vartheta_i)^G$ for $i = 1, 2$ and let ψ be an irreducible constituent of $\chi_1\chi_2$. Then

$$\psi(1)\chi_1(1) \geq \chi_2(1)\vartheta_1(1)^2.$$

Proof We have $0 \neq [\chi_1\chi_2, \psi] = [\chi_2, \psi\bar{\chi}_1]$ and thus χ_2 is a constituent of $\psi\bar{\chi}_1$. Also $(\chi_2)_N = (\chi_2(1)/\vartheta_2(1))\vartheta_2$ and it follows that

$$\chi_2(1)/\vartheta_2(1) \leq [\vartheta_2, (\psi_N)(\bar{\chi}_1)_N] = [\vartheta_2(\chi_1)_N, \psi_N].$$

However, $(\chi_1)_N = (\chi_1(1)/\vartheta_1(1))\vartheta_1$ and this yields

$$\frac{\chi_2(1)}{\vartheta_2(1)} \leq \frac{\chi_1(1)}{\vartheta_1(1)} [\vartheta_1\vartheta_2, \psi_N].$$

Now $\vartheta_1\vartheta_2$ is the unique irreducible constituent of ψ_N and thus $[\vartheta_1\vartheta_2, \psi_N] = \psi(1)/\vartheta_1(1)\vartheta_2(1)$. Substitute this in the above and simplify to obtain the result. ∎

(12.8) COROLLARY Let $N \lhd G$ and suppose $\beta \in \text{Irr}(G)$ with $N \subseteq \mathbf{Z}(\beta)$. Let $\vartheta \in \text{Irr}(N)$. Then there exists an integer b such that $b\vartheta(1) \in \text{c.d.}(G)$ and $b^2 \geq \beta(1)t$, where $t = |G : I_G(\vartheta)|$.

Proof Let $T = I_G(\vartheta)$ and let γ be an irreducible constituent of β_T. Then β is a constituent of γ^G and so $\beta(1) \leq \gamma(1)t$. Also, $N \subseteq \mathbf{Z}(\gamma)$ and we write $\gamma_N = \gamma(1)\lambda$. Let ξ be an irreducible constituent of ϑ^T and let η be an irreducible constituent of $\xi\gamma$. Since $\vartheta\lambda \in \mathrm{Irr}(N)$, Theorem 12.7, applies and so

$$\xi(1)\eta(1) \geq \gamma(1)\vartheta(1)^2.$$

Thus

$$\xi^G(1)\eta^G(1) \geq \gamma(1)t^2\vartheta(1)^2 \geq \beta(1)t\vartheta(1)^2.$$

Now $T = I_G(\vartheta) = I_G(\vartheta\lambda)$ and hence $\xi^G, \eta^G \in \mathrm{Irr}(G)$ by Theorem 6.11. Let χ be whichever of ξ^G, η^G has larger degree. Then $\chi(1) = b\vartheta(1)$ for some integer b and

$$b^2\vartheta(1)^2 = \chi(1)^2 \geq \xi^G(1)\eta^G(1) \geq \beta(1)t\vartheta(1)^2$$

and the result follows. ∎

(12.9) COROLLARY Let $\mathrm{c.d.}(G) = \{1, p\}$, where p is a prime. Then there exists abelian $A \lhd G$ with $|G : A| = p$ or p^2.

Proof By Theorem 12.5, we may assume that G is a p-group. Let $K \lhd G$ be maximal such that G/K is nonabelian and let $Z/K = \mathbf{Z}(G/K)$. By Lemma 12.3 it follows that $|G : Z| = p^2$.

Let $\beta \in \mathrm{Irr}(G/K)$ with $\beta(1) = p$ and let $\vartheta \in \mathrm{Irr}(Z)$. Then since $Z \subseteq \mathbf{Z}(\beta)$, Corollary 12.8 yields an integer b such that $b^2 \geq \beta(1) = p$ and $b\vartheta(1) \in \mathrm{c.d.}(G)$. This forces $\vartheta(1) = 1$. Thus Z is abelian and the proof is complete. ∎

(12.10) LEMMA Let $A \subseteq G$ be abelian and let $b = \max(\mathrm{c.d.}(G))$. Then

$$(1/|A|)\sum_{a \in A} |\mathbf{C}_G(a)| \geq |G|/b.$$

Proof Since $|\mathbf{C}_G(a)| = \sum_\chi |\chi(a)|^2$ for $\chi \in \mathrm{Irr}(G)$, we have

$$(1/|A|)\sum_a |\mathbf{C}_G(a)| = (1/|A|)\sum_\chi \sum_a |\chi(a)|^2 = \sum_\chi [\chi_A, \chi_A].$$

However, χ_A is the sum of $\chi(1)$ linear characters and hence $[\chi_A, \chi_A] \geq \chi(1)$. Thus

$$(1/|A|)\sum_a |\mathbf{C}_G(a)| \geq \sum_\chi \chi(1).$$

Furthermore, $|G| = \sum_\chi \chi(1)^2 \leq b \sum_\chi \chi(1)$ and thus $\sum_\chi \chi(1) \geq |G|/b$ and the result follows. ∎

(12.11) THEOREM Let G be nonabelian and let p be a prime. Then $\mathrm{c.d.}(G) = \{1, p\}$ iff one of the following holds.

(a) There exists abelian $A \lhd G$ with $|G : A| = p$.
(b) $|G : \mathbf{Z}(G)| = p^3$.

Proof If (a) holds, then $\chi(1) | p$ for every $\chi \in \mathrm{Irr}(G)$ by Ito's Theorem 6.15. Since G is nonabelian, $\mathrm{c.d.}(G) = \{1, p\}$. If $|G : \mathbf{Z}(G)| = p^3$, then $\chi(1)^2 \leq p^3$ by

Corollary 2.30 and $\chi(1)|p^3$ by 6.15 (or 3.12). Again it follows that c.d.$(G) = \{1, p\}$.

Conversely, suppose c.d.$(G) = \{1, p\}$ and assume that (a) is false. We may assume that G is a p-group. By Corollary 12.9, there exists abelian $A \lhd G$ with $|G:A| = p^2$. By Lemma 12.10, we have

$$(1/|A|)\sum_{a\in A}|\mathbf{C}_G(a)| \geq |G|/p.$$

Now A acts by conjugation on $G - A$ and by Corollary 5.15, the number of orbits of this action is

$$(1/|A|)\sum_{a\in A}(|\mathbf{C}_G(a)| - |A|) \geq (|G|/p) - |A|.$$

It follows that the average size α of these orbits satisfies

$$\alpha \leq (|G| - |A|)/((|G|/p) - |A|) = p + 1 < p^2$$

and so A has an orbit of size 1 or p on $G - A$. Since G does not have an abelian subgroup of index p, $A = \mathbf{C}_G(A)$ and thus there exists $x \in G - A$ in an orbit of size p. Let $K = \langle A, x\rangle$. Then $\mathbf{Z}(K) = \mathbf{C}_A(x)$ has index p in A and index p^3 in G. We shall show that $\mathbf{Z}(G) = \mathbf{Z}(K)$ to complete the proof.

Let $Z = \mathbf{Z}(K)$. If $\chi \in \operatorname{Irr}(G)$ and χ_K is irreducible, then $Z \subseteq \mathbf{Z}(\chi)$ and $[G, Z] \subseteq \ker \chi$. On the other hand, if χ_K reduces, then all irreducible constituents are linear and $K' \subseteq \ker \chi$. Suppose $[G, Z] > 1$. We also have $K' > 1$ but $K' \cap [G, Z] \subseteq \ker \chi$ for every $\chi \in \operatorname{Irr}(G)$ so that $K' \cap [G, Z] = 1$. Now $K'[G, Z]$ is the direct product of two nontrivial groups and so has an irreducible character ϑ with $K' \nsubseteq \ker \vartheta$ and $[G, Z] \nsubseteq \ker \vartheta$. Let χ be an irreducible constituent of ϑ^G. Then $K' \nsubseteq \ker \chi$ and $[G, Z] \nsubseteq \ker \chi$. This contradiction shows that $[G, Z] = 1$ and completes the proof. ∎

We remark that the last several sentences of the proof could be replaced by an appeal to Problem 5.26.

Next we refine Theorem 12.5 in a somewhat different direction. Namely, if c.d.$(G) = \{1, m\}$ and G has no abelian normal subgroup of index m, then the nilpotence class of G is ≤ 3.

(12.12) LEMMA Let $A \lhd G$ with A abelian and G/A cyclic. Then $|A| = |G'||A \cap \mathbf{Z}(G)|$.

Proof Let $G/A = \langle Ag\rangle$ and let $\sigma: A \to A$ be defined by $\sigma(a) = a^{-1}a^g$. Then σ is a homomorphism and $\ker \sigma = \mathbf{C}_A(g) = A \cap \mathbf{Z}(G)$. Let I be the image of σ. Then $g \in \mathbf{N}(I)$ and so $I \lhd G$. Since $G = \langle A, g\rangle$ and g centralizes A mod I, it follows that G/I is abelian and $G' \subseteq I$. Clearly, $I \subseteq G'$ and hence

$$|A| = |\ker \sigma||I| = |A \cap \mathbf{Z}(G)||G'|$$

and the proof is complete. ∎

(12.13) LEMMA Let $\text{c.d.}(G) = \{1, m\}$ and suppose $A, B \subseteq G$ are abelian of index m with $A \neq B$. Then $|G'| \leq m$ and $G' \subseteq \mathbf{Z}(G)$.

Proof If $K \subseteq G$ with $|G:K| \leq m$, then $G' \subseteq K$ and hence $K \lhd G$ by Problem 5.14(c). In particular, $A \lhd G$. Let $A < K \subseteq G$, then $K \lhd G$ and thus $K' \lhd G$. If $K' < G'$, let $1_{G'} \neq \mu \in \text{Irr}(G'/K')$ and let χ be an irreducible constituent of μ^G. Now $K' \subseteq \ker \chi$ and so χ_K has a linear constituent λ and $\chi(1) \leq \lambda^G(1) = |G:K| < m$. Thus $\chi(1) = 1$ and $G' \subseteq \ker \chi$. It follows that $G' \subseteq \ker \mu$, contradicting the choice of μ. Therefore $K' = G'$.

Since $A \neq B$, let $b \in B - A$ and let $K = \langle A, b \rangle$. Thus K/A is cyclic and $|G'| = |K'| = |A : A \cap \mathbf{Z}(K)|$ by Lemma 12.12. However,

$$A \cap \mathbf{Z}(K) \supseteq A \cap B$$

and so $|A : A \cap \mathbf{Z}(K)| \leq |A : A \cap B| \leq |G : B| = m$ and the first assertion is proved.

For $g \in G$, the conjugacy class of g is contained in the coset gG' and so has size $\leq |G'| \leq m$. Thus $|G : \mathbf{C}(g)| \leq m$ and $G' \subseteq \mathbf{C}(g)$ by the first sentence of the proof. Since g is arbitrary, we have $G' \subseteq \mathbf{Z}(G)$ and the proof is complete. ∎

(12.14) THEOREM Let $\text{c.d.}(G) = \{1, m\}$ and suppose that G has no abelian normal subgroup of index m. Then $[G, G'] \subseteq \mathbf{Z}(G)$, that is, G is nilpotent of class ≤ 3.

Proof By Theorem 12.5 we may assume that G is a p-group. If G has a faithful irreducible character χ then $\chi = \lambda^G$ for linear $\lambda \in \text{Irr}(K)$ and $K \subseteq G$ since G is an M-group. We have $|G:K| = \chi(1) = m$ and so $K \lhd G$ by Problem 5.14(c). Thus all irreducible constituents of χ_K are linear and $K' \subseteq \ker \chi = 1$, a contradiction.

Thus G has no faithful irreducible character and hence $\mathbf{Z}(G)$ is not cyclic. Let $Z_1, Z_2, Z_3 \subseteq \mathbf{Z}(G)$ be distinct subgroups of order p. If G/Z_i is of nilpotence class ≤ 3, then $[G, G, G, G] \subseteq Z_i$. If this happens for two distinct Z_i, we conclude that $[G, G, G, G] = 1$ and we are done.

Assume then that G/Z_1 and G/Z_2 (say) do not have class ≤ 3. Working by induction on $|G|$, it follows that there exist $A, B \lhd G$ with $|G:A| = m = |G:B|$ and $A' \subseteq Z_1$ and $B' \subseteq Z_2$. Let $Z = Z_1 Z_2$ so that AZ/Z and BZ/Z are abelian. Since G/Z is nonabelian, we have $m \in \text{c.d.}(G/Z)$ and so $|G : AZ| \geq m$ and $|G : BZ| \geq m$. It follows that $Z \subseteq A \cap B$.

If $A/Z = B/Z$, then $A' \subseteq Z_1 \cap Z_2 = 1$ and A is abelian, a contradiction. Thus $A/Z \neq B/Z$ and Lemma 12.13 applies to yield $G'Z/Z \subseteq \mathbf{Z}(G/Z)$. Thus $[G', G] \subseteq Z \subseteq \mathbf{Z}(G)$ and the proof is complete. ∎

We now consider groups G for which $|\text{c.d.}(G)| = 3$. Although the information obtained is not as detailed as when $|\text{c.d.}(G)| = 2$, we do prove a result analogous to Corollary 12.6.

(12.15) THEOREM Let $|\text{c.d.}(G)| = 3$. Then $G''' = 1$, that is, G is solvable of derived length ≤ 3.

Before proceeding with the proof of this result we make some general remarks. As the group A_5 shows, we cannot conclude from $|\text{c.d.}(G)| = 4$ that G is solvable. It has been conjectured by G. Seitz that for solvable G, $\text{d.l.}(G) \leq |\text{c.d.}(G)|$. (Here, $\text{d.l.}(G)$ is the *derived length* of G; the smallest integer k for which $G^{(k)}$, the kth commutator subgroup of G, is trivial.) By Theorem 5.12, this conjecture holds for M-groups. It has also been proved when $|\text{c.d.}(G)| = 4$ (S. Garrison) and when $|G|$ is odd (T. Berger). It is known that for any solvable group, $\text{d.l.}(G) \leq 3|\text{c.d.}(G)|$.

(12.16) LEMMA Let $N \lhd G$ with $(|N|, |G:N|) = 1$. Then

$$|\text{c.d.}(N)| \leq |\text{c.d.}(G)|.$$

Also if G/N is supersolvable and N is solvable with $\text{d.l.}(N) \leq |\text{c.d.}(N)|$, then $\text{d.l.}(G) \leq |\text{c.d.}(G)|$.

Proof Let π be the set of prime divisors of $|N|$. If $\chi \in \text{Irr}(G)$, let ϑ be an irreducible constituent of χ_N. By Corollary 11.29, $\chi(1)/\vartheta(1)$ divides $|G:N|$. Since $\vartheta(1)$ divides $|N|$, it follows that $\vartheta(1)$ is exactly the π-part of $\chi(1)$. Since every $\vartheta \in \text{Irr}(N)$ arises this way, we see that $\text{c.d.}(N)$ is exactly the set of π-parts of the elements of $\text{c.d.}(G)$. The first assertion follows.

Now assume that N is solvable and G/N is supersolvable. It follows that G/N' is an M-group by Theorems 6.22 and 6.23 and thus $\text{d.l.}(G/N') \leq |\text{c.d.}(G/N')|$ by Theorem 5.12. We may assume that $N' > 1$ and observe that

$$\begin{aligned}\text{d.l.}(G) \leq \text{d.l.}(G/N') + \text{d.l.}(N') &\leq |\text{c.d.}(G/N')| + \text{d.l.}(N) - 1\\ &\leq |\text{c.d.}(G/N')| + |\text{c.d.}(N)| - 1,\end{aligned}$$

where the last inequality follows from the assumption that $\text{d.l.}(N) \leq |\text{c.d.}(N)|$.

Now every $f \in \text{c.d.}(G/N')$ divides $|G:N|$ by Ito's Theorem 6.15 and thus the π-part of f is trivial. We conclude from the first part of the proof that $|\text{c.d.}(N)| \leq |\text{c.d.}(G)| - |\text{c.d.}(G/N')| + 1$ and the result follows. ∎

Proof of Theorem 12.15 Let $\text{c.d.}(G) = \{1, m, n\}$. If $(m, n) \neq 1$, then G has a proper normal p-complement N for some prime p by Corollary 12.2. By Lemma 12.16, $|\text{c.d.}(N)| \leq 3$ and so N is solvable and $\text{d.l.}(N) \leq |\text{c.d.}(N)|$ by induction if $|\text{c.d.}(N)| = 3$ and by Corollary 12.6 if $|\text{c.d.}(N)| < 3$. In this case we are done by Lemma 12.16.

Assume now that $(n, m) = 1$. Suppose there exists $K \lhd G$ with G/K solvable and nonabelian. We may assume that G/K satisfies the hypotheses of Lemma 12.3. If G/K is a p-group then (say) n is a power of p. Let $\chi \in \mathrm{Irr}(G)$ with $\chi(1) = m$. Since $p \nmid m$, we have $\chi_K \in \mathrm{Irr}(K)$ and thus by Gallagher's theorem (Corollary 6.17), $\chi\beta \in \mathrm{Irr}(G)$ for $\beta \in \mathrm{Irr}(G/K)$. This is a contradiction since $mn \notin \mathrm{c.d.}(G)$.

Therefore G/K is a Frobenius group with kernel $N/K = (G/K)'$, where N/K is a p-group and $|G:N| = n$ (say). Suppose $\psi \in \mathrm{Irr}(N)$ with $\psi(1) > 1$. Since $n\psi(1) \notin \mathrm{c.d.}(G)$, we conclude from Theorem 12.4 that $p \mid \psi(1)$. Let χ be an irreducible constituent of ψ^G. Then $p \mid \chi(1)$ and since $p \nmid n$, we have $\chi(1) = m$. Thus χ_N is irreducible since $(|G:N|, \chi(1)) = 1$. We conclude that $\chi_N = \psi$ and $\psi(1) = m$. Thus $\mathrm{c.d.}(N) = \{1, m\}$ and $\mathrm{d.l.}(N) \leq 2$ by Corollary 12.6. Since G/N is abelian, we have $\mathrm{d.l.}(G) \leq 3$ as desired.

Suppose now that no such K exists. If $\chi \in \mathrm{Irr}(G)$ with $\chi(1) \neq 1$, then $G/\ker \chi$ is nonabelian and thus is nonsolvable. However $|\mathrm{c.d.}(G/\ker \chi)| \leq 3$ and working by induction on $|G|$ we must have $\ker \chi = 1$. Therefore, every nonlinear irreducible character of G is faithful.

Now let $n < m$ and let $\chi \in \mathrm{Irr}(G)$ with $\chi(1) = n$. By Theorem 4.3, each irreducible character of G is a constituent of χ^t for suitable integers t. Let t be minimal such that χ^t has an irreducible constituent ψ of degree m. Thus $[\chi\xi, \psi] \neq 0$ for some irreducible constituent ξ of χ^{t-1}. Now $\xi(1) \neq 1$ or else $\chi\xi$ is irreducible forcing $\chi\xi = \psi$ which is not the case. Also $\xi(1) \neq m$ by the minimality of t. Thus $\xi(1) = n$.

Since $[\chi\lambda, \psi] = 0$ for linear λ, we conclude that $[\lambda, \psi\bar{\chi}] = 0$ and $\psi\bar{\chi}$ has no linear constituents. Thus

$$\psi\bar{\chi} = \sum_{i=1}^{a} \xi_i + \sum_{j=1}^{b} \eta_j,$$

where $\xi_i, \eta_j \in \mathrm{Irr}(G)$, $\xi_i(1) = n$, $\eta_j(1) = m$ and $a \geq 1$ since ξ is one of the ξ_i. Comparing degrees yields $mn = an + bm$ and since $(m, n) = 1$, we have $m \mid a$. Since $b \geq 0$ and $a > 0$, this yields $a = m$ and $b = 0$ and $\psi\bar{\chi} = \sum_{i=1}^{m} \xi_i$.

We claim that each ξ_i is of the form $\lambda_i\bar{\chi}$ for some linear character, λ_i. It suffices to find a linear constituent of $\chi\xi_i$. Suppose (for some i) that $\chi\xi_i$ has no linear constituent. Then

$$\chi\xi_i = \sum_{j=1}^{r} \chi_j + \sum_{k=1}^{s} \psi_k$$

where $\chi_j, \psi_k \in \mathrm{Irr}(G)$, $\chi_j(1) = n$ and $\psi_k(1) = m$. Thus $n^2 = rn + sm$ and $n \mid s$. However, $0 \neq [\psi\bar{\chi}, \xi_i] = [\psi, \chi\xi_i]$ and ψ is one of the ψ_k. Thus $s \geq 1$ and hence $s \geq n$. Since $r \geq 0$, this yields $n^2 \geq nm$ and $n \geq m$, a contradiction.

We now have $\xi_i = \lambda_i\bar{\chi}$ for linear λ_i and thus $\bar{\chi}\psi = \sum \xi_i = \bar{\chi} \sum \lambda_i$ and hence

$$(\psi_{G'})(\bar{\chi}_{G'}) = m\bar{\chi}_{G'}.$$

Since $\ker \psi = 1$, we have $\psi(x) \neq m$ for $x \neq 1$ and thus $\bar{\chi}(x) = 0$ for all $x \in G' - \{1\}$. It follows that $[\bar{\chi}_{G'}, 1_{G'}] \neq 0$ and thus $G' \subseteq \ker \bar{\chi}$. Thus contradiction completes the proof. ∎

Another technique for studying character degrees is based on the following lemma.

(12.17) LEMMA (*Garrison*) Let $H \subseteq G$ and let $\vartheta \in \mathrm{Irr}(H)$. Suppose for every irreducible constituent χ of ϑ^G that $\chi_H = \vartheta$. Then $\mathbf{V}(\vartheta) \lhd G$.

Proof We have $(\vartheta^G)_H$ is a multiple of ϑ and so $(\vartheta^G)_H = |G:H|\vartheta$. Let $h \in H$ with $\vartheta(h) \neq 0$ and let $S = \{x \in G \mid h^x \in H\}$. By definition of ϑ^G we have

$$(1/|H|) \sum_{x \in S} \vartheta(h^x) = \vartheta^G(h) = |G:H|\vartheta(h).$$

However, since $\vartheta^G(h^x) = \vartheta^G(h)$, it follows for $x \in S$ that $\vartheta(h^x) = \vartheta(h)$. This yields $|S|\vartheta(h) = |G|\vartheta(h)$ and since $\vartheta(h) \neq 0$, we have $|S| = |G|$ and $S = G$. Thus $h^g \in H$ for all $g \in G$ and $\vartheta(h^g) = \vartheta(h) \neq 0$. Thus G leaves the generating set for $\mathbf{V}(\vartheta)$ invariant under conjugation and the result follows. ∎

(12.18) LEMMA Let $\chi \in \mathrm{Irr}(G)$ and let $\ker \chi < N \lhd G$. Then

$$\ker \chi < N \cap \mathbf{V}(\chi).$$

Proof We may assume $\ker \chi = 1$. If $\mathbf{V}(\chi) \cap N = 1$ then χ vanishes on $N - \{1\}$ and so $[\chi_N, 1_N] \neq 0$. This forces $N \subseteq \ker \chi$, a contradiction. ∎

In the following, $\mathbf{F}(G)$ denotes the *Fitting subgroup* of G, the (unique) largest normal nilpotent subgroup.

(12.19) THEOREM (*Broline–Garrison*) Let $\chi \in \mathrm{Irr}(G)$ and $K = \ker \chi$. Either of the following conditions guarantees the existence of $\psi \in \mathrm{Irr}(G)$ with $\psi(1) > \chi(1)$ and $\ker \psi < K$:

(a) $K \nsubseteq \mathbf{F}(G)$;
(b) $K = \mathbf{F}(G)$, G/K is solvable and $K < G$.

Proof Suppose H is a maximal subgroup of G and that $KH = G$. Then $\chi_H = \vartheta \in \mathrm{Irr}(H)$. If ψ is an irreducible constituent of ϑ^G such that $\psi_H \neq \vartheta$, then ψ_H reduces and $\psi(1) > \vartheta(1) = \chi(1)$. Also, we cannot have $(\ker \psi)H = G$ or else ψ_H would be irreducible. The maximality of H thus yields $\ker \psi \subseteq H$ and since ϑ is a constituent of ψ_H, we have $\ker \psi \subseteq \ker \vartheta = H \cap K < K$. Thus the result follows in this situation and we may suppose that whenever $KH = G$ for a maximal subgroup $H \subseteq G$, the character $\vartheta = \chi_H$ satisfies the hypotheses of Lemma 12.17. In particular, $\mathbf{V}(\vartheta) \lhd G$.

Now we choose $L \lhd G$ as follows. If $K \not\subseteq \mathbf{F}(G)$, take $L = K$. If $K \subseteq \mathbf{F}(G)$, then we are in situation (b) and $\mathbf{F}(G) = K < G$. Here, G/K is solvable and we take $L > K$ with L/K an elementary abelian chief factor of G. Thus L is not nilpotent and we choose a nonnormal Sylow subgroup P of L. Note that in the case $K = \mathbf{F}(G)$ and $L > K$, we must have $L = KP$. By the Frattini argument, $G = L\mathbf{N}_G(P) = K\mathbf{N}_G(P)$ and $\mathbf{N}_G(P) < G$. Let $H \supseteq \mathbf{N}_G(P)$ be a maximal subgroup of G. Thus $G = HK$ and we let $\vartheta = \chi_H$. Then $\mathbf{V}(\vartheta) \lhd G$ and thus $\mathbf{V}(\vartheta) \cap L \lhd G$.

Now $K \cap H = \ker \vartheta \subseteq \mathbf{V}(\vartheta)$ and so in the case $K = L$, we have $P \subseteq K \cap H \subseteq L \cap \mathbf{V}(\vartheta)$. In the case that L/K is a chief factor of G, we have $(L \cap H)/(\ker \vartheta)$ is a chief factor of H and thus $L \cap H \subseteq \mathbf{V}(\vartheta)$ by Lemma 12.18. Thus in any case, $P \subseteq L \cap \mathbf{V}(\vartheta) \lhd G$ and P is Sylow in $L \cap \mathbf{V}(\vartheta)$. The Frattini argument now yields $G = (L \cap \mathbf{V}(\vartheta))\mathbf{N}_G(P) \subseteq H$. This is a contradiction and proves the theorem. ∎

(12.20) COROLLARY Let $\chi \in \operatorname{Irr}(G)$. If either $\chi(1) = \max(\text{c.d.}(G))$ or $\ker \chi$ is minimal among kernels of irreducible characters of G then $\ker \chi$ is nilpotent.

(12.21) COROLLARY (*Garrison*) Let G be solvable and let $|\text{c.d.}(G)| = n$. Then there exist $N_i \lhd G$ with

$$1 = N_0 \subseteq N_1 \subseteq \cdots \subseteq N_n = G$$

such that N_{i+1}/N_i is nilpotent for $0 \le i < n$.

Proof Let $N_1 = \mathbf{F}(G)$, the Fitting subgroup. Then $\text{c.d.}(G/N_1) \subseteq \text{c.d.}(G)$ and if $N_1 < G$, then $\text{c.d.}(G/N_1)$ does not contain the largest element of c.d.(G) by Theorem 12.19. In this case, $|\text{c.d.}(G/N_1)| < n$ and the result follows by induction on $|G|$. ∎

Suppose $A \subseteq G$ is abelian. By Problem 5.4 (or 2.9(b)), $|G : A|$ is an upper bound for c.d.(G). Conversely, suppose we know $\max(\text{c.d.}(G))$. Can we conclude that there exists an abelian subgroup with bounded index in G? We can, although it is certainly not true that there necessarily exists abelian $A \subseteq G$ with $|G : A| = \max(\text{c.d.}(G))$.

We use the notation $b(G) = \max(\text{c.d.}(G))$. Note that if $H \subseteq G$ and $\psi \in \operatorname{Irr}(H)$, then ψ is a constituent of χ_H for some $\chi \in \operatorname{Irr}(G)$, and thus

$$\psi(1) \le \chi(1) \le b(G)$$

and hence $b(H) \le b(G)$.

(12.22) LEMMA Let $b(G) = b$. Then there exists $x \in G - \{1\}$ such that $|G : \mathbf{C}_G(x)| \le b^2$.

Proof Let $k = |\mathrm{Irr}(G)| - 1$. Then

$$|G| - 1 = \sum \chi(1)^2 \leq kb^2,$$

where the sum runs over nonprincipal $\chi \in \mathrm{Irr}(G)$. Let

$$m = \min\{|G : \mathbf{C}_G(x)| \mid x \in G, x \neq 1\}.$$

Then $|G| - 1 \geq mk$ since each of the k nonidentity conjugacy classes has size $\geq m$. We now have $kb^2 \geq mk$ and the result follows. ∎

(12.23) THEOREM Let $b(G) = b$. Then G has an abelian subgroup of index $\leq (b!)^2$.

Proof Use induction on b. If $b = 1$, the result is trivial and if $b = 2$, then $\mathrm{c.d.}(G) = \{1, 2\}$ and we are done by Corollary 12.9. Assume then, that $b \geq 3$.

Let $K \lhd G$ be maximal such that G/K is nonabelian. It follows that $b(K) \leq b/2$ since if $\psi \in \mathrm{Irr}(K)$ with $\psi(1) > b/2$, then necessarily $\psi = \chi_K$ for some $\chi \in \mathrm{Irr}(G)$. By Corollary 6.17, we have $\beta\chi \in \mathrm{Irr}(G)$ for $\beta \in \mathrm{Irr}(G/K)$. Since G/K is nonabelian, we may choose β with $\beta(1) \geq 2$ and this yields $\beta(1)\chi(1) > b$, a contradiction.

By Lemma 12.3, we have three main cases to consider namely: G/K is nonsolvable, G/K is a Frobenius group and G/K is a p-group.

Case 1 G/K is nonsolvable. Since $b(G/K) \leq b$, Lemma 12.22 yields $x \in G/K$ such that $x \neq 1$ and

$$|(G/K) : \mathbf{C}_{G/K}(x)| \leq b^2.$$

Let $C/K = \mathbf{C}_{G/K}(x)$. If $b(C) \leq b - 1$, then there exists abelian $A \subseteq C$ with $|C : A| \leq ((b-1)!)^2$ and so $|G : A| = |G : C||C : A| \leq (b!)^2$ and we are done.

Assume that $b(C) = b$ and let $\psi \in \mathrm{Irr}(C)$ with $\psi(1) = b$. Then every irreducible constituent χ of ψ^G satisfies $\chi(1) = b$ and thus $\chi_C = \psi$. It follows by Lemma 12.17 that $\mathbf{V}(\psi) \lhd G$ and thus $K\mathbf{V}(\psi) \lhd G$. If $K\mathbf{V}(\psi) > K$, then $G/K\mathbf{V}(\psi)$ is abelian and so $C \lhd G$ since $K\mathbf{V}(\psi) \subseteq C$. If $Z/K = \mathbf{Z}(C/K)$ then $Z \lhd G$. Also, $x \in Z/K$ and so $Z > K$. It follows that G/Z is abelian and G/K is solvable, a contradiction. We conclude that $\mathbf{V}(\psi) \subseteq K$ and hence $|C : K| \leq \psi(1)^2 \leq b^2$ by the remarks preceding Theorem 12.4. Thus $|G : K| \leq b^4$.

By the inductive hypothesis, K has an abelian subgroup of index $\leq ([b/2]!)^2$. For $b \geq 5$, we have $([b/2]!)^2 b^4 \leq (b!)^2$ and the result follows. Assume then, that $b \leq 4$. If $2 \notin \mathrm{c.d.}(G/K)$, then $\mathrm{c.d.}(G/K) \subseteq \{1, 3, 4\}$ which contradicts the nonsolvability of G/K by Theorem 12.15. Thus we may choose $\vartheta \in \mathrm{Irr}(G/K)$ with $\vartheta(1) = 2$. We have $G/\ker\vartheta$ is nonabelian and hence $\ker\vartheta = K$. Let $M/K = (G/K)'$. Then $M' \not\subseteq K = \ker\vartheta$ and hence ϑ_M is irreducible. Pick $y \in M/K$ of order 2. Since $y \in \ker(\det(\vartheta))$, it follows that $y \in \mathbf{Z}(\vartheta) = \mathbf{Z}(G/K) = 1$, a contradiction. This completes case 1.

Case 2 G/K is a Frobenius group. Let $N/K = (G/K)'$, the Frobenius kernel. By Lemma 12.3, $|G:N| \in \text{c.d.}(G)$ and so $|G:N| \le b$. If $b(N) \le b/2$, then application of the inductive hypothesis to N yields an abelian subgroup $A \subseteq N$ with $|N:A| \le ([b/2]!)^2$. Since $b([b/2]!)^2 \le (b!)^2$ we are done.

Suppose then, that $b(N) > b/2$ and let $\psi \in \text{Irr}(N)$ with $\psi(1) > b/2$. Then $\psi(1)|G:N| \notin \text{c.d.}(G)$ and by Theorem 12.4 it follows that $\psi(1)^2 \ge |N:K|$ and hence $|G:K| \le b^3$. Since $b \ge 3$, we have $([b/2]!)^2 b^3 \le (b!)^2$ and we are done in case 2 by applying the inductive hypothesis to K.

Case 3 G/K is a p-group. Let $Z/K = \mathbf{Z}(G/K)$. By Lemma 12.3, $|G:Z| = \beta(1)^2$, where $\beta \in \text{Irr}(G)$ and $Z \subseteq \mathbf{Z}(\beta)$. In particular, $|G:Z| \le b^2$. If $b(Z) > b/2$, pick $\psi \in \text{Irr}(Z)$ with $\psi(1) > b/2$. Then $\psi = \chi_Z$ for some $\chi \in \text{Irr}(G)$. It follows from Theorem 12.7 or 6.16 that $\beta(1)\psi(1) \in \text{c.d.}(G)$ and this is a contradiction since $\beta(1)\psi(1) > b$. Thus $b(Z) \le b/2$ and the result follows as in previous cases since $([b/2]!)^2 b^2 \le (b!)^2$. The proof is now complete. ∎

It is apparent in the above proof that in cases 2 and 3, the inequalities obtained are far from being best possible. The limiting factor in this proof is the nonsolvable case 1. If one assumes that G is solvable it is possible to obtain a better bound. It is not known what the best possible bound is either in the general situation or for solvable groups.

The following result provides a tool which can be used to find an abelian subgroup in G of index $\le b(G)^4$ when G has an abelian normal subgroup with nilpotent factor group. (Compare Theorem 12.24 with part (a) of Theorem 12.19.)

(12.24) THEOREM (*Broline*) Let $\chi \in \text{Irr}(G)$ and let $K = \ker \chi$ and $F/K = \mathbf{F}(G/K)$. Suppose F is not nilpotent. Then there exists $\psi \in \text{Irr}(G)$ with $\ker \psi < K$.

Proof Let $Q \in \text{Syl}_q(F)$ with $Q \ntriangleleft F$ and let $H \supseteq \mathbf{N}_G(Q)$ be a maximal subgroup. Now $QK \triangleleft G$ since F/K is nilpotent and normal in G/K and hence $HK = G$ by the Frattini argument. Thus $\vartheta = \chi_H \in \text{Irr}(H)$.

Since $\ker(\vartheta^G) \subseteq H$, we have $\ker(\vartheta^G) \nsupseteq K$. Let ψ be an irreducible constituent of ϑ^G with $K \nsubseteq \ker \psi$ and let $L = \ker \psi$. If $L \subseteq H$, then $L \subseteq \ker \vartheta = H \cap K < K$ and we are done. Suppose then, that $L \nsubseteq H$ so that $LH = G$ and ψ_H is irreducible. Then $\psi_H = \vartheta$ and $L \cap H = \ker \vartheta = K \cap H$. Write $N = \ker \vartheta$.

Now $K \cap L \triangleleft G$ and $K > K \cap L \supseteq K \cap H$. It follows that

$$(K \cap L)H < KH = G$$

and thus $K \cap L \subseteq H$. Therefore $K \cap L = K \cap H = N$

We have
$$|F:F\cap H| = |G:H| = |L:N| = |KL:K|.$$
Since $Q \subseteq F \cap H$ and $Q \in \mathrm{Syl}_q(F)$, it follows that $q \nmid |KL/K|$. Since $KQ/K \lhd G/K$, we conclude that the commutator $[KL/K, KQ/K] = 1$ and thus $[L, Q] \subseteq K$. Since $L \lhd G$, we have $[L, Q] \subseteq K \cap L = N$ and hence $L \subseteq \mathbf{N}(NQ)$. However,
$$NQ = (H \cap K)Q = H \cap KQ \lhd H$$
since $KQ \lhd G$. Therefore, $G = LH \subseteq \mathbf{N}(NQ)$ and $NQ \lhd G$. The Frattini argument yields $G = N\mathbf{N}(Q) \subseteq H$, a contradiction. ∎

(12.25) LEMMA Let $G'' = 1$ and let $m = |G:\mathbf{F}(G)|$. Then $m \leq b(G)$ and $b(\mathbf{F}(G)) \leq b(G)/m$.

Proof Let $\chi \in \mathrm{Irr}(G)$ be such that $\ker \chi$ is minimal. Since G is a relative M-group with respect to G' which is abelian, there exists $H \supseteq G'$ and linear $\lambda \in \mathrm{Irr}(H)$ such that $\lambda^G = \chi$. Also, $H \supseteq \ker \chi$. Since $H \lhd G$, all irreducible constituents of χ_H are linear and thus $H/\ker \chi$ is abelian. Thus $H/\ker \chi \subseteq \mathbf{F}(G/\ker \chi)$ and hence H is nilpotent by Theorem 12.24. In particular, $H \subseteq \mathbf{F}(G)$ and $m \leq |G:H| = \chi(1) \leq b(G)$.

Now let $\vartheta \in \mathrm{Irr}(\mathbf{F}(G))$ and let ψ be an irreducible constituent of ϑ^G. Choose $\chi \in \mathrm{Irr}(G)$ with minimal $\ker \chi \subseteq \ker \psi$ and let H be as above with $|G:H| \leq b(G)$, $H/\ker \chi$ abelian and $H \subseteq \mathbf{F}(G)$. We have $H' \subseteq \ker \chi \subseteq \ker \psi$ and thus ψ_H has linear constituents. Thus ϑ_H has linear constituents and hence $\vartheta(1) \leq |\mathbf{F}(G):H| = |G:H|/m \leq b(G)/m$ and the proof is complete. ∎

(12.26) THEOREM Suppose $U \lhd G$ is abelian and G/U is nilpotent. Then there exists abelian $A \subseteq G$ such that $|G:A| \leq b(G)^4$.

Proof Use induction on $b = b(G)$. We may assume $b > 1$ and so G is not abelian. First, suppose that every nilpotent factor group of G is abelian. Then $G' \subseteq U$, $G'' = 1$ and Lemma 12.25 applies. Since G is not nilpotent, $\mathbf{F}(G) < G$ and $b(\mathbf{F}(G)) \leq b/m < b$, where $m = |G:\mathbf{F}(G)| \leq b$. By the inductive hypothesis, there exists abelian $A \subseteq \mathbf{F}(G)$ with $|\mathbf{F}(G):A| \leq (b/m)^4$ and hence $|G:A| \leq b^4/m^3 \leq b^4$.

Now suppose that G does have a nonabelian nilpotent factor group. Choose $K \lhd G$, maximal such that G/K is nonabelian and nilpotent. By Lemma 12.3, G/K is a p-group and $|G:Z| = f^2$, where $Z/K = \mathbf{Z}(G/K)$ and $Z = \mathbf{Z}(\beta)$ for some $\beta \in \mathrm{Irr}(G)$ with $\beta(1) = f$. By Corollary 12.8, we conclude that $b(Z) \leq b/f^{1/2} < b$. By the inductive hypothesis, there exists abelian $A \subseteq Z$, with $|Z:A| \leq b(Z)^4 \leq b^4/f^2$. Thus $|G:A| = |G:Z||Z:A| \leq b^4$ and the proof is complete. ∎

Next we discuss the "p-structure" of G, where p is a prime which is in some sense large when compared to $b(G)$. The objective here is to show that $|G:\mathbf{O}_p(G)|$ is not divisible by too large a power of p. We prove that if $b(G) < p$, then $p \nmid |G:\mathbf{O}_p(G)|$ and that if $b(G) < p^{3/2}$, then $p^2 \nmid |G:\mathbf{O}_p(G)|$.

(12.27) LEMMA Let G act transitively on a set Ω with $|\Omega| > 1$. Let $H = G_\alpha$ for some $\alpha \in \Omega$. Then the average size of the orbits of H on $\Omega - \{\alpha\}$ is $\leq b(G)$.

Proof Let $\vartheta = (1_H)^G$, the permutation character of G on Ω. Write $\vartheta = 1_G + \sum a_\chi \chi$, where the sum runs over the nonprincipal irreducible constituents of ϑ. The number of orbits t of H on $\Omega - \{\alpha\}$ is given by $t = [1_H, \vartheta_H] - 1 = [\vartheta, \vartheta] - 1 = \sum a_\chi{}^2$.

Now $|\Omega - \{\alpha\}| = \vartheta(1) - 1 = \sum a_\chi \chi(1) \leq b \sum a_\chi$, where $b = b(G)$. Thus the average orbit size s is given by

$$s = (\textstyle\sum a_\chi \chi(1))/(\sum a_\chi{}^2) \leq (b \sum a_\chi)/(\sum a_\chi) = b. \quad ∎$$

(12.28) COROLLARY If G has more than one Sylow p-subgroup, then there exist distinct $P, Q \in \mathrm{Syl}_p(G)$ such that

$$b(G) \geq |\mathbf{N}_G(P) : \mathbf{N}_G(P) \cap \mathbf{N}_G(Q)| \geq |P : P \cap Q|.$$

Proof Apply Lemma 12.27 to the conjugation action of G on $\mathrm{Syl}_p(G)$. Let $P \in \mathrm{Syl}_p(G)$ and $H = \mathbf{N}_G(P)$. Then some orbit of H on $\mathrm{Syl}_p(G) - \{P\}$ must have size $\leq b(G)$. Let Q be in such an orbit. Then $b(G) \geq |H:\mathbf{N}_H(Q)|$ and we have the first inequality. Also

$$|H:\mathbf{N}_H(Q)| \geq |P:\mathbf{N}_P(Q)|$$

and since $\mathbf{N}_p(Q) = P \cap Q$, the result follows. ∎

(12.29) THEOREM Let p be a prime and let $b(G) < p$. Then G has a normal abelian Sylow p-subgroup.

Proof Let $P \in \mathrm{Syl}_p(G)$. If $P \ntriangleleft G$, then by Corollary 12.28, we can find $Q \in \mathrm{Syl}_p(G)$ such that $Q \neq P$ and $|P:P \cap Q| \leq b(G) < p$. This is a contradiction and shows $P \triangleleft G$.

That P is abelian follows since every $f \in \mathrm{c.d.}(P)$ is a power of p satisfying $f < p$. The proof is complete. ∎

(12.30) LEMMA (*Burnside*) Let $P \in \mathrm{Syl}_p(G)$ and let $X, Y \subseteq P$ be normal subsets of P which are conjugate in G. Then X and Y are conjugate in $\mathbf{N}_G(P)$.

Proof Suppose $Y = X^g$ so that $P \subseteq \mathbf{N}(Y)$ and $P^g \subseteq \mathbf{N}(X)^g = \mathbf{N}(X^g) = \mathbf{N}(Y)$. By Sylow's theorem in $\mathbf{N}(Y)$, we have $P^{gu} = P$ for some $u \in \mathbf{N}(Y)$ and $gu \in \mathbf{N}(P)$. However, $X^{gu} = Y^u = Y$ and the proof is complete. ∎

(12.31) LEMMA Let $H \subseteq G$ with $|G:H| = p$, a prime. Then $\mathbf{O}^{p'}(H) \triangleleft G$.

Proof Let K be the kernel of the action of G on the right cosets of H (acting by right multiplication). Then $K \subseteq H$ and $K \lhd G$. Also $|G:K|$ divides $p!$ and so $|H:K|$ divides $(p-1)!$ and is prime to p. Thus $\mathbf{O}^{p'}(H) \subseteq K \lhd H$ and so $\mathbf{O}^{p'}(H) = \mathbf{O}^{p'}(K) \lhd G$ since $K \lhd G$. ∎

(12.32) THEOREM Suppose $b(G) < p^{3/2}$ for some prime p. Then $p^2 \nmid |G:\mathbf{O}_p(G)|$.

Proof We may assume that $\mathbf{O}_p(G) = 1$. Let $P \in \mathrm{Syl}_p(G)$ and assume $|P| \geq p^2$. Let $N = \mathbf{N}_G(P)$. By Corollary 12.28, choose $Q \neq P$, with $Q \in \mathrm{Syl}_p(G)$ such that

$$p^{3/2} > b(G) \geq |N : \mathbf{N}_N(Q)| \geq |P : P \cap Q|$$

and set $D = P \cap Q$. Then $|P:D| = p$. Let $M = \mathbf{N}_G(D) \supseteq P$. Now $P \nsubseteq \mathbf{N}(Q)$ and so

$$\mathbf{N}_N(Q) < P\mathbf{N}_N(Q) \subseteq N \cap M.$$

It follows that

$$p^{3/2} > |N : \mathbf{N}_N(Q)| \geq p|N : N \cap M|$$

and $|N : N \cap M| < p^{1/2}$. Note that $M < G$.

We consider the action (by right multiplication) of M on $\Omega = \{Mx \mid x \in G\}$. Suppose r of the orbits of M on $\Omega - \{M\}$ have size $< p^2$ and s have size $\geq p^2$. Let Ω_0 be the union of the r smaller orbits.

We claim

(*) If $Mx \in \Omega_0$, then $DD^x = D^xD$ and $x \in MNM$.

Assuming (*), let us complete the proof. We have $|\Omega - \{M\}| \geq p^2 s$ so that Lemma 12.27 yields

$$p^{3/2} > b(G) \geq |\Omega - \{M\}|/(r+s) \geq p^2 s/(r+s).$$

If $s = 0$, then $Mx \in \Omega_0$ for every $x \in G - M$ and hence $DD^x = D^xD$ for all $x \in G$ by (*). Thus $\langle D^x \mid x \in G \rangle$ is a p-group that is normal in G, a contradiction.

Thus $s > 0$ and $p^{1/2} < (r+s)/s \leq 1 + r$. Now if $Mx \in \Omega_0$, then $x \in MNM$ because of (*) and hence the orbit of Mx under M contains an element of the form Mn for $n \in N$. The number of distinct Mn for $n \in N$ is $|N : N \cap M| < p^{1/2}$ and thus at most $p^{1/2}$ M-orbits of Ω contain an element of the form Mn. These include the trivial orbit $\{M\}$ and so we have $1 + r \leq p^{1/2}$. This contradicts a previous inequality.

We now work to establish (*). Let $Mx \in \Omega_0$ so that $x \notin M$ and $D \neq D^x$. If either of D or D^x normalizes the other, then $DD^x = D^xD$ is a p-group properly containing D. Since $|P:D| = p$ it follows that $DD^x \in \mathrm{Syl}_p(G)$ and $|DD^x : D| = p$ so that $D \lhd DD^x$ and $DD^x \subseteq M$. Since also $P \subseteq M$ we have $P = (DD^x)^m$ for some $m \in M$ by Sylow's theorem. Now D^m, $D^{xm} \lhd P$ and hence Lemma 12.30 yields $D^m = D^{xmn}$ for some $n \in N$. Thus $xmnm^{-1} \in \mathbf{N}(D) = M$ and $x \in MNM$ as desired.

The remaining case is where $D \not\subseteq M^x$ and $D^x \not\subseteq M$. Let $W = M \cap M^x$ so that WD and WD^x are groups. Since $Mx \in \Omega_0$, we have $|M : M \cap M^x| < p^2$ and hence $|DW : W| \leq |M : W| < p^2$ and $|DW : W| = p$. Similarly $|D^xW : W| = p$. By Lemma 12.31, $\mathbf{O}^{p'}(W)$ is normal in both WD and WD^x. Write $K = \mathbf{O}^{p'}(W)$ and $L = \mathbf{N}_G(K)$.

Since $|M : W| < p^2$ and $|DW : W| = p$ it follows that $|M : DW| < p$ and DW contains a full Sylow p-subgroup of M and hence of G. It follows that DK/K and D^xK/K are Sylow subgroups of L/K. Also, $D^xK \neq DK$ since $DK \subseteq M$ and $D^x \not\subseteq M$. Thus $|\mathrm{Syl}_p(L/K)| > 1$ and $b(L/K) \geq p$ by Theorem 12.29.

By Corollary 12.8, $b(L) \geq ab(K)$ for some a with $a^2 \geq b(L/K) \geq p$. Thus $p^{3/2} > b(L) \geq p^{1/2}b(K)$ and $b(K) < p$. Thus K has a normal Sylow p-subgroup U by Theorem 12.29. Now $UD \in \mathrm{Syl}_p(M)$ and $UD \lhd WD$ since U is characteristic in $K \lhd WD$. Thus $|M : \mathbf{N}_M(UD)| \leq |M : DW| < p$ and it follows that $UD \lhd M$. Thus $|\mathrm{Syl}_p(M)| = 1$ which is a contradiction since $P, Q \in \mathrm{Syl}_p(M)$. This completes the proof. ∎

We close this chapter by considering the opposite of the stituation with which we began. Suppose no $f \in$ c.d.(G) is divisible by the prime p. A sufficient condition for this to happen is that G has a normal abelian Sylow p-subgroup. (This follows by Ito's Theorem 6.15.) It is conjectured that this condition is also necessary. The next result shows that to prove the conjecture, it would suffice to check simple groups.

(12.33) THEOREM Suppose G does not have a normal abelian Sylow p-subgroup and that no element of c.d.(G) is divisible by p. Then G has a non-abelian simple composition factor S of order divisible by p such that no element of c.d.(S) is divisible by p.

Proof If $N \lhd G$ and $\psi \in \mathrm{Irr}(N)$, choose $\chi \in \mathrm{Irr}(G)$ with $[\chi_N, \psi] \neq 0$. Then $\psi(1) | \chi(1)$ and so $p \nmid \psi(1)$. In particular, if $N \in \mathrm{Syl}_p(G)$, then N is necessarily abelian and thus G does not have a normal Sylow p-subgroup. Also, if G is simple, the result is trivial and we assume G is not simple.

Let N be a maximal normal subgroup of G. Working by induction on $|G|$, we may assume that N has a normal Sylow p-subgroup P. Then G/P does not have a normal Sylow p-subgroup and if $P > 1$ we complete the proof by applying the inductive hypothesis to G/P. Suppose then, that $P = 1$.

We must have $p \,||G : N|$ and since G/N is simple we can take $S = G/N$ unless G/N is abelian, that is, $|G/N| = p$. We suppose this is the case. Let $Q \in \mathrm{Syl}_p(G)$ so that $|Q| = p$.

If $\psi \in \text{Irr}(N)$, then ψ^G cannot be irreducible since $p \mid \psi^G(1)$. Thus $I_G(\psi) > N$ and hence $I_G(\psi) = G$ and ψ is invariant in G. Thus Q acts trivially on Irr(G). It follows by Brauer's Theorem 6.32 that Q acts trivially on the set of conjugacy classes of N.

If $\mathcal{K}$ is a class of N then Q permutes $\mathcal{K}$. Since $p \nmid |N|$, we have $p \nmid |\mathcal{K}|$ and thus Q fixes an element of $\mathcal{K}$. It follows that $C = \mathbf{C}_N(Q)$ meets every conjugacy class of N. Thus $N = \bigcup_{x \in N} C^x$ and this yields

$$|N| - 1 \le |N : \mathbf{N}_N(C)|(|C| - 1) \le |N : C|(|C| - 1) = |N| - |N : C|.$$

Thus $|N : C| \le 1$ and $C = N$. We conclude that $Q \lhd G$, a contradiction. ∎

(12.34) COROLLARY (*Ito*) Let G be solvable. Then G has a normal abelian Sylow p-subgroup iff every element of c.d.(G) is relatively prime to p.

Problems

(12.1) The normal subgroups $N_i \lhd G$ are a *Sylow tower* if

$$1 = N_0 \subseteq N_1 \subseteq \cdots \subseteq N_k = G$$

and N_{i+1}/N_i is a Sylow subgroup of G/N_i for each i, $0 \le i < k$. Suppose for every $m, n \in$ c.d.(G), either $m \mid n$ or $n \mid m$. Show that G has a Sylow tower.

(12.2) Suppose that every $f \in$ c.d.(G) is a power of the integer m. Assume that m is not a prime power. Show that there exists abelian $A \subseteq G$ with $|G : A| = b(G)$ and that such an A is necessarily normal in G.

Hint This generalizes part of Theorem 12.5. Mimic the proof of that theorem.

(12.3) Suppose that G is solvable and that for every $m, n \in$ c.d.(G) with $m \ne n$, we have $(m, n) = 1$. Show that $|\text{c.d.}(G)| \le 3$.

(12.4) Let G be solvable with $b(G) = b > 1$ and suppose that G has no factor group which is a nonabelian p-group. Show that there exists $L \subseteq G$ and an integer r with $2 \le r \le b$ such that $b(L) \le b/r$ and $|G : L| \le br$.

(12.5) Let G be solvable with $b(G) = b$. Show that G has an abelian subgroup of index $\le kb^{\log_2(b)}$ for a suitable constant k independent of b.

(12.6) Let c.d.(G) $= \{1, p^e\}$, where p is a prime and $e > 1$. If a Sylow p-subgroup of G is nonabelian, show that G is nilpotent.

Hint Use the fact that abelian Frobenius complements are cyclic to show that if G is not nilpotent, then there exists abelian $H \lhd G$ with $|G : H| = p^e$ and G/H cyclic. Now let G/K be as in Lemma 12.3(a) and consider HK.

(12.7) Let G be solvable with $b = b(G)$. Let p be a prime.

(a) Show that there exists proper $K \lhd G$ and integer $e \geq 0$ such that $p^{e/4}b(K) \leq b$ and $p^{e+1} \nmid |G:K|$.

(b) Show that if $p^f \| |G:\mathbf{O}_p(G)|$, then $p^f \leq b^4$.

(12.8) Let $G'' = 1$ and assume that all Sylow subgroups of G are abelian. Show that $b(G) = |G:A|$ for some abelian $A \lhd G$.

(12.9) Let G be solvable and suppose that all Sylow subgroups of G are abelian. Show that $|G:\mathbf{F}(G)| \leq b(G)^2$.

Hints $\mathbf{F}(G) = \mathbf{C}_G(\mathbf{F}(G))$. Let $\lambda \in \mathrm{Irr}(\mathbf{F}(G))$ be such that $|G:I_G(\lambda)|$ is maximal. Let $T = I_G(\lambda)$ and $s = |G:T|$. Show that $b(T/F(G)) \leq b/s$ and let $F/\mathbf{F}(G) = \mathbf{F}(T/\mathbf{F}(G))$. Show $b(F) \leq s$. Use Corollary 11.32.

(12.10) Suppose that every $f \in \mathrm{c.d.}(G)$ divides p^e where p is a prime. Show that there exists abelian $A \subseteq G$ with index dividing p^{4e}.

(12.11) Suppose $\beta \in \mathrm{Irr}(G)$ with $\beta(1) = f$ and that $Z = \mathbf{Z}(\beta)$ satisfies $|G:Z| = f^2$. Let $Z \subseteq H \subseteq G$.

(a) If $|G:H| > f$, show that $b(H) < b(G)$.

(b) If $|G:H| = f$ and $\chi \in \mathrm{Irr}(H)$ with $\chi(1) = b(G)$, show that $\mathbf{V}(\chi) \subseteq Z$.

(12.12) (a) Suppose $A \subseteq G$ is abelian and $|G:\mathbf{C}_G(A)| > b(G)$. Show that there exists $A_0 \subseteq A$ such that $|A:A_0| \leq b(G)^2$ and $\mathbf{C}_G(A_0) > \mathbf{C}_G(A)$.

(b) If G is a p-group, improve part (a) to read $|A:A_0| \leq b(G)$.

Hint Use Lemma 12.10.

(12.13) Show that there exists a function f defined on positive integers such that for any group G if $b(G) = b$, then there exists $H \subseteq G$ with $|G:H| \leq b$ and $|H:\mathbf{Z}(H)| \leq f(b)$.

Hint Use repeated applications of Problem 12.12.

(12.14) Let G be solvable and let p be a prime. Suppose $p^2 \nmid f$ for all $f \in \mathrm{c.d.}(G)$. Show that either $\mathbf{O}_p(G) > 1$ or a Sylow p-subgroup of G is abelian.

Hint In a minimal counterexample, let M be a minimal normal subgroup. Show that $|\mathbf{O}_p(G/M)| = p$. Now show that $\mathbf{O}_p(G/M)$ is a direct factor of a Sylow subgroup of G/M. Produce the other factor by considering $I_G(\lambda)$ for suitable $\lambda \in \mathrm{Irr}(M)$.

(12.15) Let $N \lhd G$ with G/N a p-group and $p \neq 2$. Let $\vartheta \in \mathrm{Irr}(N)$ be invariant in G. Suppose that every irreducible constituent of ϑ^G has degree $\leq p\vartheta(1)$. Show that $b(G/N) \leq p$.

Hints Extend ϑ to $\hat{\vartheta} \in \mathrm{Irr}(H)$ with $N \subseteq H$ and $|G:H| = p$. For $\varphi \in \mathrm{Irr}(H/N)$ with $\varphi(1) = p$, consider $(\varphi\hat{\vartheta})^G$. Conclude that there exists linear

$\mu \in \mathrm{Irr}(H/N)$ such that $\varphi^x = \varphi\mu$ for $x \in G$, with μ independent of the choice of φ. Use this and the hypothesis $p \neq 2$ to show that φ is invariant in G.

Note If $p = 2$, then Problem 12.15 is actually false.

(12.16) Let G be a p-group and suppose $b(G) = p^2$. If $p \neq 2$, show that $\bigcap\{\ker \chi \mid \chi(1) = p^2\} = 1$.

Hint Use Problem 12.15.

Note Passman has conjectured that if G is any p-group with $b(G) = p^e$ and $e < p$, then $\bigcap\{\ker \chi \mid \chi(1) = p^e\} = 1$. He has proved this when G has class 2.

13 Character correspondence

We have already seen several examples of the following situation (for instance, Theorems 6.11 and 6.16). We are given H, a subgroup or factor group of G. Subsets $\mathscr{S} \subseteq \text{Irr}(H)$ and $\mathscr{T} \subseteq \text{Irr}(G)$ are specified and it is proved that there exists a "natural" one-to-one correspondence between $\mathscr{S}$ and $\mathscr{T}$. Here, the word "natural" is intended to mean that the correspondence is uniquely described by some general rule and thus more is being said than merely that $|\mathscr{S}| = |\mathscr{T}|$. We shall not attempt to give a precise definition of naturalness. Most of this chapter is devoted to the study of a particular character correspondence which was discovered by G. Glauberman.

We introduce some notation. Let S and G be groups such that S acts on G. [That is, we are given a homomorphism $S \to \text{Aut}(G)$.] In this situation, we can construct the semidirect product Γ of G by S so that $G \lhd \Gamma$, $S \subseteq \Gamma$, $GS = \Gamma$, $G \cap S = 1$ and the given action of S on G is the action by conjugation in Γ. (In fact these properties characterize the semidirect product.)

If χ is a character of G and $s \in S$, then as usual we define the character χ^s of G by $\chi^s(g^s) = \chi(g)$. Then S permutes $\text{Irr}(G)$. We write

$$\text{Irr}_S(G) = \{\chi \in \text{Irr}(G) \mid \chi^s = \chi \quad \text{for all} \quad s \in S\}.$$

(13.1) THEOREM (*Glauberman*) For every pair of groups (G, S) such that S is solvable and acts on G and $(|G|, |S|) = 1$, there exists a uniquely defined one-to-one map $\pi(G, S)$: $\text{Irr}_S(G) \to \text{Irr}(\mathbf{C}_G(S))$. These maps satisfy the following properties:

(a) If $T \lhd S$ and $B = \mathbf{C}_G(T)$, then $\pi(G, T)$ maps $\text{Irr}_S(G)$ onto $\text{Irr}_S(B)$.
(b) In the situation of (a), $\pi(G, S) = \pi(G, T)\pi(B, S/T)$.

(c) Suppose S is a p-group and $C = \mathbf{C}_G(S)$. Let $\chi \in \mathrm{Irr}_S(G)$ and $\psi = (\chi)\pi(G, S)$. Then ψ is the unique irreducible constituent of χ_C such that $p \nmid [\chi_C, \psi]$.

In the situation of (a) in the theorem, note that B is S-invariant since $T \lhd S$. Thus S acts on B and in fact S/T acts on B. Therefore $\pi(B, S/T)$ is defined on $\mathrm{Irr}_{S/T}(B) = \mathrm{Irr}_S(B)$ which is the image of $\mathrm{Irr}_S(G)$ under $\pi(G, T)$. Also, $\pi(B, S/T)$ maps to the irreducible characters of $\mathbf{C}_B(S/T) = \mathbf{C}_B(S) = \mathbf{C}_G(S)$. Thus the equation in (b) makes sense.

Also note that if $T = S$ in (a), then $B = \mathbf{C}_G(S)$ and $\mathrm{Irr}_S(B) = \mathrm{Irr}(B)$. Thus this special case of (a) asserts that $\pi(G, S)$ always maps $\mathrm{Irr}_S(G)$ onto $\mathrm{Irr}(\mathbf{C}_G(S))$.

There is enough information in the statement of Theorem 13.1 to determine π uniquely. Thus suppose that $\pi_0(G, S)$ is defined whenever (G, S) satisfies the hypotheses of 13.1 and that π_0 satisfies (a), (b), and (c). We claim that $\pi_0(G, S) = \pi(G, S)$. By 13.1(c), this is certainly the case if S is a p-group so we may work by induction on $|S|$ and assume that S has composite order. Let $T \lhd S$ have prime index and let $B = \mathbf{C}_G(T)$. Then $\pi(B, S/T) = \pi_0(B, S/T)$ and $\pi(G, T) = \pi_0(G, T)$ and 13.1(b) yields $\pi(G, S) = \pi_0(G, S)$.

The preceding argument suggests how to construct the map $\pi(G, S)$; namely, prove that if S is a cyclic p-group and $\chi \in \mathrm{Irr}_S(G)$, then χ_C does have a unique irreducible constituent β such that $[\chi_C, \beta] \not\equiv 0 \bmod p$, where $C = \mathbf{C}_G(S)$. Define $\pi(G, S)$ in this case by $(\chi)\pi(G, S) = \beta$. For general solvable S, define $\pi(G, S)$ by working along a composition series for S. There are numerous technical difficulties with this approach, not the least of which is to show that the map constructed is independent of the composition series. The key to overcoming these difficulties is to find a uniform definition for $\pi(G, S)$ for all cyclic S. Following Glauberman, this is what we shall do.

We establish some notation which will be used repeatedly.

(13.2) HYPOTHESIS Let S act on G and suppose $(|G|, |S|) = 1$. Let $C = \mathbf{C}_G(S)$ and let Γ be the semidirect product $\Gamma = GS$.

(13.3) LEMMA Assume the situation in 13.2 and let $\chi \in \mathrm{Irr}_S(G)$. Then there exists a unique extension $\hat{\chi}$ of χ to Γ such that $(o(\hat{\chi}), |S|) = 1$. Also, $\hat{\chi}$ is the unique extension such that $S \subseteq \ker(\det \hat{\chi})$.

Proof The first statement is just Corollary 8.16. Since $o(\hat{\chi}_S)$ divides both $|S|$ and $o(\hat{\chi})$, we have $o(\hat{\chi}_S) = 1$ and $S \subseteq \ker(\det \hat{\chi})$. If ψ is an extension of χ with $S \subseteq \ker(\det \psi)$, then $o(\psi) = o(\det \psi)$ and divides $|\Gamma : \ker(\det \psi)|$ which is prime to $|S|$. Thus $\psi = \hat{\chi}$ and the proof is complete. ∎

We call the character $\hat{\chi}$ of Lemma 13.3 the *canonical extension* of χ.

For positive integers n, we write $\mathbb{Q}_n = \mathbb{Q}(\varepsilon)$, where ε is a primitive nth root of unity. If $(m, n) = 1$, it is well known from Galois theory that $\mathbb{Q}_n \cap \mathbb{Q}_m$

$= \mathbb{Q}$. Thus if $\alpha \in \mathbb{Q}_n$ is invariant under the Galois group $\mathscr{G}(\mathbb{Q}_{mn}/\mathbb{Q}_m)$, then $\alpha \in \mathbb{Q}$.

If $M \lhd H$ with $|M| = m$ and $|H : M| = n$ such that $(m, n) = 1$, we shall use the notation $\mathscr{G}(H/M)$ to denote $\mathscr{G}(\mathbb{Q}_{nm}/\mathbb{Q}_m)$. Note that $\mathscr{G}(H/M)$ permutes $\mathrm{Irr}(H)$ (by Problem 2.2 or Lemma 9.16). Suppose $\vartheta \in \mathrm{Irr}(M)$ is extendible to H. Since $\mathscr{G}(H/M)$ fixes ϑ, it permutes the set of extensions of ϑ to H.

In the situation of 13.2, if $\chi \in \mathrm{Irr}_S(G)$, and $\hat{\chi}$ is the canonical extension of χ to Γ, then for $\tau \in \mathscr{G}(\Gamma/G)$, $(\hat{\chi})^\tau$ is an extension of χ and $o(\hat{\chi}^\tau) = o(\hat{\chi})$. Thus $\hat{\chi}^\tau = \hat{\chi}$ and hence $\hat{\chi}$ has values in $\mathbb{Q}_{|G|}$ and $\hat{\chi}_S$ has values in $\mathbb{Q}_{|G|} \cap \mathbb{Q}_{|S|} = \mathbb{Q}$. We have thus proved the following corollary.

(13.4) COROLLARY In the notation of Lemma 13.3, $\hat{\chi}_S$ is rational valued.

The following strengthens this.

(13.5) LEMMA Assume Hypothesis 13.2. Let $\chi \in \mathrm{Irr}_S(G)$ and let $\hat{\chi}$ be the canonical extension of χ to Γ. Note that $CS = C \times S$. For each irreducible constituent β of χ_C there exists a (possibly reducible) character ψ_β of S such that $\hat{\chi}_{CS} = \sum_\beta (\beta \times \psi_\beta)$. This equation uniquely determines the ψ_β. Also, the ψ_β are rational valued.

Proof We have

$$\mathrm{Irr}(CS) = \{\beta \times \varphi \mid \beta \in \mathrm{Irr}(C), \varphi \in \mathrm{Irr}(S)\}.$$

Write $\hat{\chi}_{CS} = \sum_{\beta, \varphi} a_{\beta\varphi}(\beta \times \varphi)$. Set $\psi_\beta = \sum_\varphi a_{\beta\varphi}\varphi$ and observe that $\psi_\beta = 0$ unless $[\chi_C, \beta] \neq 0$. Thus $\hat{\chi}_{CS} = \sum_\beta (\beta \times \psi_\beta)$, where the sum runs over those $\beta \in \mathrm{Irr}(C)$, which are constituents of χ_C. Also, this equation uniquely determines the ψ_β's.

If $\tau \in \mathscr{G}(\Gamma/G)$ then $(\hat{\chi})^\tau = \hat{\chi}$ and $\beta^\tau = \beta$. Thus

$$\hat{\chi}_{CS} = ((\hat{\chi})^\tau)_{CS} = \sum_\beta (\beta \times (\psi_\beta)^\tau).$$

Thus ψ_β is invariant under $\mathscr{G}(\Gamma/G)$ and hence has values in $\mathbb{Q}_{|G|}$. Since its values also lie in $\mathbb{Q}_{|S|}$, the result follows. ∎

(13.6) THEOREM Assume Hypothesis 13.2 and that S is cyclic. Then for each $\chi \in \mathrm{Irr}_S(G)$, there exist unique $\beta \in \mathrm{Irr}(C)$ and $\varepsilon = \pm 1$ such that $\hat{\chi}(cs) = \varepsilon\beta(c)$ for all $c \in C$ and all generators s of S, where $\hat{\chi}$ is the canonical extension of χ to Γ. Also, β is a constituent of χ_C and the map $\chi \mapsto \beta$ is one-to-one.

Proof Write $\hat{\chi}_{CS} = \sum \beta \times \psi_\beta$ as in Lemma 13.5 and fix a generator s of S. Write $\vartheta(c) = \hat{\chi}(cs)$ for $c \in C$. Then $\vartheta = \sum_\beta \psi_\beta(s)\beta$. In particular, ϑ is a class function on C. We claim that $[\vartheta, \vartheta] = 1$.

Let T be a set of representatives for the right cosets of C in G. If $t_1, t_2 \in T$ and $x \in (Cs)^{t_1} \cap (Cs)^{t_2}$, then there exist $c_1, c_2 \in C$ with

$$c_1^{t_1}s^{t_1} = (c_1 s)^{t_1} = x = (c_2 s)^{t_2} = c_2^{t_2}s^{t_2}.$$

We thus have two factorizations of x into products of commuting elements of orders dividing $|G|$ and $|S|$. By Lemma 8.18, $s^{t_1} = s^{t_2}$ and thus $t_1 t_2{}^{-1} \in \mathbf{C}(s) = C$ since $S = \langle s \rangle$. Thus $t_1 = t_2$. Hence the sets $(Cs)^t$ are disjoint for distinct $t \in T$ and $|\bigcup_t (Cs)^t| = |T||C| = |G| = |Gs|$. Since $s^t s^{-1} \in G$, we have $(Cs)^t \subseteq Gs$ and hence $Gs = \bigcup_t (Cs)^t$ is a disjoint union.

We have

$$|C|[\vartheta, \vartheta] = \sum_{x \in Cs} |\hat{\chi}(x)|^2 = \sum_{x \in Cs} |\hat{\chi}(x^t)|^2$$

for all $t \in T$. It follows that

$$|T||C|[\vartheta, \vartheta] = \sum_{x \in Gs} |\hat{\chi}(x)|^2.$$

By Lemma 8.14(c), the latter sum equals $|G| = |T||C|$ and thus $[\vartheta, \vartheta] = 1$ as claimed.

We now have

$$1 = [\vartheta, \vartheta] = \sum_{\beta} |\psi_\beta(s)|^2.$$

However, $\psi_\beta(s)$ is a rational algebraic integer by Lemma 13.5 and so lies in $\mathbb{Z}$. Thus $\psi_\beta(s)$ is nonzero for some unique β and for that β, $\psi_\beta(s) = \varepsilon = \pm 1$. This yields $\hat{\chi}(cs) = \varepsilon\beta(c)$ for $c \in C$. This equation clearly determines ε and β uniquely.

We now show that β is independent of the choice of the generator s of S. If $S = \langle s_0 \rangle$, then $s_0 = s^m$ for some m with $(m, |S|) = 1$. Thus there exists an automorphism σ of the field $\mathbb{Q}_{|S|}$ such that $\lambda^\sigma(s) = \lambda(s)^m = \lambda(s_0)$ for all $\lambda \in \mathrm{Irr}(S)$. Therefore $0 \neq \psi_\beta(s)^\sigma = \psi_\beta(s_0)$ and hence replacing s by s_0 yields the same β.

Finally, suppose $\chi_1, \chi_2 \in \mathrm{Irr}_S(G)$ determine the same character β so that $\hat{\chi}_i(cs) = \varepsilon_i \beta(c)$ for $c \in C$ and $i = 1, 2$. Since $Gs = \bigcup (Cs)^t$, it follows that $\hat{\chi}_1(gs) = \varepsilon_1 \varepsilon_2 \hat{\chi}_2(gs)$ for all $g \in G$ and thus by Lemma 8.14(b), it follows that

$$\chi_1 = (\hat{\chi}_1)_G = (\hat{\chi}_2)_G = \chi_2$$

and the map $\chi \mapsto \beta$ is one-to-one. The proof is complete. ▮

(13.7) DEFINITION Assume Hypothesis 13.2 and that S is cyclic. Construct maps $\gamma(G, S)$: $\mathrm{Irr}_S(G) \to \mathrm{Irr}(C)$ and $\varepsilon(G, S)$: $\mathrm{Irr}_S(G) \to \{-1, 1\}$ by $(\chi)\gamma(G, S) = \beta$ and $(\chi)\varepsilon(G, S) = \varepsilon = \pm 1$ where $\hat{\chi}(cs) = \varepsilon\beta(c)$ for $c \in C$, $\langle s \rangle = S$, and $\hat{\chi}$ is the canonical extension of χ to Γ.

It will turn out that the map $\pi(G, S)$ of Theorem 13.1 equals $\gamma(G, S)$ when $\gamma(G, S)$ is defined, that is, for cyclic S.

The map $\gamma(G, S)$: $\mathrm{Irr}_S(G) \to \mathrm{Irr}(C)$ of Definition 13.7 is one-to-one by 13.6. It is also onto. One way to prove this is to show that $|\mathrm{Irr}_S(G)| = |\mathrm{Irr}(C)|$.

Since S is cyclic, it follows from Brauer's Theorem 6.32 that $|\mathrm{Irr}_S(G)|$ is equal to the number of S-invariant conjugacy classes of G. It is true that each S-invariant class of G intersects C nontrivially and, in fact, the interesction is a single conjugacy class of C. It follows that the number of S-invariant classes of G is equal to the total number of conjugacy classes of C, and hence equals $|\mathrm{Irr}(C)|$. Thus $\gamma(G. S)$ maps onto.

The assertion about S-invariant classes of G in the preceding paragraph follows from the Schur–Zassenhaus theorem. We digress from the discussion of characters in order to give a proof.

(13.8) LEMMA (*Glauberman*) Let S act on G with $(|S|, |G|) = 1$. Assume that one of S or G is solvable. Let S and G both act on a set Ω such that

(a) $(\alpha \cdot g) \cdot s = (\alpha \cdot s) \cdot g^s$ for all $\alpha \in \Omega$, $g \in G$, and $s \in S$.
(b) G is transitive on Ω.

Then S fixes a point of Ω.

Proof Let $\Gamma = GS$, the semidirect product. For $gs \in \Gamma$ and $\alpha \in \Omega$, define $\alpha \cdot (gs) = (\alpha \cdot g) \cdot s$. Condition (a) above guarantees that this is an action. Pick $\alpha \in \Omega$ and let $H = \Gamma_\alpha$. Since $|\Omega| = |G : G \cap H| = |\Gamma : H|$ by (b), it follows that $|H : G \cap H| = |S|$. By the existence part of the Schur–Zassenhaus theorem, let T be a complement for $G \cap H$ in H.

Then $|T| = |S|$ and T is a complement for G in Γ. Now the conjugacy part of the Schur–Zassenhaus theorem yields $S = T^x$ for some $x \in \Gamma$. Thus $S \subseteq H^x$ and S fixes $\alpha \cdot x \in \Omega$. The proof is complete. ∎

(13.9) COROLLARY In the situation of Lemma 13.8, the set of S-fixed points of Ω is an orbit under the action of $\mathbf{C}_G(S)$.

Proof If $\alpha \in \Omega$ is fixed by S and $c \in \mathbf{C}_G(S)$, then $(\alpha \cdot c) \cdot s = (\alpha \cdot s) \cdot c^s = \alpha \cdot c$ and $\alpha \cdot c$ is S-fixed. Now suppose $\alpha, \beta \in \Omega$ are S-fixed. Let $X = \{g \in G \mid \alpha \cdot g = \beta\}$. Then X is a left coset of G_β and is S-invariant. Let G_β act on X by right multiplication. Note that G_β is S-invariant and is transitive on X. For $x \in X$, $g \in G_\beta$ and $s \in S$, we have $(x \cdot g) \cdot s = (xg)^s = x^s g^s = (x \cdot s) \cdot g^s$ and Lemma 13.8 applies to the actions of S on G_β and S and G_β on X. Thus S fixes a point $x \in X$. Then $x \in \mathbf{C}_G(S)$ and $\alpha \cdot x = \beta$. The proof is complete. ∎

(13.10) COROLLARY Assume Hypothesis 13.2 and that at least one of G or S is solvable. Then $\mathscr{K} \mapsto \mathscr{K} \cap C$ defines a bijection from the set of S-invariant conjugacy classes of G onto the set of conjugacy classes of C.

Proof Let $\mathscr{K}$ be an S-invariant class of G. The conjugation action of G on $\mathscr{K}$ is transitive. For $k \in \mathscr{K}$, $g \in G$, and $s \in S$, we have

$$(k \cdot g) \cdot s = (g^{-1}kg)^s = (g^{-1})^s k^s g^s = (k \cdot s) \cdot g^s.$$

By Lemma 13.8 we conclude that $\mathscr{K} \cap C \neq \varnothing$ and by 13.9, $\mathscr{K} \cap C$ is a class of C.

Since the classes of G are disjoint, the map $\mathscr{K} \mapsto \mathscr{K} \cap C$ is one-to-one. If $c \in C$, then the G-class of c is S-invariant. It follows that the map $\mathscr{K} \mapsto \mathscr{K} \cap C$ is onto. ∎

(13.11) COROLLARY The map $\gamma(G, S)$ of Definition 13.7 maps $\mathrm{Irr}_S(G)$ onto $\mathrm{Irr}(C)$.

Assume Hypothesis 13.2 and that S is solvable. Let $\mathscr{S}$ be a composition series for S with

$$\mathscr{S} : 1 = S_0 \lhd S_1 \lhd \cdots \lhd S_k = S.$$

Let $C_i = \mathbf{C}_G(S_i)$ so that

$$G = C_0 \supseteq C_1 \supseteq \cdots \supseteq C_k = C.$$

Also, $S_{i+1} \subseteq \mathbf{N}_\Gamma(C_i)$ for $0 \leq i < k$ and we view S_{i+1}/S_i as acting on C_i. Since S_{i+1}/S_i is cyclic,

$$\gamma_i = \gamma(C_i, S_{i+1}/S_i)$$

is defined for each i, $0 \leq i < k$. We have

$$\gamma_i\colon \mathrm{Irr}_{S_{i+1}}(C_i) \to \mathrm{Irr}(C_{i+1}).$$

Each γ_i is one-to-one and maps onto $\mathrm{Irr}(C_{i+1})$ and so we can define $\gamma_i^{-1}\colon \mathrm{Irr}(C_{i+1}) \to \mathrm{Irr}(C_i)$.

(13.12) DEFINITION Assume Hypothesis 13.2 with S solvable. Let $\mathscr{S}$ be a composition series for S and use the above notation. Put $\mathscr{X}(G, \mathscr{S}) = (\mathrm{Irr}(C))\gamma_{k-1}^{-1}\gamma_{k-2}^{-1} \cdots \gamma_1^{-1}\gamma_0^{-1} \subseteq \mathrm{Irr}(G)$ and let $\pi(G, \mathscr{S})\colon \mathscr{X}(G, \mathscr{S}) \to \mathrm{Irr}(C)$ be given by $\pi(G, \mathscr{S}) = \gamma_0\gamma_1 \cdots \gamma_{k-1}$.

Thus $\mathscr{X}(G, \mathscr{S})$ is the largest set on which the composite function $\gamma_0\gamma_1 \cdots \gamma_{k-1}$ is defined. Since each γ_i is one-to-one, $\pi(G, \mathscr{S})$ is one-to-one and by construction of $\mathscr{X}(G, \mathscr{S})$, we see that $\pi(G, \mathscr{S})$ maps $\mathscr{X}(G, \mathscr{S})$ onto $\mathrm{Irr}(C)$. Our objective now is to show that $\mathscr{X}(G, \mathscr{S}) = \mathrm{Irr}_S(G)$, that $\pi(G, \mathscr{S})$ is independent of the choice of $\mathscr{S}$ and that for cyclic S, $\pi(G, \mathscr{S}) = \gamma(G, \mathscr{S})$. We obtain these results by considering the case that S is a p-group.

(13.13) LEMMA Let $\mathbb{Z} \subseteq R \subseteq \mathbb{C}$, where R is a ring containing the values of all $\xi \in \mathrm{Irr}(G)$. Let ϑ be a generalized character of G with values in an ideal I of R. Assume $I \cap \mathbb{Z} \subseteq p\mathbb{Z}$ for some prime p not dividing $|G|$. Then p divides $[\vartheta, \xi]$ for all $\xi \in \mathrm{Irr}(G)$.

Proof We have $|G|[\vartheta, \xi] = \sum \vartheta(g)\overline{\xi(g)} \in I \cap \mathbb{Z}$. Thus $p \,||\, G|[\vartheta, \xi]$ and the result follows since $p \nmid |G|$. ∎

(13.14) THEOREM Assume Hypothesis 13.2 and that S is a p-group. Let $\chi \in \mathrm{Irr}_S(G)$. Then there exists a unique $\beta \in \mathrm{Irr}(C)$, with $[\chi_C, \beta] \not\equiv 0 \bmod p$. Furthermore

(a) $[\chi_C, \beta] \equiv \pm 1 \bmod p$;

(b) if S is cyclic, then $\beta = (\chi)\gamma(G, S)$ and $(\chi)\varepsilon(G, S) \equiv [\chi_C, \beta] \bmod p$.

(c) If $\mathscr{S}$ is a composition series for S, then $\mathscr{X}(G, \mathscr{S}) = \mathrm{Irr}_S(G)$ and $(\chi)\pi(G, \mathscr{S}) = \beta$. In particular, $\pi(G, \mathscr{S})$ is independent of the choice of $\mathscr{S}$.

Proof First assume that S is cyclic and let $\beta = (\chi)\gamma(G, S)$ and $\varepsilon = (\chi)\varepsilon(G, S)$ so that $\hat{\chi}(cs) = \varepsilon\beta(c)$ for $c \in S$ and $S = \langle s \rangle$, where $\hat{\chi}$ is the canonical extension of χ to Γ.

Let R be the ring of algebraic integers in $\mathbb{Q}_{|\Gamma|}$ and let I be a maximal ideal of R with $p \in I$. In the notation of Theorem 8.20, we have $(cs)_{p'} = c$ and thus that theorem yields $\chi(c) = \hat{\chi}(c) \equiv \hat{\chi}(cs) \bmod I$.

We therefore have $\chi(c) \equiv \varepsilon\beta(c) \bmod I$ for all $c \in C$ and thus $\chi_C - \varepsilon\beta$ is a generalized character of C with values in I. Since $1 \notin I$ and $p \in I$, we have $I \cap \mathbb{Z} = p\mathbb{Z}$ and since $p \nmid |C|$, Lemma 13.13 yields $[\chi_c - \varepsilon\beta, \xi] \equiv 0 \bmod p$ for all $\xi \in \mathrm{Irr}(C)$. It follows that $[\chi_C, \xi] \equiv 0 \bmod p$ for $\xi \neq \beta$ and $[\chi_C, \beta] \equiv \varepsilon \bmod p$. When S is cyclic, this proves everything but (c).

If $|S| = p$, then $\mathscr{S} : 1 \lhd S$ and $\mathscr{X}(G, \mathscr{S}) = \mathrm{Irr}_S(G)$ and $\pi(G, S) = \gamma(G, S)$. In particular, part (c) of the theorem holds when $|S| = p$.

Now assume $|S| > p$ and drop the assumption that S is cyclic. Work by induction on $|S|$. Let

$$\mathscr{S} : 1 = S_0 \lhd \cdots \lhd S_k = S$$

and write $T = S_{k-1}$ and

$$\mathscr{T} : 1 = S_0 \lhd \cdots \lhd S_{k-1} = T.$$

Let $B = \mathbf{C}_G(T)$. Then $\pi(G, \mathscr{S}) = \pi(G, \mathscr{T})\gamma(B, S/T)$. Also, $\mathscr{X}(G, \mathscr{S})$ is the inverse image in $\mathscr{X}(G, \mathscr{T})$ of $\mathrm{Irr}_{S/T}(B) = \mathrm{Irr}_S(B)$ under the map

$$\pi(G, \mathscr{T}) \colon \mathscr{X}(G, \mathscr{T}) \to \mathrm{Irr}(B).$$

By the inductive hypothesis applied to T, we have $\chi \in \mathrm{Irr}_S(G) \subseteq \mathrm{Irr}_T(G) = \mathscr{X}(G, \mathscr{T})$ and $\chi_B = p\vartheta \pm \xi$, where ϑ is a character of B or is zero and $\xi = (\chi)\pi(G, \mathscr{T})$. If $s \in S$, we have $[\chi_B, \xi] = [(\chi^s)_B, \xi^s] = [\chi_B, \xi^s]$ and thus $[\chi_B, \xi^s] \not\equiv 0 \bmod p$. Thus $\xi = \xi^s$ and $\xi \in \mathrm{Irr}_S(B)$. In particular, $\chi \in \mathscr{X}(G, \mathscr{S})$.

If $\beta \in \mathrm{Irr}(C)$, we have

$$[\chi_C, \beta] = [(p\vartheta \pm \xi)_C, \beta] \equiv \pm[\xi_C, \beta] \bmod p$$

and since $\xi \in \mathrm{Irr}_{S/T}(B)$ and S/T is cyclic, it follows from the first part of the proof that there is a unique $\beta \in \mathrm{Irr}(C)$ with $[\xi_C, \beta] \not\equiv 0 \bmod p$, namely $\beta = (\xi)\gamma(B, S/T)$. Thus $\beta = (\chi)\pi(G, \mathscr{S})$ is unique in $\mathrm{Irr}(C)$ such that $[\chi_C, \beta] \not\equiv 0 \bmod p$ and in fact $[\chi_C, \beta] \equiv \pm[\xi_C, \beta] \equiv \pm 1 \bmod p$.

We already have $\mathrm{Irr}_S(G) \subseteq \mathscr{X}(G, \mathscr{S})$. Now suppose

$$\psi \in \mathscr{X}(G, \mathscr{S}) \subseteq \mathscr{X}(G, \mathscr{T}) = \mathrm{Irr}_T(G).$$

Let $\eta = (\psi)\pi(G, \mathscr{T})$ so that $\eta \in \mathrm{Irr}_S(B)$. Let $s \in S$. Then $0 \not\equiv [\psi_B, \eta] = [(\psi^s)_B, \eta^s] = [(\psi^s)_B, \eta] \bmod p$. Since $T \lhd S$ and $\psi \in \mathrm{Irr}_T(G)$, it follows that $\psi^s \in \mathrm{Irr}_T(G) = \mathscr{X}(G, \mathscr{T})$ and thus $\eta = (\psi^s)\pi(G, \mathscr{T})$ since $[(\psi^s)_B, \eta] \not\equiv 0 \bmod p$. Since $\pi(G, \mathscr{T})$ is one-to-one, we have $\psi = \psi^s$ and thus $\psi \in \mathrm{Irr}_S(G)$.

We have now shown that $\mathscr{X}(G, \mathscr{S}) = \mathrm{Irr}_S(G)$ and have given a description of the map $\pi(G, \mathscr{S})$ which is independent of $\mathscr{S}$. This completes the proof. ∎

Assume Hypothesis 13.2 and that S is cyclic so that

$$\gamma(G, S)\colon \mathrm{Irr}_S(G) \to \mathrm{Irr}(C)$$

is defined. Suppose $\sigma\colon \Gamma \to \Gamma_1$ is an isomorphism and that $\sigma(G) = G_1$ and $\sigma(S) = S_1$. Then $\sigma(C) = C_1 = \mathbf{C}_{G_1}(S_1)$ and $\gamma(G_1, S_1)\colon \mathrm{Irr}_{S_1}(G_1) \to \mathrm{Irr}(C_1)$ is defined. Because $\gamma(G, S)$ is uniquely defined, independently of any arbitrary choices, it is clear that if

$$(\chi)\gamma(G, S) = \beta,$$

then

$$(\chi_1)\gamma(G_1, S_1) = \beta_1,$$

where χ_1 and β_1 correspond to χ and β via the isomorphism, σ. That is, $\chi_1(g^\sigma) = \chi(g)$ and $\beta_1(c^\sigma) = \beta(c)$ for $g \in G$ and $c \in C$. (Recall that the computation of $(\chi)\gamma(G, S)$ requires choosing a generator s of S, but that the result is independent of this choice.)

An important special case of this invariance under isomorphism of $\gamma(G, S)$ is when $\sigma \in \mathrm{Aut}(\Gamma)$ and $G^\sigma = G$ and $S^\sigma = S$. In that case we have

$$(\chi^\sigma)\gamma(G, S) = ((\chi)\gamma(G, S))^\sigma.$$

(13.15) LEMMA Assume Hypothesis 13.2 and let $T \lhd S$ with T cyclic and $B = \mathbf{C}_G(T)$. Then $\gamma(G, T)$ maps $\mathrm{Irr}_S(G)$ onto $\mathrm{Irr}_S(B)$.

Proof Since $\mathrm{Irr}_S(G) \subseteq \mathrm{Irr}_T(G)$, $\gamma(G, T)$ is defined on $\mathrm{Irr}_S(G)$. Let $H = GT \lhd \Gamma$. If $s \in S$, then s defines an automorphism of H with $G^s = G$ and $T^s = T$. Therefore, by the above discussion, we have

$$(\chi^s)\gamma(G, T) = ((\chi)\gamma(G, T))^s$$

for all $\chi \in \mathrm{Irr}_T(G)$. It is immediate that $\gamma(G, T)$ maps $\mathrm{Irr}_S(G)$ into $\mathrm{Irr}_S(B)$. Since $\gamma(G, T)$ maps onto $\mathrm{Irr}(B)$ and is one-to-one, the result follows. ∎

(13.16) THEOREM Assume Hypothesis 13.2 and that S is cyclic. Let p be a prime and let T be the p-complement in S. Let $B = \mathbf{C}_G(T)$. Then $\gamma(G, S) = \gamma(G, T)\gamma(B, S/T)$ on $\mathrm{Irr}_S(G)$.

Proof If $p \nmid |S|$, then $T = S$, $B = C$, and $\gamma(C, 1)$ is the identity map on $\mathrm{Irr}(C)$ (as is clear from Definition 13.7). The result is thus trivial in this case and we assume $p \mid |S|$. In particular, $p \nmid |G|$.

Let $\chi \in \mathrm{Irr}_S(G) \subseteq \mathrm{Irr}_T(G)$ and let $\xi = (\chi)\gamma(G, T) \in \mathrm{Irr}(B)$. Let $\beta = (\chi)\gamma(G, S)$. By Lemma 13.15, $\xi \in \mathrm{Irr}_S(B) = \mathrm{Irr}_{S/T}(B)$. We must show that $\beta = (\xi)\gamma(B, S/T)$.

Let $\beta_0 = (\xi)\gamma(B, S/T)$. Since S/T is a p-group, Theorem 13.14 yields $\xi_C = p\varphi + \varepsilon_0 \beta_0$, where φ is a character (or is zero) and $\varepsilon_0 = \pm 1$. Let R be the ring of algebraic integers in $\mathbb{Q}_{|\Gamma|}$ and let I be a maximal ideal containing p. We have then $\xi(c) \equiv \varepsilon_0 \beta_0(c) \bmod I$ for $c \in C$.

By definition of $\gamma(G, T)$, we have $\hat{\chi}(bt) = \varepsilon\xi(b)$, where $T = \langle t \rangle$, $b \in B$, $\varepsilon = \pm 1$ and $\hat{\chi}$ is the canonical extension of χ to Γ. Applying this to $c \in C \subseteq B$ yields

$$\hat{\chi}(ct) = \varepsilon\xi(c) \equiv \varepsilon\varepsilon_0 \beta_0(c) \bmod I$$

for any generator, t of T.

Now let $S = \langle s \rangle$ and let $t = (s)_{p'}$ in the notation of Theorem 8.20. Then $\langle t \rangle = T$ and $(cs)_{p'} = ct$ for $c \in C$. Thus by 8.20, we have

$$\hat{\chi}(cs) \equiv \hat{\chi}(ct) \bmod I$$

and thus

$$\hat{\chi}(cs) \equiv \varepsilon\varepsilon_0 \beta_0(c) \bmod I.$$

By definition of $\gamma(G, S)$, we have $\hat{\chi}(cs) = \delta\beta(c)$ for $c \in C$ with $\delta = \pm 1$. Thus $\beta(c) \equiv \delta\varepsilon\varepsilon_0 \beta_0(c) \bmod I$ for all $c \in C$. By Lemma 13.13, $[\beta - \delta\varepsilon\varepsilon_0 \beta_0, \beta]$ is divisible by p and thus $\beta = \beta_0$ as desired. ∎

(13.17) COROLLARY Assume Hypothesis 13.2 and that S is cyclic with $|S| = pq$, where p and q are primes. Let $\mathscr{S}$ be a composition series for S. Then $\mathscr{X}(G, \mathscr{S}) = \mathrm{Irr}_S(G)$ and $\pi(G, \mathscr{S}) = \gamma(G, S)$.

Proof If $p = q$, the result is immediate from Theorem 13.14(b) and (c). Assume then that $p \neq q$. Write $\mathscr{S}: 1 \lhd T \lhd S$. We may assume $|T| = q$. Now Lemma 13.15 yields that $\mathscr{X}(G, \mathscr{S}) = \mathrm{Irr}_S(G)$. Theorem 13.16 asserts that

$$\pi(G, \mathscr{S}) = \gamma(G, T)\gamma(B, S/T) = \gamma(G, S)$$

where $B = \mathbf{C}_G(T)$. The proof is complete. ∎

(13.18) THEOREM Assume Hypothesis 13.2 with S solvable. Let $\mathscr{S}$ and $\mathscr{T}$ be composition series for S. Then $\mathscr{X}(G, \mathscr{S}) = \mathscr{X}(G, \mathscr{T})$ and $\pi(G, \mathscr{S}) = \pi(G, \mathscr{T})$.

Proof Use induction on the composition length k of S. If $k = 1$ then $\mathscr{S} = \mathscr{T}$ and there is nothing to prove so assume $k > 1$. Write

$$\mathscr{S}: 1 = S_0 \lhd \cdots \lhd S_k = S,$$
$$\mathscr{T}: 1 = T_0 \lhd \cdots \lhd T_k = S.$$

Let $\mathscr{S}^*$ and $\mathscr{T}^*$ be the composition series for S_{k-1} and T_{k-1} respectively, obtained by deleting S from $\mathscr{S}$ and $\mathscr{T}$.

Consider the case that $S_{k-1} = T_{k-1}$. Let $B = \mathbf{C}_G(S_{k-1})$. Then $\mathrm{Irr}_S(B) = (\mathrm{Irr}(C))\gamma(B, S/S_{k-1})^{-1}$ and so

$$\mathscr{X}(G, \mathscr{S}) = (\mathrm{Irr}_S(B))\pi(G, \mathscr{S}^*)^{-1}$$

and

$$\mathscr{X}(G, \mathscr{T}) = (\mathrm{Irr}_S(B))\pi(G, \mathscr{T}^*)^{-1}.$$

By the inductive hypothesis, $\pi(G, \mathscr{S}^*) = \pi(G, \mathscr{T}^*)$ and hence $\mathscr{X}(G, \mathscr{S}) = \mathscr{X}(G, \mathscr{T})$. Also

$$\pi(G, \mathscr{S}) = \pi(G, \mathscr{S}^*)\gamma(B, S/S_{k-1}) = \pi(G, \mathscr{T}^*)\gamma(B, S/T_{k-1}) = \pi(G, \mathscr{T}).$$

Assume now that $S_{k-1} \neq T_{k-1}$ and let $M = S_{k-1} \cap T_{k-1}$. Let $\mathscr{M}$ be a composition series for M and extend $\mathscr{M}$ to composition series $\mathscr{S}^\circ$ and $\mathscr{T}^\circ$ for S which run through S_{k-1} and T_{k-1}, respectively. By the preceding paragraph,

$$\mathscr{X}(G, \mathscr{S}^\circ) = \mathscr{X}(G, \mathscr{S}) \qquad \text{and} \qquad \pi(G, \mathscr{S}^\circ) = \pi(G, \mathscr{S}).$$

We may thus replace $\mathscr{S}$ by $\mathscr{S}^\circ$ and similarly replace $\mathscr{T}$ by $\mathscr{T}^\circ$. We may now assume that $S_i = T_i$ for $i \leq k - 2$ and that

$$\mathscr{M}: 1 = S_0 \lhd \cdots \lhd S_{k-2} = M.$$

Let

$$\bar{\mathscr{S}}: 1 \lhd S_{k-1}/M \lhd S/M, \qquad \bar{\mathscr{T}}: 1 \lhd T_{k-1}/M \lhd S/M$$

and let $D = \mathbf{C}_G(M)$. It follows that

$$\mathscr{X}(G, \mathscr{S}) = (\mathscr{X}(D, \bar{\mathscr{S}}))\pi(G, \mathscr{M})^{-1}$$

and

$$\mathscr{X}(G, \mathscr{T}) = (\mathscr{X}(D, \bar{\mathscr{T}}))\pi(G, \mathscr{M})^{-1}.$$

Also

$$\pi(G, \mathscr{S}) = \pi(G, \mathscr{M})\pi(D, \bar{\mathscr{S}}) \qquad \text{and} \qquad \pi(G, \mathscr{T}) = \pi(G, \mathscr{M})\pi(D, \bar{\mathscr{T}}).$$

Therefore, it suffices to prove that

$$\mathscr{X}(D, \bar{\mathscr{S}}) = \mathscr{X}(D, \bar{\mathscr{T}}) \qquad \text{and} \qquad \pi(D, \bar{\mathscr{S}}) = \pi(D, \bar{\mathscr{T}}).$$

We may therefore assume that $M = 1$ and $D = G$. Thus $S = S_{k-1} \times T_{k-1}$ is abelian of order pq for primes p and q. If $p = q$, then

$$\mathscr{X}(G, \mathscr{S}) = \mathrm{Irr}_S(G) = \mathscr{X}(G, \mathscr{T}) \qquad \text{and} \qquad \pi(G, \mathscr{S}) = \pi(G, \mathscr{T})$$

by Theorem 13.14. If $p \neq q$, then G is cyclic and Corollary 13.17 yields the result. The proof is complete. ∎

Assume Hypothesis 13.2 with solvable S and choose a composition series $\mathscr{S}$ for S. By Theorem 13.18, $\mathscr{X}(G, \mathscr{S})$ is independent of the particular composition series and so we can write $\mathscr{X}(G, S) = \mathscr{X}(G, \mathscr{S})$. Similarly, we can write $\pi(G, S) = \pi(G, \mathscr{S})$. Then $\mathscr{X}(G, S)$ and $\pi(G, S)$ are unambiguously defined. We have $\mathscr{X}(G, S) \subseteq \mathrm{Irr}(G)$ and $\pi(G, S)$ is a bijection of $\mathscr{X}(G, S)$ onto $\mathrm{Irr}(C)$.

As in the discussion preceding Lemma 13.15, suppose $\sigma\colon \Gamma \to \Gamma_1$ is an isomorphism and let $G_1 = \sigma(G)$, $S_1 = \sigma(S)$, and $C_1 = \sigma(C) = \mathbf{C}_{G_1}(S_1)$. Then σ induces bijections $\mathrm{Irr}(G) \to \mathrm{Irr}(G_1)$ and $\mathrm{Irr}(C) \to \mathrm{Irr}(C_1)$. Since $\mathscr{X}(G, S)$ is uniquely defined, we have $\mathscr{X}(G_1, S_1)$ is the image of $\mathscr{X}(G, S)$ in $\mathrm{Irr}(G_1)$ and $\beta = (\chi)\pi(G, S)$ implies that $\beta_1 = (\chi_1)\pi(G_1, S_1)$, where $\chi_1(g^\sigma) = \chi(g)$ and $\beta_1(c^\sigma) = \beta(c)$. In particular, if $\sigma \in \mathrm{Aut}(\Gamma)$ and $G^\sigma = G$ and $S^\sigma = S$, then σ leaves $\mathscr{X}(G, S)$ setwise invariant and $(\chi^\sigma)\pi(G, S) = ((\chi)\pi(G, S))^\sigma$ for $\chi \in \mathscr{X}(G, S)$.

(13.19) COROLLARY Assume Hypothesis 13.2, with S solvable, and let $\mathscr{X}(G, S)$ and $\pi(G, S)$ be as above. Then $\mathscr{X}(G, S) = \mathrm{Irr}_S(G)$. Also, if $T \lhd S$ and $B = \mathbf{C}_G(T)$, then

(a) $\pi(G, T)$ maps $\mathrm{Irr}_S(G)$ onto $\mathrm{Irr}_S(B)$;
(b) $\pi(G, S) = \pi(G, T)\pi(B, S/T)$.

Proof If $T \lhd S$ and $B = \mathbf{C}_G(T)$, use a composition series for S, which runs through T in order to construct $\pi(G, S)$. Then (b) is immediate and

$$\mathscr{X}(G, S) = (\mathscr{X}(B, S/T))\pi(G, T)^{-1}.$$

Thus (a) will follow once we prove the first statement.

We show that $\mathscr{X}(G, S) = \mathrm{Irr}_S(G)$ by induction on the composition length k of S. If $k = 1$ then $\mathscr{X}(G, S) = (\mathrm{Irr}(C))\gamma(G, S)^{-1} = \mathrm{Irr}_S(G)$. Suppose then, that $k > 1$ and let $T \lhd S$ with $1 < T < S$. Let $H = GT \lhd \Gamma$. For $s \in S$, we have $G^s = G$ and $T^s = T$ and hence by the discussion preceding the statement of the corollary, we see that

$$(\chi^s)\pi(G, T) = ((\chi)\pi(G, T))^s$$

for $\chi \in \mathscr{X}(G, T) = \mathrm{Irr}_T(G)$. Since $\pi(G, T)$ is a bijection from $\mathrm{Irr}_T(G)$ onto $\mathrm{Irr}(B)$, where $B = \mathbf{C}_G(T)$, we see that $\pi(G, T)$ carries the S-invariant characters in $\mathrm{Irr}_T(G)$ onto $\mathrm{Irr}_S(B)$. Since $\mathrm{Irr}_S(G) \subseteq \mathrm{Irr}_T(G)$, it follows that

$$\mathrm{Irr}_S(G) = (\mathrm{Irr}_S(B))\pi(G, T)^{-1} = \mathscr{X}(B, S/T)\pi(G, T)^{-1} = \mathscr{X}(G, S),$$

where the second equality is by the inductive hypothesis applied to S/T. ∎

(13.20) DEFINITION Assume Hypothesis 13.2 with S solvable. Then the *Glauberman map* is the map $\pi(G, S)\colon \mathrm{Irr}_S(G) \to \mathrm{Irr}(C)$ constructed above.

We have now completed the proof of Theorem 13.1. The Glauberman map satisfies conditions (a) and (b) of 13.1 by Corollary 13.19. It satisfies condition (c) by Theorem 13.14(c). Also, by Theorem 13.14(a), we have the following.

(13.21) COROLLARY Let $\pi(G, S)$ be the Glauberman map with S a p-group. Let $C = \mathbf{C}_G(S)$ and let $\chi \in \mathrm{Irr}_S(G)$ and $\beta = (\chi)\pi(G, S)$. Then $[\chi_C, \beta] \equiv \pm 1 \bmod p$.

There is one further loose end.

(13.22) COROLLARY Assume Hypothesis 13.2 with S cyclic. Then $\pi(G, S) = \gamma(G, S)$.

Proof Use induction on $|S|$. Let $p \mid |S|$. If S is a p-group, the result follows from Theorem 13.14(b). Assume that S is not a p-group and let T be the p-complement in S and $B = \mathbf{C}_G(T)$. Then $\pi(G, T) = \gamma(G, T)$ and $\pi(B, S/T) = \gamma(B, S/T)$ by the inductive hypothesis. The result now follows from Corollary 13.19(b) (or Theorem 13.1(b)) and Theorem 13.16. ∎

Let S act on G. Then S permutes $\mathrm{Irr}(G)$ and S permutes the set $\mathrm{Cl}(G)$ of conjugacy classes of G. By Brauer's Theorem 6.32, the permutation characters of S on $\mathrm{Irr}(G)$ and $\mathrm{Cl}(G)$ are equal and it is natural to ask if these actions are permutation isomorphic. That is, does there exist a bijection $\alpha: \mathrm{Irr}(G) \to \mathrm{Cl}(G)$ such that $\alpha(\chi^s) = \alpha(\chi)^s$ for all $\chi \in \mathrm{Irr}(G)$ and $s \in S$? In general, the answer is no. However, if $(|G|, |S|) = 1$ and S is solvable, it follows via Glauberman's Theorem 13.1 that the actions of S on $\mathrm{Irr}(G)$ and $\mathrm{Cl}(G)$ are permutation isomorphic.

(13.23) LEMMA Let the group S permute two sets Ω and Λ. Suppose that for every $T \subseteq S$, the number of fixed points of T on Ω equals the number on Λ. Then Ω and Λ are permutation isomorphic.

Proof We prove the existence of a bijection $\alpha: \Omega \to \Lambda$ such that $\alpha(\omega \cdot s) = \alpha(\omega) \cdot s$ for all $\omega \in \Omega$ and $s \in S$ by induction on $|\Omega|$. (Note that taking $T = 1$ yields $|\Omega| = |\Lambda|$.)

Let $T \subseteq S$ be maximal such that T has a fixed point on Ω. (Possibly $T = S$.) Let T fix $\omega \in \Omega$ and $\lambda \in \Lambda$. By the maximality of T, we have $S_\omega = T = S_\lambda$. Let $\mathcal{O}_\omega$ be the orbit of ω and $\mathcal{O}_\lambda$ the orbit of λ under S. Write $\Omega = \mathcal{O}_\omega \cup \Omega_1$ and $\Lambda = \mathcal{O}_\lambda \cup \Lambda_1$ where the unions are disjoint. Map $\alpha_0: \mathcal{O}_\omega \to \mathcal{O}_\lambda$ by $\alpha_0(\omega \cdot s) = \lambda \cdot s$ and check that α_0 is well-defined, one-to-one, onto and that $\alpha_0(v \cdot s) = \alpha_0(v) \cdot s$ for all $v \in \mathcal{O}_\omega$ and $s \in S$.

Since every $H \subseteq S$ has equal numbers of fixed points on $\mathcal{O}_\omega$ and $\mathcal{O}_\lambda$, it follows that H has equal numbers of fixed points on Ω_1 and Λ_1. By the inductive hypothesis, there exists a permutation isomorphism $\alpha_1: \Omega_1 \to \Lambda_1$. Now define α on Ω by combining α_0 and α_1. ∎

(13.24) THEOREM Let S act on G with S solvable and $(|G|, |S|) = 1$. Then

(a) S fixes the same numbers of irreducible characters and conjugacy classes of G.
(b) The actions of S on $\mathrm{Irr}(G)$ and $\mathrm{Cl}(G)$ are permutation isomorphic.

Proof It suffices to prove (a) since (b) follows via Lemma 13.23 by application of (a) to all subgroups of S.

Since $\pi(G, S)$ maps $\mathrm{Irr}_S(G)$ one-to-one and onto $\mathrm{Irr}(C)$, it follows that

$$|\mathrm{Irr}_S(G)| = |\mathrm{Irr}(C)| = |\mathrm{Cl}(C)|.$$

By Corollary 13.10, intersection defines a bijection from the set of S-fixed conjugacy classes of G onto $\mathrm{Cl}(C)$. The result now follows. ∎

Theorem 13.24 becomes false if the hypothesis that $(|G|, |S|) = 1$ is dropped. (See Problem 13.16 for an example.)

Suppose we continue to assume that $(|G|, |S|) = 1$, but drop the assumption that S is solvable. If S is nonsolvable, then $2 \mid |S|$ by the Feit–Thompson theorem and thus $2 \nmid |G|$. Thus G is solvable, again by Feit–Thompson. In this situation, where Hypothesis 13.2 is satisfied with G solvable of odd order, it is possible to construct a natural character correspondence from $\mathrm{Irr}_S(G)$ onto $\mathrm{Irr}(C)$ by a method entirely different from Glauberman's. (And thus Theorem 13.24 remains valid without the hypothesis that S is solvable.) We shall describe the map $\mathrm{Irr}_S(G) \to \mathrm{Irr}(C)$ in this case but without giving the proof since the only known proofs are too long to include. The key step is the following.

(13.25) THEOREM Assume Hypothesis 13.2 and that G is solvable of odd order. Let $C \subseteq H \subseteq G$. Suppose that there exist S-invariant normal subgroups, K and L of G such that

(a) $L \subseteq K$ and K/L is abelian;
(b) $G = KC$;
(c) $H = LC$.

Then for each $\chi \in \mathrm{Irr}_S(G)$, there exists a unique $\psi \in \mathrm{Irr}_S(H)$ such that $[\chi_H, \psi]$ is odd. The map $\chi \mapsto \psi$ is a bijection from $\mathrm{Irr}_S(G)$ onto $\mathrm{Irr}_S(H)$.

Proof Omitted. ∎

To construct the correspondence between $\mathrm{Irr}_S(G)$ and $\mathrm{Irr}(C)$ we construct a chain of subgroups

$$G = C_0 > C_1 > \cdots > C_k = C$$

and apply Theorem 13.25 to obtain maps $\mathrm{Irr}_S(C_i) \to \mathrm{Irr}_S(C_{i+1})$. The composition of these maps is the desired correspondence.

Assume Hypothesis 13.2 and that G is solvable of odd order. We consider the set $\mathscr{H}$ of S-invariant subgroups H with $C \subseteq H \subseteq G$. For $H \in \mathscr{H}$, define $H^* = [H, S]'C$. Since $[H, S] \lhd HS$ it follows that $[H, S]' \lhd HS$ and thus $H^* \in \mathscr{H}$. If $H > C$, then $[H, S] > 1$ and $[H, S] > [H, S]'$ by solvability. Since $[H^*, S] \subseteq [H, S]'$, it follows that $H^* < H$. We now define the subgroups $C_i \in \mathscr{H}$ by $C_0 = G$ and $C_{i+1} = (C_i)^*$ and we have the desired chain of subgroups.

We must now check that the hypotheses of Theorem 13.25 are satisfied by taking $H = G^*$. Let $K = [G, S]$ and $L = [G, S]'$. Then K and L are normal S-invariant subgroups of G and K/L is abelian. That $G = KC$ follows fairly easily from Glauberman's Lemma 13.8. Condition (c) of Theorem 13.25 is automatic from the definition of G^*.

It has been conjectured for every group G and prime p that if $N = \mathbf{N}_G(P)$ for $P \in \mathrm{Syl}_p(G)$, then the numbers of irreducible characters of p'-degree of G and of N are equal. (For simple groups G, this conjecture is due to McKay.) Using Glauberman's Theorem 13.1, we prove a result which includes the special case of this conjecture when G has a normal p-complement.

(13.26) THEOREM Let $G = KH$ with $K \lhd G$, H solvable, and $(|H|, |K|) = 1$. Let $N = \mathbf{N}_G(H)$ and put

$$\mathscr{X} = \{\chi \in \mathrm{Irr}(G) \mid (|H|, \chi(1)) = 1\}$$

and

$$\mathscr{Y} = \{\eta \in \mathrm{Irr}(N) \mid (|H|, \eta(1)) = 1\}.$$

Then there exists a uniquely defined bijection of $\mathscr{X}$ onto $\mathscr{Y}$.

Proof We have $N = (N \cap K)H$ and the commutator

$$[N \cap K, H] \subseteq H \cap K = 1$$

so that $N \cap K = C = \mathbf{C}_K(H)$ and $N = C \times H$. It follows that

$$\mathscr{Y} = \{\beta \times \lambda \mid \beta \in \mathrm{Irr}(C), \lambda \in \mathrm{Irr}(H), \lambda(1) = 1\}.$$

Now if $\chi \in \mathscr{X}$, let ϑ be an irreducible constituent of χ_K. Then $|G : I_G(\vartheta)|$ divides both $\chi(1)$ and $|G : K| = |H|$. Thus $I_G(\vartheta) = G$ and $\vartheta \in \mathrm{Irr}_H(K)$. Let

$\hat{\vartheta}$ be the canonical extension of ϑ to G. Then $\chi = \hat{\vartheta}\xi$ for some unique $\xi \in \mathrm{Irr}(G/K)$ by Gallagher's theorem (Corollary 6.17). Since $(\chi(1), |H|) = 1$, we must have $\xi(1) = 1$. Also, χ uniquely determines ϑ by $\chi_K = \vartheta$ and thus determines $\hat{\vartheta}$ and ξ. We can now map $\mathscr{X}$ to $\mathscr{Y}$ by $\chi \mapsto ((\vartheta)\pi(K, H)) \times \xi_H$. Since restriction defines a bijection of $\mathrm{Irr}(G/K)$ onto $\mathrm{Irr}(H)$ and since $\hat{\vartheta}\xi \in \mathscr{X}$ for every $\vartheta \in \mathrm{Irr}_H(K)$ and linear $\xi \in \mathrm{Irr}(G/K)$, it follows that we have mapped $\mathscr{X}$ onto $\mathscr{Y}$. The map is one-to-one since the $\eta \in \mathscr{Y}$ are uniquely of the form $\beta \times \lambda$. ∎

Suppose S acts on G and that $N \lhd G$ is S-invariant. Also, assume that $(|G:N|, |S|) = 1$. We consider a pair of "dual" questions:

(a) Let $\chi \in \mathrm{Irr}_S(G)$. Does there exist $\vartheta \in \mathrm{Irr}_S(N)$ with $[\chi_N, \vartheta] \neq 0$?
(b) Let $\vartheta \in \mathrm{Irr}_S(N)$. Does there exist $\chi \in \mathrm{Irr}_S(G)$ with $[\vartheta^G, \chi] \neq 0$?

Both questions can be answered in the affirmative. By the Feit–Thompson theorem, at least one of S and G/N is solvable and we assume this. Also, note that if S is a p-group, both facts can be proved relatively easily by counting arguments.

(13.27) THEOREM Let S act on G and leave $N \lhd G$ invariant. Assume that $(|S|, |G:N|) = 1$ and that one of S or G/N is solvable. Let $\chi \in \mathrm{Irr}_S(G)$. Then χ_N has an S-invariant irreducible constituent.

Proof Let $\Omega = \{\vartheta \in \mathrm{Irr}(N) \mid [\chi_N, \vartheta] \neq 0\}$. Then G/N permutes Ω transivively and since χ is S-invariant, S permutes Ω. Also, the action of S on G induces an action on G/N. Let $\vartheta \in \Omega$, $s \in S$, and $g \in G$. For $x \in N$, we have

$$(\vartheta^g)^s(x^s) = \vartheta^g(x) = \vartheta(gxg^{-1})$$

and

$$(\vartheta^s)^{g^s}(x^s) = \vartheta^s(g^s x^s (g^s)^{-1}) = \vartheta(gxg^{-1}).$$

The hypotheses of Glauberman's Lemma 13.8 are thus satisfied and the result follows. ∎

The situation of question (b) is more difficult and interesting. First we consider the case where G/N is solvable.

(13.28) THEOREM Let S act on G and leave $N \lhd G$ invariant. Assume that $(|S|, |G:N|) = 1$ and that G/N is solvable. Let $\vartheta \in \mathrm{Irr}_S(N)$. Then ϑ^G has an S-invariant irreducible constituent.

Proof First assume that G/N is abelian, and let A be the group of linear characters of G/N. Let $\Omega = \{\chi \in \mathrm{Irr}(G) \mid [\vartheta^G, \chi] \neq 0\}$. If $\chi \in \Omega$ and $\lambda \in A$, then

$\chi\lambda \in \Omega$ since $(\chi\lambda)_N = \chi_N$. This defines an action of A on Ω. We claim that this action is transitive.

Let $\chi, \psi \in \Omega$. Then $0 \neq [\chi_N, \psi_N] = [(\chi_N)^G, \psi]$ and ψ is an irreducible constituent of $(\chi_N)^G = (\chi_N 1_N)^G = \chi(1_N)^G$. However, $(1_N)^G = \sum_{\lambda \in A} \lambda$ and thus $(\chi_N)^G = \sum \chi\lambda$. Since $\chi\lambda \in \mathrm{Irr}(G)$ for each $\lambda \in A$, it follows that $\psi = \chi\lambda$ for some λ and A is transitive on Ω as claimed.

Since ϑ is S-invariant, it follows that S permutes Ω and also S acts on the group A since $(\lambda\mu)^s = \lambda^s\mu^s$. If $\chi \in \Omega$, $\lambda \in A$, and $s \in S$, we clearly have

$$(\chi\lambda)^s = (\chi^s)\lambda^s$$

and the hypotheses of Glauberman's Lemma 13.8 are satisfied since $|A| = |G:N|$ and thus $(|A|, |S|) = 1$. The result follows in this case.

We now assume that G/N is nonabelian and work by induction on $|G:N|$. Let $M/N = (G/N)'$. Then $M \lhd G$ is S-invariant and $M < G$ since G/N is solvable. By the inductive hypothesis, ϑ^M has an S-invariant irreducible constituent ψ and by the first part of the proof, ψ^G has an S-invariant irreducible constituent. The result now follows since $\vartheta^G = (\vartheta^M)^G$. ∎

To handle the case that G/N is not solvable, we appeal to the Glauberman correspondence, Theorem 13.1. We first restrict attention to the situation where $(|G|, |S|) = 1$.

(13.29) THEOREM Assume Hypothesis 13.2 and that S is solvable. Let $N \lhd \Gamma$ with $N \subseteq G$. Let $\chi \in \mathrm{Irr}_S(G)$ and $\vartheta \in \mathrm{Irr}_S(N)$. Write $\xi = (\chi)\pi(G, S)$ and $\varphi = (\vartheta)\pi(N, S)$. Then $[\vartheta^G, \chi] \neq 0$ iff $[\varphi^C, \xi] \neq 0$.

Proof We first consider the case that S is a p-group. Then $[\chi_C, \xi] \not\equiv 0 \bmod p$ and every irreducible constituent of χ_C other than ξ occurs with multiplicity divisible by p. Since $\chi_{C \cap N} = (\chi_C)_{C \cap N}$, it follows that

(1) $$[\chi_{C \cap N}, \varphi] \equiv [\chi_C, \xi][\xi_{C \cap N}, \varphi] \bmod p.$$

Now write $\chi_N = \sum b_\Delta \Delta$, where Δ runs over sums of orbits of the action of S on $\mathrm{Irr}(N)$. If $\Delta = \sum \mathcal{O}$ where $\mathcal{O}$ is such an orbit, then $[\Delta_{C \cap N}, \varphi] = |\mathcal{O}|[\eta_{C \cap N}, \varphi]$ for $\eta \in \mathcal{O}$. If $|\mathcal{O}| > 1$, then $p \mid |\mathcal{O}|$ and $[\Delta_{C \cap N}, \varphi] \equiv 0 \bmod p$. Thus

$$[\chi_{C \cap N}, \varphi] \equiv \sum_{\eta \in \mathrm{Irr}_S(N)} [\chi_N, \eta][\eta_{C \cap N}, \varphi] \bmod p.$$

However, for $\eta \in \mathrm{Irr}_S(N)$, we have $[\eta_{C \cap N}, \varphi] \equiv 0 \bmod p$ unless $\varphi = (\eta)\pi(N, S)$, that is, unless $\eta = \vartheta$. Thus

(2) $$[\chi_{C \cap N}, \varphi] \equiv [\chi_N, \vartheta][\vartheta_{C \cap N}, \varphi] \bmod p.$$

Since $[\chi_C, \xi] \not\equiv 0 \not\equiv [\vartheta_{C \cap N}, \varphi]$, comparison of Equations (1) and (2) yields that $[\chi_N, \vartheta] \equiv 0 \bmod p$ iff $[\xi_{C \cap N}, \varphi] \equiv 0 \bmod p$. Since $N \lhd G$ and

$p \nmid |G|$ it follows that $[\chi_N, \vartheta] \equiv 0 \bmod p$ iff $[\chi_N, \vartheta] = 0$. Similarly, $[\xi_{C \cap N}, \varphi] \equiv 0 \bmod p$ iff $[\xi_{C \cap N}, \varphi] = 0$. The result thus follows if S is a p-group.

To complete the proof, we use induction on $|S|$. We may choose $T \lhd S$ with S/T a nontrivial p-group. Let $B = \mathbf{C}_G(T)$. By Theorem 13.1, we have $\pi(G, S) = \pi(G, T)\pi(B, S/T)$ and similarly, $\pi(N, S) = \pi(N, T)\pi(B \cap N, S/T)$. Now $[\vartheta^G, \chi] \neq 0$ iff $[((\vartheta)\pi(N, T))^B, (\chi)\pi(G, T)] \neq 0$ by the inductive hypothesis. The result now follows by the first part of the proof. ∎

(13.30) COROLLARY Let S act on G with S solvable and $(|S|, |G|) = 1$. Let $N \lhd G$ be S-invariant. Let $\vartheta \in \mathrm{Irr}_S(N)$. Then ϑ^G has an S-invariant irreducible constituent.

Proof Let $C = \mathbf{C}_G(S)$ and $\varphi = (\vartheta)\pi(N, S)$ so that $\varphi \in \mathrm{Irr}(C \cap N)$. Let ξ be an irreducible constituent of φ^C and let $\chi = (\varphi)\pi(G, S)^{-1}$. Then $\chi \in \mathrm{Irr}_S(G)$ and $[\vartheta^G, \chi] \neq 0$ by Theorem 13.29. The proof is complete. ∎

To complete our analysis of question (b), we need to consider the case where S is solvable and $(|S|, |G|) \neq 1$. Of course, we continue to assume that $(|S|, |G:N|) = 1$. First, we observe that it is no loss to assume that ϑ is invariant in G since if $U = I_G(\vartheta)$, then S leaves U invariant. Now, if we can find an S-invariant irreducible constituent ψ of ϑ^U, then $\psi^G \in \mathrm{Irr}_S(G)$ as desired. We shall finish the proof by an appeal to the theory of projective representations of Chapter 11. The following proves slightly more than we need.

(13.31) THEOREM Let $N \subseteq G \lhd \Gamma$ with $N \lhd \Gamma$ and $(|\Gamma : G|, |G : N|) = 1$. Assume that one of Γ/G or G/N is solvable. Let $\vartheta \in \mathrm{Irr}(N)$ be invariant in Γ. Then ϑ^G has some Γ-invariant irreducible constituent.

Proof By Theorem 11.28, we can find a character triple $(\Gamma_1, N_1, \vartheta_1)$ and an isomorphism (τ, σ): $(\Gamma, N, \vartheta) \to (\Gamma_1, N_1, \vartheta_1)$ with $\vartheta_1(1) = 1$. Let $G_1 = G^\tau$ so that the isomorphism τ: $\Gamma/N \to \Gamma_1/N_1$ carries G/N to G_1/N_1. Suppose we can find $\psi_1 \in \mathrm{Irr}(G_1 | \vartheta_1)$ with ψ_1 invariant in Γ_1. Let $\psi \in \mathrm{Irr}(G | \vartheta)$ with $\sigma_G(\psi) = \psi_1$. We claim that ψ is invariant in Γ. To see this, let χ be an irreducible constituent of ψ^Γ so that $\chi \in \mathrm{Irr}(\Gamma | \vartheta)$. Let $\chi_1 = \sigma_\Gamma(\chi)$. Then $(\chi_1)_{G_1} = e\psi_1$ for some integer e and it follows that $\chi_G = e\psi$. Thus ψ is invariant as desired.

The argument of the preceding paragraph shows that it is no loss to assume that ϑ is linear. We may thus factor $\vartheta = \lambda\mu$ where $(o(\lambda), |G : N|) = 1$ and $(o(\mu), |\Gamma : G|) = 1$ and λ and μ are powers of ϑ. Thus λ and μ are invariant in Γ. By Corollary 6.27, λ has a unique extension, $\hat{\lambda} \in \mathrm{Irr}(G)$ such that $o(\hat{\lambda}) = o(\lambda)$. Because of the uniqueness, it follows that $\hat{\lambda}$ is invariant in Γ. Suppose we can find a Γ-invariant irreducible constituent ψ of μ^G. Then $\psi\hat{\lambda} \in \mathrm{Irr}(G)$ is Γ-invariant and $\vartheta = \mu\lambda$ is a constituent of $(\psi\hat{\lambda})_N$.

We may therefore assume that $\vartheta = \mu$, that is, that ϑ is linear and $(o(\vartheta), |\Gamma : G|) = 1$. Also, we may replace Γ by $\Gamma/\ker \vartheta$ and thus assume that ϑ is faithful. Thus $|N| = o(\vartheta)$ is relatively prime to $|\Gamma : G|$ and hence $(|G|, |\Gamma : G|) = 1$. By the Schur–Zassenhaus theorem, we can find $S \subseteq \Gamma$ with $SG = \Gamma$ and $S \cap G = 1$. Thus S acts on G and leaves N and ϑ invariant. Also, one of S or G/N is solvable. Now Corollary 13.30 or Theorem 13.28 yields an S-invariant irreducible constituent of ϑ^G. The result follows. ∎

To close the chapter, we obtain some further information in a special case of Theorem 13.6.

(13.32) THEOREM Assume Hypothesis 13.2 and that S is cyclic. Also suppose that $C = \mathbf{C}_G(x)$ for all $x \in S - \{1\}$. Let $\chi \in \mathrm{Irr}_S(G)$ and $\beta = (\chi)\pi(G, S)$. Let $\hat{\chi}$ be the canonical extension of χ to Γ. Then there exists $\varepsilon = \pm 1$ and $\mu \in \mathrm{Irr}(S)$ with $\mu^2 = 1_S$ such that

(a) $\hat{\chi}(cx) = \varepsilon\beta(c)\mu(x)$ for all $x \in S - \{1\}$;
(b) $\chi_C = |S|\vartheta + \varepsilon\beta$, where ϑ is a character of C or is zero;
(c) $(\chi(1) - \varepsilon\beta(1))/|S| = k \in \mathbb{Z}$.
(d) $\hat{\chi}_S = k\rho_S + \varepsilon\beta(1)\mu$, where ρ_S is the regular character.
(e) $\mu \neq 1_S$ iff $|S|$ is even and k is odd.

Proof If $1 < T \subseteq S$ then $\pi(G, S) = \pi(G, T) = \gamma(G, T)$ and hence $\hat{\chi}(cx) = \pm\beta(c)$ for all $x \in S - \{1\}$ where the sign is independent of c. It follows that in the notation of Lemma 13.5, we have $\psi_\beta(x) = \pm 1$ for all $x \neq 1$ and $\psi_\varphi(x) = 0$ for $x \neq 1$ and $\beta \neq \varphi \in \mathrm{Irr}(C)$.

Write $\psi = \psi_\beta$. We work to express ψ in terms of $\mathrm{Irr}(S)$. Let $\lambda_1, \lambda_2 \in \mathrm{Irr}(S)$ and compute

$$[(\lambda_1 - \lambda_2), \psi] = (1/|S|) \sum_{x \in S;\, x \neq 1} \pm(\lambda_1(x) - \lambda_2(x)).$$

Since $|\lambda_1(x) - \lambda_2(x)| \leq 2$ for $x \in S$, this yields

$$|[(\lambda_1 - \lambda_2), \psi]| \leq 2(|S| - 1)/|S| < 2.$$

Therefore, the multiplicities with which λ_1 and λ_2 occur in ψ differ by at most 1. It follows that $\psi = a\rho_S \pm \mu$ where $a \in \mathbb{Z}$ and μ is a sum of distinct linear characters of S with $\mu(1) \leq |S|/2$. Also, $\mu(x) = \pm\psi(x) = \pm 1$ for $1 \neq x \in S$ and hence

$$\mu(1) = [\mu, \mu] = (1/|S|)(\mu(1)^2 + |S| - 1)$$

and $\mu(1)^2 - |S|\mu(1) + |S| - 1 = 0$. It follows that $\mu(1) = 1$ and $\mu \in \mathrm{Irr}(S)$. Define $\varepsilon = \pm 1$ by the equation $\psi = a\rho_S + \varepsilon\mu$. (Note that if $|S| = 2$, then neither ε nor μ is uniquely defined.) Now $\hat{\chi}(cx) = \beta(c)\psi(x) = \varepsilon\beta(c)\mu(x)$ for $1 \neq x \in S$ and (a) is proved. Also $\mu(x) = \pm 1$ for $x \in S$ so that $\mu^2 = 1_S$.

Since all ψ_φ for $\varphi \neq \beta$ vanish on $S - \{1\}$, each is a multiple of ρ_S and hence $(\varphi \times \psi_\varphi)_C$ is a multiple of $|S|\varphi$. Since $(\beta \times \psi_\beta)_C = a|S|\beta + \varepsilon\beta$, statement (b) follows and (c) is immediate from (b). Also (d) now follows.

We now prove (e). Since $\mu^2 = 1_S$, it follows that $\mu = 1_S$ if $|S|$ is odd. Suppose then, that $|S|$ is even and let $\tau \in \operatorname{Irr}(S)$ with $\tau^2 = 1_S$ but $\tau \neq 1_S$. (This uniquely defines τ.) Now $\det(\rho_S) = \prod_{\lambda \in \operatorname{Irr}(S)} \lambda = \tau$. By (d) it follows that

$$1_S = \det(\hat{\chi}_S) = \tau^k \mu^{\varepsilon\beta(1)}.$$

Since $2 \mid |S|$, we have $2 \nmid \beta(1)$ and therefore $\mu = \tau^k$. Statement (e) now follows and the proof is complete. ∎

(13.33) COROLLARY Let (Γ, N, ϑ) be a character triple and let $N \subseteq G \lhd \Gamma$. Suppose Γ/G is cyclic and that Γ/N is a Frobenius group with kernel G/N. Assume that $\vartheta^G = e\chi$ for $\chi \in \operatorname{Irr}(G)$ (and thus $e^2 = |G:N|$). Then $|\Gamma:G|$ divides $e - \varepsilon$ for some $\varepsilon = \pm 1$.

Proof We may replace (Γ, N, ϑ) by an isomorphic character triple and assume $\vartheta(1) = 1$. Write $\vartheta = \mu\lambda$, where $(o(\lambda), |G:N|) = 1$ and $(o(\mu), |\Gamma:G|) = 1$ and λ and μ are powers of ϑ and thus invariant in Γ and λ is extendible to G by Corollary 6.27. Let ν be an extension of $\bar{\lambda}$. Then

$$\mu^G = (\bar{\lambda}\vartheta)^G = (\nu_N\vartheta)^G = \nu\vartheta^G = e(\nu\chi).$$

Since $\nu\chi \in \operatorname{Irr}(G)$ we may replace ϑ by μ and assume $(o(\vartheta), |\Gamma:G|) = 1$. We may also assume that $\ker \vartheta = 1$ so that $N \subseteq \mathbf{Z}(\Gamma)$ and $(|G|, |\Gamma:G|) = 1$.

Let S be a complement for G in Γ. If $1 \neq x \in S$, then $\mathbf{C}_{G/N}(x) = 1$ and it follows that $N = \mathbf{C}_G(x)$ and we are in the situation of Theorem 13.32. (Note that $\chi \in \operatorname{Irr}_S(G)$ since χ is the unique irreducible constituent of ϑ^G.) The result now follows from 13.32(c). ∎

Problems

(13.1) Assume Hypothesis 13.2 with S solvable. Prove the following facts by using the statement of Theorem 13.1, but do not appeal to any of the results used in constructing the Glauberman map.

(a) If $\chi \in \operatorname{Irr}_S(G)$ then $(\chi)\pi(G, S)$ is a constituent of χ_C.
(b) If $\beta = (\chi)\pi(G, S)$, then $\mathbb{Q}(\chi) = \mathbb{Q}(\beta)$.
(c) If $C = G$, then $\pi(G, S)$: $\operatorname{Irr}(G) \to \operatorname{Irr}(G)$ is the identity map.

(13.2) In the situation of Theorem 13.1, let $\beta = (\chi)\pi(G, S)$ for $\chi \in \operatorname{Irr}_S(G)$. Show that $\chi(1)$ divides $|G:C|\beta(1)$.

Hint In the case that S is cyclic, consider $\omega = \omega_{\hat{\chi}}$ as in Chapter 3.

(13.3) In the situation of Theorem 13.1, let $N \lhd G$ be S-invariant with $NC = G$. Let $\vartheta \in \mathrm{Irr}_S(N)$. Show that

$$I_C((\vartheta)\pi(N, S)) = I_G(\vartheta) \cap C.$$

(13.4) Let S act on G and leave $N \lhd G$ invariant. Assume that $(|S|, |G:N|) = 1$ and that one of S or G/N is solvable. Suppose $\mathbf{C}_{G/N}(S) = 1$. If $\chi \in \mathrm{Irr}_S(G)$, show that there exists a unique $\psi \in \mathrm{Irr}_S(N)$ such that $[\chi_N, \psi] \neq 0$. Write $(\chi)\delta = \psi$. Show that δ maps $\mathrm{Irr}_S(G)$ onto $\mathrm{Irr}_S(N)$.

(13.5) Assume Hypothesis 13.2 and that S is solvable. Let $C \subseteq N \subseteq G$ with $N \lhd \Gamma$.

(a) Show that $\mathbf{C}_{G/N}(S) = 1$.
(b) Let δ: $\mathrm{Irr}_S(G) \to \mathrm{Irr}_S(N)$ be as in Problem 13.4. Show that δ is one-to-one.
(c) Show that $\delta\pi(N, S) = \pi(G, S)$.

Hint (For (a)) Use Glauberman's Lemma 13.8.

(13.6) In the situation of Theorem 13.1, assume that G is solvable. Let $\chi \in \mathrm{Irr}_S(G)$ and $\beta = (\chi)\pi(G, S)$. Show that $\beta(1)$ divides $\chi(1)$.

Hints Let $N < G$ be normal, S-invariant and maximal such. Then either $NC = G$ or $N \supseteq C$. Use Problem 13.3 or 13.5(c) and induction on $|G|$.

(13.7) In Problem 13.4, assume that G/N is solvable. Show that δ is one-to-one.

Hint Use induction on $|G:N|$.

(13.8) Assume Hypothesis 13.2 and that G is nilpotent. Show that $|\mathrm{Irr}_S(G)| = |\mathrm{Irr}(C)|$. Do not assume Theorem 13.25.

Hint Use Problem 13.7.

(13.9) Assume Hypothesis 13.2 and that G is solvable. Show that $C > 1$ iff $|\mathrm{Irr}_S(G)| > 1$. Do not assume Theorem 13.25.

(13.10) Let $N \lhd \Gamma$ with $N \subseteq G \lhd \Gamma$ and $(|\Gamma:G|, |G:N|) = 1$. Assume that one of Γ/G or G/N is solvable. Let K/N be a complement for G/N in Γ/N and assume $\mathbf{C}_{G/N}(K/N) = 1$. Let $\vartheta \in \mathrm{Irr}(N)$ be invariant in K. Show that there exists a unique Γ-invariant $\chi \in \mathrm{Irr}(G)$ with $[\vartheta^G, \chi] \neq 0$.

Hints Use the argument of Theorem 13.31 to reduce to the case $(|G|, |\Gamma:G|) = 1$. Use Problem 13.5(b) if Γ/G is solvable.

(13.11) Let S be solvable and let H be a group with $(|S|, |H|) = 1$. Let $G = H \times H \times \cdots \times H$ where there are $|S|$ factors and let S act on G by permuting the factors regularly. Let $C = \mathbf{C}_G(S)$ and note that

$$C = \{(h, h, \ldots, h) \mid h \in H\} \cong H.$$

Also, if $\chi \in \mathrm{Irr}_S(G)$, then $\chi = \vartheta \times \vartheta \times \cdots \times \vartheta$ for some $\vartheta \in \mathrm{Irr}(H)$. If $(\chi)\pi(G, S) = \beta$, show that $\beta((h, h, \ldots, h)) = \vartheta(h^{|S|})$.

(13.12) Let $N \lhd G$ and suppose $\vartheta \in \mathrm{Irr}(N)$ is invariant in G. Suppose $\vartheta^G = e\chi$ for some $\chi \in \mathrm{Irr}(G)$. (Thus (G, N, ϑ) is a fully ramified character triple and $e^2 = |G:N|$.) Assume that S is solvable, acts on G, leaves N and ϑ invariant and that $(|G:N|, |S|) = 1$. Let $C/N = \mathbf{C}_{G/N}(S)$. Show that $\chi_C = f\xi$ with $\xi \in \mathrm{Irr}(C)$ and $f^2 = |G:C|$.

Hint Use the technique of the proof of Theorem 13.31 to reduce to the case that $(|G|, |S|) = 1$ and $\vartheta(1) = 1$ with ϑ faithful. In this case, $C = \mathbf{C}_G(S)$. Use Theorem 13.1.

(13.13) Let S act on G and leave $N \lhd G$ invariant. Assume that $(|S|, |G:N|) = 1$ and that S acts trivially on G/N. Let $\vartheta \in \mathrm{Irr}(N)$ and $\chi \in \mathrm{Irr}(G)$ with $[\chi_N, \vartheta] \neq 0$. Show that ϑ is S-invariant iff χ is S-invariant. Do not assume the Feit–Thompson theorem.

Hint In showing that $\vartheta \in \mathrm{Irr}_S(N)$ implies $\chi \in \mathrm{Irr}_S(G)$, it suffices to assume that G/N is cyclic.

(13.14) In the situation of Theorem 13.1, let $N \lhd G$ be S-invariant with $NC = G$. Let $\vartheta \in \mathrm{Irr}_S(N)$ and $\varphi = (\vartheta)\pi(N, S)$. Let $I = I_G(\vartheta)$ (so that $I \cap C = I_C(\varphi)$ by Problem 13.3). If ψ is an irreducible constituent of ϑ^I, show that

$$(\psi^G)\pi(G, S) = ((\psi)\pi(I, S))^C.$$

(13.15) Let E be elementary abelian of order p^2 for $p \neq 2$ and let S be dihedral of order $2p$.

(a) Define an action of S on E such that $\mathrm{Irr}_S(E) = \{1_E\}$ but $\mathbf{C}_E(S) > 1$.
(b) Define an action of S on E such that $\mathbf{C}_E(S) = 1$ but $|\mathrm{Irr}_S(E)| > 1$.

14 Linear groups

Suppose χ is a faithful character of G. In this chapter we are concerned with drawing conclusions about G when given information about χ. For instance, we already know that if χ is irreducible, then $\mathbf{Z}(G)$ is cyclic and that if all irreducible constituents of χ are linear, then G is abelian. Another, less trivial example which we have seen is Theorem 3.13.

A faithful F-representation of G with degree n is an isomorphism of G with a *linear group* of degree n over F; in other words, a subgroup of $\mathrm{GL}(n, F)$. For our purposes, we will restrict attention to finite linear groups. (It should be pointed out, however, that stating that an infinite group is a linear group imposes a type of finiteness condition on it, that is, it guarantees that the group is not "too badly" infinite.)

We shall also restrict attention to complex linear groups. Thus from now on, a "linear group" is a finite subgroup of $\mathrm{GL}(n, \mathbb{C})$ for some n. A group is thus isomorphic to a linear group of degree n iff it has a faithful character of degree n. We say that a linear group is *irreducible* if the identity map is an irreducible representation.

(14.1) THEOREM (*Blichfeldt*) Let G be a linear group of degree n and let $\pi = \{p \mid p \text{ is prime}, p > n + 1\}$. Then G has an abelian Hall π-subgroup.

We need a lemma. Recall that a character χ is p-rational (where p is prime) if its values lie in $\mathbb{Q}_r$ for some r with $p \nmid r$. (This is Definition 6.29.)

(14.2) LEMMA Let $p \neq q$ be primes such that G has no element of order pq. Then each $\chi \in \mathrm{Irr}(G)$ is either p-rational or q-rational.

Proof Let $\chi \in \mathrm{Irr}(G)$ and suppose that χ is neither p-rational nor q-rational. Then in the notation of the discussion following Definition 6.29, there exists $\sigma \in \mathscr{G}_p(G)$ and $\tau \in \mathscr{G}_q(G)$ with $\chi^\sigma \neq \chi \neq \chi^\tau$. Also, by Problem 2.2(b), $\chi^\sigma, \chi^\tau \in \mathrm{Irr}(G)$.

If $g \in G$ and $p \nmid o(g)$, then $\chi(g) \in \mathbb{Q}_m$, where $|G| = p^a m$ and $p \nmid m$. Thus by definition of $\mathscr{G}_p(G)$, we have $\chi(g)^\sigma = \chi(g)$. Similarly, if $q \nmid o(g)$, then $\chi(g)^\tau = \chi(g)$.

Since G has no element of order pq, it follows for every $g \in G$ that either $p \nmid o(g)$ or $q \nmid o(g)$ and we conclude that

$$[(\chi - \chi^\sigma), (\chi - \chi^\tau)] = 0.$$

Since $[\chi, \chi^\tau] = 0 = [\chi^\sigma, \chi]$, we obtain

$$0 = [\chi, \chi] + [\chi^\sigma, \chi^\tau] = 1 + [\chi^\sigma, \chi^\tau] \geq 1$$

and this contradiction proves the result. ∎

The following easy lemma is a special case of a more general result due to Schur which we will prove later.

(14.3) LEMMA Let χ be a faithful p-rational character of G and assume that p divides $|G|$. Then $\chi(1) \geq p - 1$.

Proof Since χ is p-rational, it has values in $\mathbb{Q}_r$ for some r with $p \nmid r$. Let $P \subseteq G$ with $|P| = p$. Then χ_P has values in $\mathbb{Q}_r \cap \mathbb{Q}_p = \mathbb{Q}$. Let $\mathscr{G} = \mathscr{G}(\mathbb{Q}_p/\mathbb{Q})$. Then $\mathscr{G}$ fixes χ_P and thus permutes the linear constituents of χ_P. Since χ is faithful, χ_P has a nonprincipal linear constituent λ and $\lambda(x) \neq 1$, where $1 \neq x \in P$. Since $\mathscr{G}$ is transitive on the $p - 1$ primitive pth roots of 1, it follows that the images of λ under $\mathscr{G}$ take on $p - 1$ different values at x and thus there are at least $p - 1$ different characters in the orbit of λ under $\mathscr{G}$. Since each is a constituent of χ_P, the result follows. ∎

An observation that is often useful when working with linear groups is that if $\{K_i\}$ is a family of normal subgroups of G with $\bigcap K_i = 1$, then G can be isomorphically embedded in the direct product

$$\prod (G/K_i) \quad \text{via} \quad g \mapsto (\ldots, gK_i, \ldots).$$

Proof of Theorem 14.1 Use induction on n and for groups with degree n, induct on $|G|$. Let χ be a faithful character of G with $\chi(1) = n$. Suppose that χ is reducible and write $\chi = \chi_1 + \chi_2$. Let $K_i = \ker \chi_i$ so that G/K_i is isomorphic to a linear group of degree $\chi_i(1) < n$. By the inductive hypothesis, G/K_i has an abelian Hall π_i-subgroup where $\pi_i = \{p \mid p > 1 + \chi_i(1)\}$. Since $\pi \subseteq \pi_i$, it follows that G/K_i has an abelian Hall π-subgroup H_i/K_i. If $H_i < G$, then the inductive hypothesis yields an abelian Hall π-subgroup H of H_i. Since no prime in π divides $|(G/K_i):(H_i/K_i)| = |G:H_i|$, it follows that H is a

Hall π-subgroup of G and we are done if $H_i < G$. We may thus assume that $H_i = G$ and G/K_i is an abelian π-group for $i = 1, 2$. However, $K_1 \cap K_2 = \ker \chi_1 \cap \ker \chi_2 = \ker \chi = 1$ and thus G is isomorphic to a subgroup of the abelian π-group, $(G/K_1) \times (G/K_2)$. The result follows in this case. We now assume that χ is irreducible.

If $H \subseteq G$ is a π-subgroup, let ϑ be an irreducible constituent of χ_H. Then $\vartheta(1)$ divides $|H|$ and so a prime divisor p of $\vartheta(1)$ satisfies $n < p \leq \vartheta(1) \leq n$ and this contradiction shows that $\vartheta(1) = 1$. Thus χ_H is a sum of linear characters and since χ is faithful, it follows that H is abelian.

Suppose there exists $M \lhd G$ with $|G:M| = p \in \pi$. Let H be a Hall π-subgroup of M, which exists by the inductive hypothesis. If $H \lhd G$, let $P \in \mathrm{Syl}_p(G)$. Then HP is a Hall π-subgroup of G and we are done. Suppose then, $H \ntriangleleft G$ and choose a Sylow subgroup Q of H with $Q \ntriangleleft G$. Then Q is Sylow in M and thus $G = M\mathbf{N}_G(Q)$ by the Frattini argument. Now H is abelian and so $H \subseteq \mathbf{N}_G(Q)$. We have $|G:\mathbf{N}_G(Q)| = |M:\mathbf{N}_M(Q)|$ which divides $|M:H|$. Thus $|G:\mathbf{N}_G(Q)|$ involves no primes from π. Since $\mathbf{N}_G(Q) < G$, it has a Hall π-subgroup which is one for G. We may thus assume that no such M exists.

Next, suppose $Z \subseteq \mathbf{Z}(G)$ with $|Z| = p \in \pi$. Then $\chi_Z = \chi(1)\lambda$ with $o(\lambda) = p$. Thus $(\det(\chi))_Z = \lambda^{\chi(1)} \neq 1_Z$ since $\chi(1) < p$. Therefore, $p \mid o(\chi)$ and it follows that $p \mid |G:\ker(\det \chi)|$. This yields a normal subgroup of index p, a contradiction. (Note that we have reproved part of Theorem 5.6 here.) Thus no such Z exists.

Now let $H \subseteq G$ be a π-subgroup of maximum possible order. Suppose H is not a Hall π-subgroup. Then there exists $q \in \pi$ with $q \mid |G:H|$. In particular, $H \neq 1$ or else $H < Q \in \mathrm{Syl}_q(G)$ which violates the maximality of H. Let $x \in H$ have prime order $p \in \pi$ and let $C = \mathbf{C}_G(x)$. Then $C < G$ by the previous paragraph and thus C has a Hall π-subgroup K by the inductive hypothesis. Since H is abelian, we have $H \subseteq C$ and thus $|H| \leq |K|$. The maximality of $|H|$ yields $|H| = |K|$ and H is a Hall π-subgroup of C. In particular, $q \nmid |C:H|$ and it follows that $q \mid |G:C|$.

Let $x \in P \in \mathrm{Syl}_p(G)$. Then P is abelian and so $P \subseteq \mathbf{C}(x) = C$ and $p \nmid |G:C|$. Thus $p \neq q$. Since p can be any prime divisor of $|H|$, we have $q \nmid |H|$. Therefore, $q \nmid |C|$ since H is a Hall π-subgroup of C. We conclude that x centralizes no element of order q in G.

We claim that G contains no element of order pq. Otherwise, there exist commuting $y, z \in G$ with $o(y) = p$ and $o(z) = q$. However, H contains a full Sylow p-subgroup of G since C does and H is a Hall π-subgroup of C. Thus H contains a conjugate of y which we may suppose to be x. Since x centralizes no element of order q, this is a contradiction and proves the claim.

Lemma 14.2 now yields that χ is r-rational for $r = p$ or q. Thus $\chi(1) \geq r - 1$ by Lemma 14.3. This is a contradiction since $r \in \pi$ and the proof is complete. ∎

(14.4) COROLLARY Suppose $\chi(1)$ is prime for every $\chi \in \mathrm{Irr}(G)$ with $\chi(1) > 1$. Then G is solvable.

Proof The hypothesis is inherited by factor groups and by normal subgroups and so we may assume that G is simple. Let $\chi \in \mathrm{Irr}(G)$ with minimal $\chi(1) > 1$. By Problem 3.3, $\chi(1) = p > 2$. Let $\pi = \{q \mid q$ is prime, $q > p + 1\}$. Since G is simple, χ is faithful and Theorem 14.1 yields an abelian Hall π-subgroup $A \subseteq G$.

Let $1 \neq a \in A$. Then $A \subseteq \mathbf{C}_G(a)$ and the only prime divisors of $|G : \mathbf{C}_G(a)|$ are $\leq p + 1$ and hence $\leq p$. If $\psi \in \mathrm{Irr}(G)$ and $\psi(1) \notin \{1, p\}$, then

$$(\psi(1), |G : \mathbf{C}_G(a)|) = 1$$

and by Burnside's Theorem 3.8, we have $\psi(a) = 0$ or $a \in \mathbf{Z}(\psi)$. Since G is simple, $\mathbf{Z}(\psi) = 1$ and hence $\psi(a) = 0$. Now

$$0 = \sum_{\xi \in \mathrm{Irr}(G)} \xi(1)\xi(a) = 1 + p \sum_{\xi \in \mathrm{Irr}(G);\, \xi(1) = p} \xi(a)$$

and hence $1/p$ is an algebraic integer. This contradiction completes the proof. ∎

In the situation of Corollary 14.4, Problem 12.3 applies since G is solvable. It follows that $|\mathrm{c.d.}(G)| \leq 3$ and thus $G''' = 1$ by Theorem 12.15.

Our next results concern finding *normal* subgroups of a linear group of degree n whose order is divisible by a prime which is large when compared with n.

(14.5) THEOREM (*D. L. Winter*) Let G be a solvable irreducible linear group of degree n. Suppose that a Sylow p-subgroup of G is not normal. Then n is divisible by a prime power $q > 1$ such that $q \equiv -1, 0$, or $1 \bmod p$.

Proof Let $\chi \in \mathrm{Irr}(G)$ be faithful with $\chi(1) = n$. Suppose G is a counterexample to the theorem with minimum possible order. We argue first that every proper normal subgroup of G has index divisible by p and has a normal Sylow p-subgroup.

Let $M \lhd G$ be proper and let $\vartheta_1, \ldots, \vartheta_r$ be the distinct irreducible constituents of χ_M. Let m be the common degree of the ϑ_i. Then $m \mid n$ and hence no prime power $q > 1$ with $q \equiv -1, 0$, or 1 can divide m. Since $|M| < |G|$, it follows that $M/\ker \vartheta_i$ has a normal Sylow p-subgroup. Now $\bigcap \ker \vartheta_i \subseteq \ker \chi = 1$ and thus M is isomorphic to a subgroup of

$$(M/\ker \vartheta_1) \times \cdots \times (M/\ker \vartheta_r).$$

It follows that M has a normal Sylow p-subgroup which is necessarily normal in G. Since G does not have a normal Sylow p-subgroup, we conclude that p divides $|G:M|$ as claimed.

Now let K be a maximal normal subgroup of G and let $P \in \text{Syl}_p(K)$. Then $P \lhd G$ and $|G:K| = p$. Since G is not a p-group, we have $P < K$ and we choose $L \supseteq P$ such that K/L is a chief factor of G. Write $|K:L| = a$ and let $S \in \text{Syl}_p(G)$.

Now $LS < G$ and $p \nmid |G:LS|$. Thus $LS \ntriangleleft G$. Since $K(LS) = G$, it follows that $\mathbf{C}_{K/L}(S) < K/L$. However, $\mathbf{C}_{K/L}(S) \lhd G/L$ and we conclude that $\mathbf{C}_{K/L}(S) = 1$. Since S is a p-group, it follows that $a \equiv 1 \bmod p$.

We claim that χ_L reduces. Otherwise, χ_{LS} is irreducible and the minimality of G yields $S \lhd LS$. Since $p \nmid \chi(1)$ it follows that χ_S is a sum of linear constituents and hence S is abelian. Thus $S \subseteq \mathbf{C}_G(P) \lhd G$ and hence $p \nmid |G:\mathbf{C}_G(P)|$. By the first part of the proof, we conclude that $\mathbf{C}_G(P) = G$ and $P \subseteq \mathbf{Z}(G)$. Therefore, $L = P \times N$ where $N = \mathbf{O}_{p'}(L) \lhd G$. Since $S \lhd LS$ and $S \cap N = 1$, it follows that $S \subseteq \mathbf{C}_G(N) \lhd G$ and thus $p \nmid |G:\mathbf{C}_G(N)|$. Therefore, $\mathbf{C}_G(N) = G$ and $N \subseteq \mathbf{Z}(G)$. Thus $L = NP \subseteq \mathbf{Z}(G)$ and since $\chi_L \in \text{Irr}(L)$, we have $n = 1$ and G is abelian. This is a contradiction and proves that χ_L reduces, as claimed.

Since $p \nmid \chi(1)$, we have $\chi_K \in \text{Irr}(K)$. Since $\chi_L \notin \text{Irr}(L)$ there are two possibilities by Theorem 6.18. Either χ_L is a sum of $|K:L| = a$ distinct irreducible constituents or else $\chi_L = e\psi$, with $\psi \in \text{Irr}(L)$, and $e^2 = a$. In the first case, $a > 1$ is a prime power dividing n and $a \equiv 1 \bmod p$, a contradiction. In the second case, $e > 1$ is a prime power dividing n and $e^2 = a \equiv 1 \bmod p$. Thus $e \equiv \pm 1 \bmod p$ and the proof is complete. ∎

(14.6) COROLLARY (*Ito*) Let G be a solvable linear group of degree n and let $p \geq n + 1$ be a prime. Suppose that a Sylow p-subgroup of G is not normal. Then G is irreducible, $p = n + 1$, and n is a power of 2.

Proof If G is irreducible, then by Winter's Theorem 14.5, there exists a prime power $q > 1$ that divides n such that $q \equiv -1$, 0, or 1 mod p. Thus $p - 1 \geq n \geq q \geq p - 1$ and hence $n = p - 1$ is a prime power. We conclude that n is a power of 2 and the proof is complete in this case.

If G is reducible, write $\chi = \chi_1 + \chi_2$, where χ is a faithful character of G of degree n. Let $K_i = \ker \chi_i$ so that $K_1 \cap K_2 = 1$ and G is isomorphic to a subgroup of $(G/K_1) \times (G/K_2)$. Each G/K_i is isomorphic to a solvable linear group of degree $\chi_i(1) < n$. Working by induction on n, it follows that each G/K_i has a normal Sylow p-subgroup and hence so does their direct product. Since G does not have a normal Sylow subgroup we have a contradiction and the proof is complete. ∎

The hypothesis that G is solvable in Corollary 14.6 can easily be relaxed to the assumption that G is "p-solvable." We say that G is *p-solvable* if there exist subgroups $N_i \lhd G$ with

$$1 = N_0 \subseteq N_1 \subseteq \cdots \subseteq N_k = G$$

and such that each factor N_{i+1}/N_i is either a p-group or of order prime to p.

(14.7) COROLLARY Let G be a p-solvable linear group of degree $n \le p - 1$ and suppose that a Sylow p-subgroup of G is not normal. Then $n = p - 1$ is a power of 2 and G is irreducible.

Proof Assume either that G is reducible, $n < p - 1$ or that n is not a power of 2. Use induction on $|G|$. Let M be the next to last term in a p-solvable series for G so that $M < G$, G/M is either a p-group or a p'-group and M is p-solvable. By the inductive hypothesis, M has a normal Sylow p-subgroup P and thus we may assume that G/M is a p-group.

Let $Q/P \in \mathrm{Syl}_q(M/P)$ for some prime $q \ne p$. The Frattini argument yields $M\mathbf{N}_G(Q) = G$ and thus $|G:\mathbf{N}_G(Q)| = |M:\mathbf{N}_M(Q)|$. Since $P \subseteq Q \subseteq \mathbf{N}_M(Q)$ and $P \in \mathrm{Syl}_p(M)$, we conclude that $p \nmid |G:\mathbf{N}_G(Q)|$. Let $S \in \mathrm{Syl}_p(\mathbf{N}_G(Q))$. Thus $S \in \mathrm{Syl}_p(G)$.

Now SQ is solvable and hence $S \lhd SQ$ by Corollary 14.6. Since $|G:Q|$ is not divisible by q, we conclude that $q \nmid |G:\mathbf{N}_G(S)|$. Since $|G:\mathbf{N}_G(S)|$ is independent of the choice of $S \in \mathrm{Syl}_p(G)$ and $q \ne p$ is arbitrary, the result follows. ∎

It is conjectured that Winter's Theorem 14.5 holds for all p-solvable groups. The conjecture is known to hold for $n \le 2p + 1$. Unlike the ease with which Corollary 14.7 follows from Corollary 14.6, the facts for $n > p$ seem quite deep. Even the special case where G has a normal p-complement is open if n is significantly larger than $2p$. The results which have been obtained all seem to depend heavily on the Glauberman correspondence, and in particular on Theorem 13.14. The proofs are too complicated to give here.

Now we drop all hypotheses of solvability or p-solvability. Suppose G is a linear group of degree n and a Sylow p-subgroup of G is not normal. How big can p be? The first bound was established by Blichfeldt and the best possible bound, $p \le 2n + 1$, was proved by Feit and Thompson. The Feit–Thompson proof depends on a very deep result of Brauer which gives the bound under the additional assumption that $p^2 \nmid |G|$. Here we shall prove $p < (n + 1)^2$ and also show how the Feit–Thompson reduction to the case $p^2 \nmid |G|$ works.

We need some preliminary results.

(14.8) LEMMA Let $H \subseteq G$ be abelian and let $\mathscr{S}$ be the set of nonprincipal irreducible constituents of $(1_H)^G$. Compute $\alpha = \min\{\chi(1)/[\chi, (1_H)^G] \mid \chi \in \mathscr{S}\}$. Then $\psi(1) \ge \alpha - 1$ for every nonlinear $\psi \in \mathrm{Irr}(G)$.

Proof Let $\psi \in \text{Irr}(G)$ with $\psi(1) > 1$. Write $\psi\bar{\psi} = 1_G + \sum a_\chi \chi$, where the sum runs over nonprincipal $\chi \in \text{Irr}(G)$. Since H is abelian, we have

$$\psi(1) \leq [\psi_H, \psi_H] = [\psi_H \bar{\psi}_H, 1_H] = [\psi\bar{\psi}, (1_H)^G] = 1 + \sum_{\chi \in \mathscr{S}} a_\chi [\chi, (1_H)^G].$$

However, $[\chi, (1_H)^G] \leq \chi(1)/\alpha$ for $\chi \in \mathscr{S}$ and this yields

$$\psi(1) \leq 1 + (1/\alpha) \sum a_\chi \chi(1) \leq 1 + (1/\alpha)(\psi(1)^2 - 1).$$

Thus $\alpha(\psi(1) - 1) \leq \psi(1)^2 - 1$ and since $\psi(1) > 1$ we obtain $\alpha \leq \psi(1) + 1$ as desired. ∎

(14.9) LEMMA Let $N \lhd G$ and suppose $g \in G$ with $N \cap \mathbf{C}(g) = 1$. Let χ be a character of G such that $[\chi_N, 1_N] = 0$. Then $\chi(g) = 0$.

Proof Let $\bar{g} = gN \in G/N$. Then $\mathbf{C}_{G/N}(\bar{g}) \supseteq \mathbf{C}_G(g)N/N$ and hence $|\mathbf{C}_{G/N}(\bar{g})| \geq |\mathbf{C}_G(g)N/N| = |\mathbf{C}_G(g)|$. Thus

$$\sum_{\psi \in \text{Irr}(G/N)} |\psi(\bar{g})|^2 \geq \sum_{\xi \in \text{Irr}(G)} |\xi(g)|^2.$$

Viewing $\text{Irr}(G/N) \subseteq \text{Irr}(G)$, we conclude that $\xi(g) = 0$ for $\xi \in \text{Irr}(G)$ with $N \nsubseteq \ker \xi$. The result follows. ∎

Note that the above proof is essentially a repetition of the proof of Corollary 2.24. Also, if g and N are as in the Lemma 14.9, then all elements of the coset Ng are conjugate in G and thus the result follows from Problem 2.1(b).

(14.10) THEOREM (*Feit–Thompson*) Let $H \subseteq G$ be abelian with $H \cap \mathbf{Z}(G) = 1$. Assume for every $g \in G - \mathbf{Z}(G)$ that $\mathbf{C}_G(g)$ has nontrivial intersection with at most one conjugate of H in G. Let $\chi \in \text{Irr}(G)$ with $H \nsubseteq \ker \chi$. Then

(a) $\chi(1)^2 > |H|[\chi_H, 1_H]^2$.

Also, if $\chi(1) > 1$ and H is contained in no proper normal subgroup of G, then

(b) $(1 + \chi(1))^2 > |H|$.

Proof Let $N = \mathbf{N}_G(H)$ and

$$X = \{x \in G - \mathbf{Z}(G) \mid \mathbf{C}(x) \cap H > 1\}.$$

Note that $H - \{1\} \subseteq X$ and that $N \subseteq \mathbf{N}(X)$. We claim that X is a T.I. set and that $N = \mathbf{N}(X)$. In particular, $X \subseteq N$.

Let $g \in G$ be such that $X \cap X^g \neq \varnothing$. Choose $x \in X$ with $x^g \in X$. Then $\mathbf{C}(x) \cap H > 1$ and so $\mathbf{C}(x^g) \cap H^g > 1$. Also, $\mathbf{C}(x^g) \cap H > 1$ and hence $H = H^g$ by hypothesis. Thus $g \in N$ and $X = X^g$ and the claim is established.

We have

$$|G| = |G|[\chi, \chi] = \sum_{g \in G} |\chi(g)|^2 \geq \chi(1)^2 + |G : N| \sum_{x \in X} |\chi(x)|^2$$

and thus

(1) $$\sum_{x \in X} |\chi(x)|^2 < |N|.$$

Now write $\chi_N = \alpha + \beta$, where $[\alpha_H, 1_H] = 0$ and $\beta_H = \beta(1)1_H$. (This is possible since $H \lhd N$.) Since $H \nsubseteq \ker \chi$, we have $\alpha \neq 0$ but we allow the possibility that $\beta = 0$.

Write $Z = \mathbf{Z}(G)$. Then $\sum_{z \in Z} |\chi(z)|^2 = |Z|\chi(1)^2$ and since $Z \cap X = \varnothing$, we conclude from (1) that

(2) $$|Z|\chi(1)^2 + |N| > \sum_{x \in X \cup Z} |\chi(x)|^2 = \sum_{x \in X \cup Z} |\alpha(x) + \beta(x)|^2.$$

By Lemma 14.9, we have $\alpha(y) = 0$ for $y \in N - (X \cup Z)$ and thus

(3) $$\sum_{x \in X \cup Z} |\alpha(x) + \beta(x)|^2 = |N|[\alpha, \alpha] + 2|N|[\alpha, \beta] + \sum_{x \in X \cup Z} |\beta(x)|^2.$$

Since $[\alpha, \beta] = 0$ and $[\alpha, \alpha] \geq 1$, (2) and (3) yield

$$|Z|\chi(1)^2 + |N| > |N| + \sum_{x \in X \cup Z} |\beta(x)|^2.$$

Because $HZ \subseteq X \cup Z$, we obtain

(4) $$|Z|\chi(1)^2 > \sum_{x \in HZ} |\beta(x)|^2.$$

Write $\chi_Z = \chi(1)\lambda$. Then $\beta_Z = \beta(1)\lambda$ and since $\beta_H = \beta(1)1_H$, we have $\beta(hz) = \beta(1)\lambda(z)$ for $h \in H$ and $z \in Z$. Thus $|\beta(x)|^2 = \beta(1)^2$ for $x \in HZ$ and (4) yields

$$\beta(1)^2|H||Z| < |Z|\chi(1)^2.$$

Thus $\chi(1)^2 > \beta(1)^2|H|$ and since $\beta(1) = [\chi_H, 1_H]$, the proof of (a) is complete.

Now assume that no proper normal subgroup of G contains H. Let $\mathscr{S}$ be the set of nonprincipal irreducible constituents of $(1_H)^G$. If $\psi \in \mathscr{S}$, then $H \nsubseteq \ker \psi$ and (a) yields $\psi(1)^2 > [\psi_H, 1_H]^2|H|$ and

$$|H|^{1/2} < \alpha = \min\{\psi(1)/[\psi, (1_H)^G] \mid \psi \in \mathscr{S}\}.$$

Lemma 14.8 then gives for $\chi(1) > 1$ that

$$\chi(1) + 1 \geq \alpha > |H|^{1/2}$$

and (b) follows. ∎

(14.11) THEOREM (*Feit–Thompson–Blichfeldt*). Let G be a linear group of degree n and let p be a prime. Assume one of the following.

(a) Every subgroup of G of order not divisible by p^2 has a normal Sylow p-subgroup and $p > n + 1$.

(b) $p \geq (n + 1)^2$.

Then G has a normal Sylow subgroup.

Proof Use induction on $|G|$. We may thus suppose that every proper subgroup of G has a normal Sylow p-subgroup. Let $P \in \mathrm{Syl}_p(G)$ and assume $P \ntriangleleft G$. Since P has a faithful character of degree $n < p$ and this character can have no nonlinear irreducible constituents, we conclude that P is abelian.

If $P \cap \mathbf{Z}(G) \neq 1$, then by Theorem 5.6, we cannot have $P \cap \mathbf{Z}(G) \subseteq G'$ and thus $p \mid |G : G'|$. It follows that there exists $N \triangleleft G$ with $|G : N| = p$. Since N has a normal Sylow p-subgroup, we conclude that G is p-solvable. Then G violates Corollary 14.7 since $n < p - 1$ and $P \ntriangleleft G$. We conclude that $P \cap \mathbf{Z}(G) = 1$.

Now if $1 \neq y \in P \cap P^g$, then $P, P^g \in \mathrm{Syl}_p(\mathbf{C}(y))$ and this forces $P = P^g$ since $\mathbf{C}(y) < G$. Thus P is a T.I. set. Now suppose that $x \in G - \mathbf{Z}(G)$. Then $\mathbf{C}(x)$ has a unique Sylow p-subgroup S and if $S > 1$, then S is contained in a unique conjugate of P. It follows that $\mathbf{C}(x)$ has nontrivial intersection with at most one conjugate of P.

If $P \subseteq M \triangleleft G$ with $M < G$, then $P \triangleleft M$ and thus $P \triangleleft G$, a contradiction. We have now shown that P satisfies the hypotheses of Theorem 14.10(b) and thus $|P| < (\psi(1) + 1)^2$ for every nonlinear $\psi \in \mathrm{Irr}(G)$.

Since G is nonabelian, the given faithful character must have some nonlinear irreducible constituent and we conclude that $|P| < (n + 1)^2$. Thus $p < (n + 1)^2$, contradicting hypothesis (b). We are therefore in the situation of hypothesis (a) and in particular, $p^2 \mid |G|$. Thus $p^2 \leq |P| < (n + 1)^2$ and $p < n + 1$ which is our final contradiction. ∎

As was mentioned before, Brauer proved that if G is a linear group of degree n and $p^2 \nmid |G|$ where $p > 2n + 1$, then a Sylow p-subgroup of G is normal. It then follows from Theorem 14.11(a) that the hypothesis $p^2 \nmid |G|$ is unnecessary.

There is a great deal of information known about linear groups of degree n in which a Sylow p-subgroup is not normal and $n < p \leq 2n + 1$. This all requires deep results from Brauer's "modular character" theory and we will not discuss it further here.

There are many ways in which the structure of a group is limited in terms of its degree as a linear group. For example, there exist integer valued functions f_i such that for linear groups G of degree n we have:

(a) If G is solvable, then $\mathrm{d.l.}(G) \leq f_1(n)$.
(b) If G is p-solvable and p^e divides $|G : \mathbf{O}_p(G)|$, then $e \leq f_2(n)$.
(c) If G is a p-group and $|G : \Phi(G)| = p^e$, then $e \leq f_3(n)$.

A much more general result of this type is Jordan's theorem.

(14.12) THEOREM (*C. Jordan*) Let G be a linear group of degree n. Then there exists abelian $A \lhd G$ such that

$$|G:A| \leq (n!)12^{n(\pi(n+1)+1)}$$

where $\pi(k)$ denotes the number of primes $\leq k$.

In fact, the existence of the functions f_1, f_2, and f_3 mentioned above follows easily from Jordan's theorem, although the best known bounds for f_1, f_2, and f_3 are very much better than those which can be derived from the inequality in Theorem 14.12 (or from any other known bound for Jordan's theorem). There is no reason to suppose that the index of an abelian normal subgroup of maximum possible order in a linear group of degree n can be anywhere nearly as large as $(n!)12^{n(\pi(n+1)+1)}$. (For instance, if $n = 2$, the "correct" bound is 60 rather than $2 \cdot 12^6$.) The good bounds for the functions f_i are proved by methods independent of Jordan's theorem.

We begin work now on a proof of Theorem 14.12 which is due to Frobenius. As the reader will see, this proof has an unusual geometric flavor.

Recall that a square complex matrix U is said to be "unitary" if $\bar{U}^{\mathsf{T}} = U^{-1}$. Note that the unitary $n \times n$ matrices form a subgroup of $GL(n, \mathbb{C})$. Also, if U is unitary, then there exists unitary V such that $V^{-1}UV$ is diagonal. (More generally, this holds for all normal matrices U, that is, those which satisfy $UU^* = U^*U$, where $U^* = \bar{U}^{\mathsf{T}}$.) Also, the diagonal matrix $V^{-1}UV$ is unitary since both U and V are. Write $V^{-1}UV = \operatorname{diag}(\lambda_1, \ldots, \lambda_n)$ so that the λ_j are the eigenvalues of U. Since $V^{-1}UV$ is unitary, it follows that $\lambda_j^{-1} = \bar{\lambda}_j$ and this proves that the eigenvalues of a unitary matrix lie on the unit circle.

(14.13) LEMMA Let A and B be $n \times n$ complex unitary matrices. Assume that the eigenvalues of B lie in the interior of some arc of length π on the unit circle. Suppose that A commutes with $A^{-1}B^{-1}AB$. Then A commutes with B.

Proof We may conjugate A and B by a unitary matrix so as to diagonalize B and hence we may assume that $B = \operatorname{diag}(b_1, \ldots, b_n)$. Since a permutation matrix is unitary, we may rearrange the b_j in any desired order by conjugating both A and B by an appropriate permuation matrix. We may thus write $b_j = e^{i\vartheta_j}$ where $\vartheta_1 \leq \vartheta_2 \leq \cdots \leq \vartheta_n < \vartheta_1 + \pi$.

Write $A = (a_{\mu\nu})$. We shall show that $a_{\mu\nu} = 0$ if $b_\mu \neq b_\nu$. This immediately yields that $AB = BA$. Now write $C = A^{-1}B^{-1}AB$. We have $A^{-1} = \bar{A}^{\mathsf{T}}$ and $B^{-1} = \bar{B}$ so that $C = \bar{A}^{\mathsf{T}}\bar{B}AB$. Also, since $AC = CA$ we have

$$\bar{A}^{\mathsf{T}}\bar{B}AB = C = ACA^{-1} = B^{-1}ABA^{-1} = \bar{B}AB\bar{A}^{\mathsf{T}}$$

Now evaluate the diagonal entry, c_{mm} of C. We have

$$\sum_\mu \bar{a}_{\mu m}\bar{b}_\mu a_{\mu m} b_m = c_{mm} = \sum_\nu \bar{b}_m a_{m\nu} b_\nu \bar{a}_{m\nu}$$

and thus

$$\sum_{\mu} |a_{\mu m}|^2 \bar{b}_{\mu} b_m = \sum_{\nu} |a_{m\nu}|^2 \bar{b}_m b_{\nu}.$$

We compare imaginary parts and obtain

$$\sum_{\mu} |a_{\mu m}|^2 \sin(\vartheta_m - \vartheta_{\mu}) = \sum_{\nu} |a_{m\nu}|^2 \sin(\vartheta_{\nu} - \vartheta_m).$$

Thus for all m, $1 \leq m \leq n$, we have

$$(*) \qquad \sum_{\nu} (|a_{m\nu}|^2 + |a_{\nu m}|^2) \sin(\vartheta_{\nu} - \vartheta_m) = 0.$$

We now use $(*)$ to prove that if $\vartheta_j \neq \vartheta_k$, then $a_{jk} = 0$. Suppose this is false and choose j minimal such that there exists k with $\vartheta_j < \vartheta_k$ and either $a_{jk} \neq 0$ or $a_{kj} \neq 0$. Take $m = j$ in $(*)$. If $\nu \leq j$, then either $\vartheta_{\nu} = \vartheta_j$ in which case $\sin(\vartheta_{\nu} - \vartheta_j) = 0$ or $\vartheta_{\nu} < \vartheta_j$ in which case $a_{\nu j} = 0 = a_{j\nu}$ by the minimality of j. Thus $(*)$ yields

$$\sum_{\nu > j} (|a_{j\nu}|^2 + |a_{\nu j}|^2) \sin(\vartheta_{\nu} - \vartheta_j) = 0.$$

Now for $\nu > j$ we have $\vartheta_j \leq \vartheta_{\nu} < \vartheta_j + \pi$ and thus $\sin(\vartheta_{\nu} - \vartheta_j) \geq 0$. Also, $|a_{j\nu}|^2 + |a_{\nu j}|^2 \geq 0$. We conclude that for each value $\nu > j$, either

$$|a_{j\nu}|^2 + |a_{\nu j}|^2 = 0$$

or $\sin(\vartheta_{\nu} - \vartheta_j) = 0$. Now $0 < \vartheta_k - \vartheta_j < \pi$ and hence $\sin(\vartheta_k - \vartheta_j) > 0$. Therefore $|a_{jk}|^2 + |a_{kj}|^2 = 0$ and thus $a_{jk} = 0 = a_{kj}$. This is a contradiction and completes the proof. ∎

(14.14) LEMMA Let A and B be $n \times n$ complex unitary matrices and suppose that the eigenvalues of A lie in an arc of length $\sigma < \pi$ on the unit circle. Then the eigenvalues of $A^{-1}B^{-1}AB$ lie on the unit circle between $-\sigma$ and σ.

Proof Certainly $A^{-1}B^{-1}AB$ is unitary and so its eigenvalues lie on the unit circle. Let $\mathfrak{a}$ be an arc of length σ which contains the eigenvalues of A and write $C = B^{-1}AB$. We need to find the eigenvalues of $A^{-1}C$.

Let λ be an eigenvalue of $A^{-1}C$ and let the column vector x be a corresponding eigenvector so that $A^{-1}Cx = \lambda x$ and thus $Cx = \lambda Ax$ and $\bar{x}^{\mathsf{T}}Cx = \lambda(\bar{x}^{\mathsf{T}}Ax)$. Thus it suffices to show that $\arg(\bar{x}^{\mathsf{T}}Ax) \in \mathfrak{a}$ and $\arg(\bar{x}^{\mathsf{T}}Cx) \in \mathfrak{a}$, where $\arg(\alpha) = \alpha/|\alpha|$ for $0 \neq \alpha \in \mathbb{C}$. Since both A and C are unitary with eigenvalues in $\mathfrak{a}$ it suffices to show that

$$\arg(\bar{y}^{\mathsf{T}}Uy) \in \mathfrak{a}$$

whenever $y \neq 0$ and U is unitary with eigenvalues in $\mathfrak{a}$. Let U and y be such and let V be unitary with $V^{-1}UV = D$, a diagonal matrix. Let $z = V^{-1}y$ so that

$$\bar{y}^{\mathsf{T}}Uy = \bar{z}^{\mathsf{T}}\bar{V}^{\mathsf{T}}UVz = \bar{z}^{\mathsf{T}}Dz$$

since $\bar{V}^{\mathsf{T}} = V^{-1}$. Now if $D = \operatorname{diag}(d_1, \ldots, d_n)$ and $z = \operatorname{col}(z_1, \ldots, z_n)$, we have

$$\bar{z}^{\mathsf{T}} D z = \sum |z_i|^2 d_i.$$

Since the d_i lie in $\mathfrak{a}$ and $|z_i|^2 \geq 0$, it follows that $\sum |z_i|^2 d_i$ lies in the infinite wedge which is the union of all rays from the origin through points of $\mathfrak{a}$. Since $\sigma < \pi$ and some $|z_i|^2$ is nonzero, it follows that $\sum |z_i|^2 d_i \neq 0$ and $\arg(\sum |z_i|^2 d_i) \in \mathfrak{a}$ as desired. ∎

(14.15) THEOREM (*Frobenius*) Let G be a linear group, let $A, B \in G$ and suppose that the eigenvalues of A and B lie in arcs of length α and β respectively, on the unit circle. Then

(a) If $\beta < \pi$ and A commutes with the commutator $[A, B]$, then $[A, B] = 1$.
(b) If $\alpha < \pi/3$ and $\beta < \pi$, then $[A, B] = 1$.

Proof By Theorem 4.17 we can conjugate all of the elements of G by a matrix such that the resulting matrices are all unitary. We may thus suppose that A and B are unitary matrices. Now (a) is just a restatement of Lemma 14.13.

To prove (b), conjugate by a unitary matrix which diagonalizes A. We may thus assume that A is diagonal. Construct elements $B_i \in G$ by setting $B_0 = B$ and $B_i = [A, B_{i-1}]$ for $i \geq 1$.

By Lemma 14.14, the eigenvalues of B_i lie between $-\pi/3$ and $\pi/3$ on the unit circle for $i \geq 1$ and thus for each $i \geq 0$, they lie in the interior of some arc of length π. Thus by part (a) we see that if $B_{i+1} = 1$ for any $i \geq 0$, then $B_i = 1$. It thus suffices to prove that $B_i = 1$ for some i.

If M is any complex matrix, we define $\vartheta(M) = \operatorname{tr}(\bar{M}^{\mathsf{T}} M)$ so that if $M = (m_{\mu\nu})$, we have $\vartheta(M) = \sum_{\mu\nu} |m_{\mu\nu}|^2$. Thus $\vartheta(M) \geq 0$ and $\vartheta(M) = 0$ iff M is a zero matrix. If U is unitary, we have

$$\vartheta(UM) = \operatorname{tr}(\bar{M}^{\mathsf{T}} \bar{U}^{\mathsf{T}} U M) = \operatorname{tr}(\bar{M}^{\mathsf{T}} M) = \vartheta(M).$$

Now we compute

$$(*) \qquad \begin{aligned} \vartheta(B_{i+1} - 1) &= \vartheta(B_i A(B_{i+1} - 1)) = \vartheta(B_i A(A^{-1} B_i^{-1} A B_i - 1)) \\ &= \vartheta(AB_i - B_i A) = \vartheta(A(B_i - 1) - (B_i - 1)A). \end{aligned}$$

Now write $B_i - 1 = (b_{\mu\nu})$. Since A is diagonal, we can write $A = \operatorname{diag}(a_1, a_2, \ldots, a_n)$, where all of the a_j lie inside some arc of length $\pi/3$. The (μ, ν) entry of $A(B_i - 1) - (B_i - 1)A$ is $a_\mu b_{\mu\nu} - b_{\mu\nu} a_\nu = b_{\mu\nu}(a_\mu - a_\nu)$. Note that $|a_\mu - a_\nu| < 1$. Equation $(*)$ now yields that

$$\vartheta(B_{i+1} - 1) = \sum_{\mu, \nu} |b_{\mu\nu}|^2 |a_\mu - a_\nu|^2 \leq \sum_{\mu, \nu} |b_{\mu\nu}|^2 = \vartheta(B_i - 1),$$

where in fact the inequality is strict unless all $b_{\mu\nu} = 0$.

If no $B_i = 1$, we therefore have

$$\vartheta(B_0 - 1) > \vartheta(B_1 - 1) > \vartheta(B_2 - 1) > \cdots.$$

Since G is a finite group, there are only finitely many B_i and this is a contradiction. We conclude that $B_i = 1$ for some i and the proof is complete. ∎

Observe that we have used our standing assumption that the groups we consider are finite twice in the above proof. The first time was in the appeal to Theorem 4.17 which is false for infinite groups. This use of finiteness could have been avoided by assuming that G was contained in the group of unitary matrices. The second application of finiteness in the last paragraph of the proof is more fundamental. For instance, part (b) of the theorem is not valid for the full unitary group.

(14.16) THEOREM Let G be a linear group of degree n. Then there exists abelian $A \lhd G$ such that

$$|H : H \cap A| \leq 12^n$$

for every abelian subgroup $H \subseteq G$.

Proof Since G is a finite group, the eigenvalues of every element of G lie on the unit circle. If $g \in G$, let $\alpha(g)$ be the length of the shortest closed arc which contains the eigenvalues of g. Note that α is a class function on G. Put $A = \langle a \in G \mid \alpha(a) < \pi/3 \rangle$. Then $A \lhd G$ and A is abelian since the generators of A commute by Theorem 14.15(b).

Now let $H \subseteq G$ with H abelian. By Maschke's Theorem 1.9 we may conjugate all of the matrices in G so as to assume that all $h \in H$ are diagonal matrices.

Partition the unit circle into twelve half-open arcs, $\mathfrak{a}_1, \mathfrak{a}_2, \ldots, \mathfrak{a}_{12}$, each of length $\pi/6$. Each element $h \in H$ determines a function f_h from $\{i \mid 1 \leq i \leq n\}$ into $\{j \mid 1 \leq j \leq 12\}$ by writing $h = \mathrm{diag}(\alpha_1, \ldots, \alpha_n)$ and setting $f_h(i) = j$ if $\alpha_i \in \mathfrak{a}_j$. Note that there are at most 12^n different functions f_h that can be obtained this way.

Suppose $x, y \in H$ with $f_x = f_y$. Write $x = \mathrm{diag}(\alpha_1, \ldots, \alpha_n)$ and $y = \mathrm{diag}(\beta_1, \ldots, \beta_n)$. Then for each i, α_i and β_i lie in the same $\mathfrak{a}_j$ and hence $\alpha_i\overline{\beta}_i$ lies strictly between $-\pi/6$ and $\pi/6$ on the unit circle. Thus $xy^{-1} \in A$ and it follows that elements in distinct cosets of $A \cap H$ in H determine different functions. Thus $|H : A \cap H| \leq 12^n$ and the proof is complete. ∎

(14.17) LEMMA Let P be a linear p-group of degree n. Then there exists abelian $A \lhd P$ such that $|P : A|$ divides $n!$

Proof By Corollary 6.14, P is an M-group and hence if $\psi \in \mathrm{Irr}(P)$, then there exists $H \subseteq P$ with $|P : H| = \psi(1)$ and such that ψ_H has a linear constituent. Choose such a subgroup H_ψ for each $\psi \in \mathrm{Irr}(P)$.

Let χ be a faithful character of P of degree n and let Ω be the set of all right cosets of all H_ψ for irreducible constituents ψ of χ. Then $|\Omega| \le n$ and G acts on Ω by right multiplication. Let A be the kernel of this action. Thus $|P:A|$ divides $n!$

To show that A is abelian, it suffices to show that χ_A is a sum of linear characters. However, for each irreducible constituent ψ of χ we have $A \subseteq H_\psi$ and thus ψ_A has a linear constituent. Since $A \lhd G$, it follows that all irreducible constituents of ψ_A are linear and the result follows. ∎

Proof of Theorem 14.12 Let $A \lhd G$ be as in Theorem 14.16 so that A is abelian and $|H : H \cap A| \le 12^n$ for every abelian subgroup $H \subseteq G$. We claim that $|G:A| \le (n!)12^{n(\pi(n+1)+1)}$. Factor $|G:A|$ into prime powers. For each prime p let $r_p = |G:A|_p$.

By Blichfeldt's Theorem 14.1, there exists an abelian Hall subgroup H_0 of G for the set of primes $> n+1$. We have $|G:A| = |G:H_0A||H_0A:A|$ and it follows that

$$\prod_{p>n+1} r_p = |H_0A:A| = |H_0:H_0 \cap A| \le 12^n.$$

For $p \le n+1$, let $P \in \mathrm{Syl}_p(G)$ and let $H_p \subseteq P$ be abelian with $|P:H_p|$ dividing $n!$ (using Lemma 14.17). Then $|P:H_p| \le (n!)_p$ and

$$\begin{aligned} r_p = |PA:A| = |PA:H_pA||H_pA:A| &= |P:P \cap H_pA||H_p:H_p \cap A| \\ &\le |P:H_p|12^n \le (n!)_p 12^n. \end{aligned}$$

Thus

$$\prod_{p \le n+1} r_p \le \left(\prod_p (n!)_p\right)12^{n\pi(n+1)} = (n!)12^{n\pi(n+1)}.$$

The result now follows. ∎

We can also use Frobenius' Theorem 14.15 in a different way.

(14.18) THEOREM Let $p \ge 7$ and let G be an irreducible linear group of degree $n < 2p$ with $n \ne p$. If the Sylow p-subgroups of G are nonabelian, then $p^2 \nmid |G:\mathbf{O}_p(G)|$. Thus in any case, $G/\mathbf{O}_p(G)$ has abelian Sylow p-subgroups.

Proof Let $P \in \mathrm{Syl}_p(G)$ be nonabelian so that $P' \cap \mathbf{Z}(P) \ne 1$. Let $U \subseteq P' \cap \mathbf{Z}(P)$ with $|U| = p$.

Let $\chi \in \mathrm{Irr}(G)$ be the character of the given faithful representation so that $\chi(1) = n$. Since P is not abelian, χ_P has some nonlinear irreducible constituent ψ. We must have $\psi(1) = p$. Since $\chi(1) < 2p$, we conclude that $\chi_P = \psi + \Lambda$, where Λ is a sum of linear characters.

Since $U \subseteq P'$, we have $U \subseteq \ker \Lambda$ and since $U \subseteq \mathbf{Z}(P)$, we also have $\psi_U = p\mu$ for some linear $\mu \in \mathrm{Irr}(U)$. Thus $\chi_U = p\mu + (n-p)1_U$. Since χ is faithful, we have $\mu \ne 1_U$.

Now pick $u \in U$, with $\mu(u) = e^{2\pi i/p}$, and note that $2\pi/p < \pi/3$ since $p \geq 7$. Therefore, u has only two distinct eigenvalues, namely 1 and $e^{2\pi i/p}$, and these lie in an arc of length $< \pi/3$ on the unit circle. By Frobenius' Theorem 14.15(b) it follows that u commutes with all of its conjugates in G and thus $A = \langle u^g \mid g \in G \rangle \lhd G$ is abelian.

Write $\chi_A = e \sum_{i=1}^{t} \lambda_i$, where the λ_i are distinct linear characters of A and $et = n$. Since $U = \langle u \rangle \subseteq A$, we conclude that e divides $[\chi_U, \mu] = p$. Since also $e \mid n$, we have $e = 1$ and $t = n$. Now let K be the kernel of the permutation action of G on $\{\lambda_i\}$. Then $|G : K|$ divides $n!$ and hence $p^2 \nmid |G : K|$. If ϑ is any irreducible constituent of χ_K, then ϑ_A is the sum of an orbit of the action of K on $\{\lambda_i\}$. By definition of K, it follows that $\vartheta(1) = 1$. We conclude that K is abelian and so the Sylow p-subgroup of K is contained in $\mathbf{O}_p(G)$. The result follows. ∎

We now present the promised generalization of Lemma 14.3.

(14.19) THEOREM (*Schur*) Let G have a faithful p-rational character of degree n and let $P \in \mathrm{Syl}_p(G)$ with $|P| = p^a$. Then

$$a \leq \sum_{i=0}^{\infty} [n/(p-1)p^i] < np/(p-1)^2.$$

The brackets above denote the greatest integer function and thus the "infinite" sum has only finitely many nonzero terms. If $n < p - 1$ then all terms in the sum are zero and the result that $a = 0$ is Lemma 14.3.

Note that G is really irrelevant in Theorem 14.19; the result is really about P. Also, a p-rational character of a p-group is necessarily rational valued since the values lie in $\mathbb{Q}_{p^a} \cap \mathbb{Q}_r$ where $p \nmid r$.

Suppose $\chi \in \mathrm{Irr}(P)$ is faithful and we wish to bound $|P|$. Obviously, knowledge of $\chi(1)$ is not sufficient since even if $\chi(1) = 1$, P could be an unboundedly large cyclic group. However, $|P|$ is bounded in terms of $\chi(1)$ and the degree of the field extension $\mathbb{Q}(\chi) \supseteq \mathbb{Q}$. If $z \in \mathbf{Z}(P)$ with $o(z) = p$, then $\chi(z) = \chi(1)\varepsilon$, where ε is a primitive pth root of 1. Thus $\varepsilon \in \mathbb{Q}(\chi)$ and hence $p - 1$ divides $|\mathbb{Q}(\chi) : \mathbb{Q}|$. Since $|\mathbb{Q}_{p^a} : \mathbb{Q}| = (p-1)p^{a-1}$, it follows that if $P \neq 1$, then $|\mathbb{Q}(\chi) : \mathbb{Q}|$ is of the form $(p-1)p^t$ for some integer $t \geq 0$.

(14.20) THEOREM Let $|P| = p^a$, where p is a prime and $a > 0$. Let $\chi \in \mathrm{Irr}(P)$ be faithful with $\chi(1) = f$ and $|\mathbb{Q}(\chi) : \mathbb{Q}| = (p-1)p^t$. Then

$$a \leq (f-1)/(p-1) + (t+1)f.$$

Proof We use induction on f. If $f = 1$, then P is cyclic and since χ is faithful, it follows that $\mathbb{Q}(\chi) = \mathbb{Q}_{p^a}$ and thus $t = a - 1$. In this case, the inequality is actually equality.

Now suppose $f > 1$. Since P is an M-group, there exists a maximal subgroup H and $\vartheta \in \mathrm{Irr}(H)$ with $\vartheta^P = \chi$. Thus $|P:H| = p$ and $\chi_H = \sum_{i=1}^{p} \vartheta_i$, where the ϑ_i are the distinct conjugates of ϑ under P. Note that the fields $\mathbb{Q}(\vartheta_i)$ are all equal and that since $\vartheta^P = \chi$, we have $\mathbb{Q}(\chi) \subseteq \mathbb{Q}(\vartheta)$.

Suppose $\mathbb{Q}(\chi) = \mathbb{Q}(\vartheta)$. Let $K_i = \ker \vartheta_i$. Thus $\bigcap K_i = 1$ and H is isomorphically contained in $(H/K_1) \times \cdots \times (H/K_p)$. The $|H:K_i|$ are all equal, say to p^b. Thus $a \le pb + 1$. Since $\vartheta(1) = f/p$, the inductive hypothesis yields

$$b \le ((f/p) - 1)/(p - 1) + (t + 1)f/p$$

and hence

$$a \le pb + 1 \le (f - p)/(p - 1) + (t + 1)f + 1 = (f - 1)/(p - 1) + (t + 1)f$$

as required.

The remaining case is where $\mathbb{Q}(\chi) < \mathbb{Q}(\vartheta)$. Consider the Galois group $\mathcal{G} = \mathcal{G}(\mathbb{Q}(\vartheta)/\mathbb{Q}(\chi))$ so that $|\mathcal{G}| = |\mathbb{Q}(\vartheta):\mathbb{Q}(\chi)|$. In particular, $\mathcal{G} \ne 1$ is a p-group. Since $\mathcal{G}$ fixes χ, it permutes the irreducible constituents ϑ_i of χ_H. Since the stabilizer of ϑ in $\mathcal{G}$ is trivial, the orbit of ϑ under $\mathcal{G}$ has size $|\mathcal{G}|$. It follows that $|\mathcal{G}| = p$ and $|\mathbb{Q}(\vartheta):\mathbb{Q}| = (p - 1)p^{t+1}$. Also, the ϑ_i are $\mathcal{G}$-conjugate and hence all $\ker \vartheta_i$ are equal. Since $\bigcap \ker \vartheta_i = 1$, we conclude that $\ker \vartheta = 1$. We can now apply the inductive hypothesis to $\vartheta \in \mathrm{Irr}(H)$ and obtain

$$\begin{aligned} a - 1 \le ((f/p) - 1)/(p - 1) + (t + 2)f/p &= (f - 1)/(p - 1) + (t + 1)f/p \\ &\le (f - 1)/(p - 1) + (t + 1)f - 1 \end{aligned}$$

and the proof is complete. ∎

We note that if $p \ne 2$, it is always possible to choose H in the above proof so that $\mathbb{Q}(\chi) = \mathbb{Q}(\vartheta)$. The inequality in Theorem 14.20 is best possible.

Proof of Theorem 14.19 Define the function α on positive integers by

$$\alpha(k) = \sum_{i=0}^{\infty} [k/(p - 1)p^i].$$

Since $[k/(p - 1)p^i] \le k/(p - 1)p^i$ and this inequality is strict for large i, we have

$$\alpha(k) < (k/(p - 1)) \sum_{i=0}^{\infty} p^{-i} = kp/(p - 1)^2$$

and the second inequality in the statement of the theorem follows.

Since $[x + y] \ge [x] + [y]$, we have $\alpha(k + l) \ge \alpha(k) + \alpha(l)$. We know that χ_P is rational valued, where χ is the given character of G. If $\chi_P = \chi_1 + \chi_2$, where the χ_i are rational valued, then working by induction on n we have $a_i \le \alpha(\chi_i(1))$, where $|P/\ker \chi_i| = p^{a_i}$. By the usual argument, $a \le a_1 + a_2$

and so

$$a \le \alpha(\chi_1(1)) + \alpha(\chi_2(1)) \le \alpha(\chi_1(1) + \chi_2(1)) = \alpha(n)$$

as desired.

We may now suppose that χ_P is not the sum of two rational characters. It follows that $\chi_P = \sum_{i=1}^r \xi_i$, where the $\xi_i \in \mathrm{Irr}(P)$ constitute an orbit under the Galois group $\mathcal{G} = \mathcal{G}(\mathbb{Q}_{p^a}/\mathbb{Q})$. Let $\xi = \xi_1$. Then $\ker \xi_i = \ker \xi$ for all i and thus $1 = \bigcap \ker \xi_i = \ker \xi$ and ξ is faithful. The stabilizer in $\mathcal{G}$ of ξ is $\mathcal{G}(\mathbb{Q}_{p^a}/\mathbb{Q}(\xi))$ and it follows that the orbit size $r = |\mathbb{Q}(\xi):\mathbb{Q}| = (p-1)p^t$ for some $t \ge 0$. Let $f = \xi(1) = \xi_i(1)$ for all i so that $n = (p-1)p^t f$. Now

$$\begin{aligned}\alpha(n) &= 1 + p + p^2 + \cdots + p^t f \\ &= (f-1)/(p-1) + f + pf + \cdots + p^t f \\ &\ge (f-1)/(p-1) + (t+1)f \ge a,\end{aligned}$$

where the last inequality is by Theorem 14.20. ∎

As an application of Schur's Theorem 14.19, we prove the following.

(14.21) THEOREM Let G be a p-solvable linear group of degree n. Let $P \in \mathrm{Syl}_p(G)$ and $|P : \mathbf{O}_p(G)| = p^a$. Then

$$a \le \sum_{i=0}^{\infty} [n/(p-1)p^i].$$

The proof of Theorem 14.21 depends on some standard facts about p-solvable groups. The first is nearly trivial, namely that subgroups and factor groups of p-solvable groups are p-solvable. The second fact is a special case of what is often called "Lemma 1.2.3" since that was the designation in the paper of P. Hall and G. Higman where it first appeared.

(14.22) LEMMA (*Hall–Higman*) Let G be p-solvable with $\mathbf{O}_p(G) = 1$. Then $\mathbf{C}_G(\mathbf{O}_{p'}(G)) \subseteq \mathbf{O}_{p'}(G)$.

Proof Let $H = \mathbf{O}_{p'}(G)$ and $C = \mathbf{C}_G(\mathbf{O}_{p'}(G))$. Then $C \lhd G$ and C is p-solvable. Let $D = \mathbf{O}_{p'}(C)$ so that $D \lhd G$ and thus $D \subseteq H$. If $H \not\supseteq C$, then C/D is a nontrivial p-solvable group and $\mathbf{O}_{p'}(C/D) = 1$. Thus $\mathbf{O}_p(C/D) > 1$ and we let $E/D = \mathbf{O}_p(C/D)$ and $P \in \mathrm{Syl}_p(E)$ so that $P > 1$.

Now $D \subseteq H \subseteq \mathbf{C}(C) \subseteq \mathbf{N}(P)$ and thus $P \lhd PD = E \lhd G$. It follows that $P \lhd G$. Since $P > 1$ and $\mathbf{O}_p(G) = 1$, this is a contradiction and the proof is complete. ∎

Proof of Theorem 14.21 Let $U = \mathbf{O}_p(G)$ and $H/U = \mathbf{O}_{p'}(G/U)$. By the Schur–Zassenhaus Theorem, U is complemented in H and we may choose a complement K. Let $N = \mathbf{N}_G(K)$. Since G permutes the set of complements for U in H and H is transitive on this set, it follows that $G = NH = NU$. Let $S \in \mathrm{Syl}_p(N)$ and put $G_0 = KS$. We have $N \cap U \subseteq S \subseteq G_0$ and

$$|G:U|_p = |N:N \cap U|_p = |G_0 : N \cap U|_p.$$

We claim that $N \cap U = \mathbf{O}_p(G_0)$ and thus it suffices to assume $G = G_0$. To prove the claim, note that $N \cap U \lhd N$ and so $N \cap U \lhd G_0$ and thus $N \cap U \subseteq V = \mathbf{O}_p(G_0)$. Since $V \lhd G_0$, $K \lhd G_0$ and $V \cap K = 1$, we have $V \subseteq \mathbf{C}(K)$ and thus $VU/U \subseteq \mathbf{C}(KU/U) = \mathbf{C}(H/U)$. Since $\mathbf{O}_p(G/U) = 1$, Lemma 14.22 yields $\mathbf{C}(H/U) \subseteq H/U$ and hence $VU/U \subseteq H/U$ and $VU \subseteq H$. Since VU is a p-group and $U \in \mathrm{Syl}_p(H)$, we have $VU = U$ and thus $V \subseteq U$. Therefore, $V \subseteq N \cap U$ and the claim is established.

We may now assume that $G_0 = G$ so that K is a normal p-complement for G. Let χ be a faithful character of G with $\chi(1) = n$. For each irreducible constituent ϑ of χ_K, let $\hat{\vartheta}$ be the canonical extension of ϑ to $I_G(\vartheta)$ and let $\vartheta^* = (\hat{\vartheta})^G \in \mathrm{Irr}(G)$. Note that any field automorphism of $\mathbb{Q}_{|G|}$ which fixes ϑ also fixes ϑ^*, and it follows that ϑ^* is p-rational.

Choose a set $\mathscr{S}$ of representatives for the G-orbits of irreducible constituents of χ_K and define $\psi = \sum_{\vartheta \in \mathscr{S}} \vartheta^*$. Then

$$n = \chi(1) \geq \sum_{\vartheta \in \mathscr{S}} \vartheta(1)|G : I_G(\vartheta)| = \sum_{\vartheta \in \mathscr{S}} \vartheta^*(1) = \psi(1).$$

Also χ_K and ψ_K have the same sets of irreducible constituents and so $\ker \psi \cap K = \ker \chi \cap K = 1$. Thus $\ker \psi$ is a p-group and $\ker \psi \subseteq \mathbf{O}_p(G)$. Now application of Schur's Theorem 14.19 to $G/\ker \psi$ yields the desired result. ∎

Write $\alpha_p(k) = \sum_{i=0}^{\infty} [k/(p-1)p^i]$. With further work one can replace the inequality $a \leq \alpha_p(n)$ in Theorem 14.21 by $a \leq \alpha_p(\mu n)$, where $\mu = \frac{2}{3}$ if $p = 2$ and $\mu = (p-1)/p$ if $p \neq 2$ is not of the form $2^e + 1$. With these improvements, equality can be obtained for all n. If G is solvable, this improvement is due to J. D. Dixon. The general p-solvable case was done by D. L. Winter and depends on the results of Chapter 13.

Suppose G is a primitive linear group of degree n. In other words, the identity map is a primitive representation in the sense of Chapter 5. By Corollary 6.13, it follows that every abelian normal subgroup of G is central and thus by Jordan's Theorem 14.12 there are only finitely many possiblities for the group $G/\mathbf{Z}(G)$. These have been explicitly enumerated for certain small values of n. We give a sample of this type of result although this proof is not typical.

(14.23) THEOREM Let G be a primitive linear group of degree 2. Then $|G : \mathbf{Z}(G)| = 12, 24$, or 60.

Proof Write $Z = \mathbf{Z}(G)$ and let $\chi \in \mathrm{Irr}(G)$ be faithful with $\chi(1) = 2$. If $H \subseteq G$ is nonabelian, then $\chi_H \in \mathrm{Irr}(H)$ and it follows that $\mathbf{Z}(H) \subseteq \mathbf{Z}(\chi) = Z$. Thus if $g \in G - Z$, then $\mathbf{C}(g)$ is abelian.

Let $\mathscr{S}$ be the set of maximal abelian subgroups of G. Then clearly $G = \bigcup \mathscr{S}$ and $Z = \bigcap \mathscr{S}$. If $A, B \in \mathscr{S}$ with $A \cap B > Z$, then $\mathbf{C}(A \cap B)$ is abelian and contains both A and B. Therefore $A = B$ and thus G is a disjoint union:

$$G = Z \cup \bigcup_{A \in \mathscr{S}} (A - Z).$$

For $A \in \mathscr{S}$ we have $[\chi_A, \chi_A] = 2$ and $[\chi_Z, \chi_Z] = 4$. Thus

$$\sum_{x \in A - Z} |\chi(x)|^2 = 2|A| - 4|Z| = 2|A - Z| - 2|Z|.$$

This yields

$$\begin{aligned} |G| = \sum_{x \in G} |\chi(x)|^2 &= \sum_{x \in Z} |\chi(x)|^2 + \sum_{A \in \mathscr{S}} \sum_{x \in A - Z} |\chi(x)|^2 \\ &= 4|Z| + 2 \sum_{A \in \mathscr{S}} |A - Z| - 2|\mathscr{S}||Z| \\ &= 4|Z| + 2(|G| - |Z|) - 2s|Z| \end{aligned}$$

where $s = |\mathscr{S}|$. We thus obtain $|G : Z| = 2s - 2$.

Now $\mathscr{S}$ is a union of conjugacy classes of subgroups. If $A \in \mathscr{S}$, we want to compute $N = \mathbf{N}_G(A)$. If $A < N$ then N is nonabelian and $\chi_N \in \operatorname{Irr}(N)$. However, $\chi_A = \lambda + \mu$ with $\lambda \neq \mu$. It follows that $|N : I_N(\lambda)| = 2$ and the restriction of χ to $I_N(\lambda)$ is reducible. Thus $I_N(\lambda)$ is abelian and hence $A = I_N(\lambda)$. We conclude that for all $A \in \mathscr{S}$, we have $|\mathbf{N}(A) : A| \leq 2$. Also $A \ntriangleleft G$.

We now consider the group $G/Z = \bar{G}$ of order $2s - 2$. Let $\mathscr{T} = \{A/Z \mid A \in \mathscr{S}\}$ so that $\mathscr{T}$ partitions $\bar{G}$ and $|\mathscr{T}| = s$. Also, $\mathscr{T}$ is a union of conjugacy classes of subgroups of $\bar{G}$. Write $\mathscr{T} = \mathscr{T}_1 \cup \mathscr{T}_2 \cup \cdots \cup \mathscr{T}_r$, where the $\mathscr{T}_i$ are the distinct conjugacy classes. Let a_i be the common size of the subgroups in $\mathscr{T}_i$ and let $t_i = |\mathscr{T}_i|$.

If $B \in \mathscr{T}_i$, then $|\bar{G} : \mathbf{N}(B)| = |\bar{G} : B|/2$ or $|\bar{G} : \mathbf{N}(B)| = |\bar{G} : B|$ and hence

(1) $$t_i = (s - 1)/a_i \qquad \text{or} \qquad t_i = 2(s - 1)/a_i.$$

We also have

(2) $$\sum t_i = s,$$

(3) $$\sum t_i(a_i - 1) = 2(s - 1) - 1,$$

and all $t_i > 1$ and $a_i > 1$. We may assume that $a_1 \leq a_2 \leq \cdots \leq a_r$ and that if $a_i = a_{i+1}$ then $t_i \geq t_{i+1}$.

If $a_1 > 2$ then $\sum t_i(a_i - 1) \geq 2 \sum t_i = 2s$ which contradicts (3). Thus $a_1 = 2$ and $t_1 = s - 1$ or $(s - 1)/2$. However, if $t_1 = s - 1$, then (2) yields

$t_2 = 1$, a contradiction. Thus $t_1 = (s-1)/2$. If $a_2 = 2$, then $t_2 = (s-1)/2$ forcing $t_3 = 1$, which is impossible. Thus $a_2 \geq 3$. If $a_2 \geq 4$, then

$$2(s-1) - 1 = \sum t_i(a_i - 1) \geq (s-1)/2 + 3\sum_{i=2}^{r} t_i$$
$$= (s-1)/2 + 3(s - (s-1)/2)$$
$$= 2(s-1) + 3,$$

a contradiction. Thus $a_2 = 3$ and $t_2 = (s-1)/3$ or $2(s-1)/3$.

Suppose $t_2 = 2(s-1)/3$. Then

$$s \geq t_1 + t_2 = (s-1)/2 + 2(s-1)/3 = 7(s-1)/6 > s - 1$$

and hence $s = 7(s-1)/6$ which yields $s = 7$ and $|G:Z| = 12$. We now suppose that $t_2 = (s-1)/3$. Since $t_1 + t_2 = 5(s-1)/6 < s$, we have $r \geq 3$.

If $a_3 \geq 6$, Equation (3) yields

$$2(s-1) - 1 = \sum t_i(a_i - 1) = (s-1)/2 + 2(s-1)/3 + \sum_{i=3}^{r} t_i(a_i - 1)$$
$$\geq 7(s-1)/6 + 5(s - t_1 - t_2)$$
$$= 7(s-1)/6 + 5(s-1)/6 + 5,$$

a contradiction. Thus $a_3 = 3, 4$, or 5.

If $a_3 = 3 = a_2$ then $t_3 \leq t_2$ and so $t_3 = (s-1)/3$. This yields

$$s \geq t_1 + t_2 + t_3 = (s-1)/2 + (s-1)/3 + (s-1)/3$$

and again $s = 7(s-1)/6$ and $|G:Z| = 12$. Suppose $a_3 = 4$. If $t_3 = (s-1)/2$, we have

$$s \geq (s-1)/2 + (s-1)/3 + (s-1)/2 = 4(s-1)/3 > s - 1$$

and thus $s = 4$. This yields $|\overline{G}| = 6$, which is impossible since a_3 must divide $|\overline{G}|$ by Lagrange's theorem. Thus $t_3 = (s-1)/4$ which yields

$$s \geq (s-1)/2 + (s-1)/3 + (s-1)/4 = 13(s-1)/12 > s - 1.$$

Thus $s = 13(s-1)/12$ and $s = 13$ and $|G:Z| = 24$ in this case.

The remaining case is $a_3 = 5$. If $t_3 = 2(s-1)/5$, this yields

$$s \geq (s-1)/2 + (s-1)/3 + 2(s-1)/5 = 37(s-1)/30 > s - 1.$$

since this has no integer solution, we have $t_3 = (s-1)/5$ and

$$s \geq 31(s-1)/30 > s - 1$$

and $s = 31$. This gives $|G:Z| = 60$ and the proof is complete. ∎

We mention that all three cases of Theorem 14.23 can occur. If $G =$ SL(2, 3), we get $G/Z \cong A_4$ of order 12. If $G =$ GL(2, 3), we get $G/Z \cong \Sigma_4$ of order 24 and if $G =$ SL(2, 5), we get $G/Z \cong A_5$ of order 60. In fact, A_4, Σ_4, and A_5 are the only possibilities for G/Z.

Problems

(14.1) Let G be a linear group of degree n. Show that there exists another linear group G^* of degree n such that $G/\mathbf{Z}(G) \cong G^*/\mathbf{Z}(G^*)$ and $\det(g^*) = 1$ for $g^* \in G^*$. Do this in such a way that G^* is irreducible iff G is irreducible.

Hint Let I be the $n \times n$ identity matrix and let $S = \{\alpha I \mid \alpha \in \mathbb{C}, \alpha^n = \det(g)$ for some $g \in G\}$. Consider $GS \subseteq \mathrm{GL}(n, \mathbb{C})$.

(14.2) Let $\chi \in \mathrm{Irr}(G)$ be p-rational and faithful and assume that

$$\chi(1) < k(p - 1)$$

for some $k \leq p$. Show that the Sylow p-subgroups of G are elementary abelian of order $< p^k$.

(14.3) Let G be an irreducible p-solvable linear group of degree $p^a > 1$. Let $N = \mathbf{O}_{p'}(G)$ and $U/N = \mathbf{O}_p(G/N)$. Show that U is nonabelian.

(14.4) Let G be a linear group of degree n and suppose that n is not divisible by any prime power $q > 1$ such that $q \equiv 1, 0$, or $-1 \bmod p$,where p is some prime. Assume that G has a solvable irreducible normal subgroup. Show that a Sylow p-subgroup of G is normal.

Hint Let $P_0 \in \mathrm{Syl}_p(S)$ where S is normal, irreducible, and solvable and consider $C = \mathbf{C}_G(P_0) \lhd G$.

(14.5) Let G have a normal p-complement N and assume that a Sylow p-subgroup of G is not normal. Suppose that G is a linear group of degree $p - 1$. Show that $N/(N \cap \mathbf{Z}(G))$ is a 2-group and thus N is nilpotent.

Hint Let $P \in \mathrm{Syl}_p(G)$. Show that $\mathbf{C}_N(P) \subseteq \mathbf{Z}(N)$.

(14.6) Let G be a linear group of degree $p - 1$, where p is a prime. Suppose there exists p-solvable $M \lhd G$ such that M does not have a normal Sylow p-subgroup. Show that G is solvable.

(14.7) Let G be an irreducible linear group of degree $p + 1$ where p is an odd prime. Let $P \in \mathrm{Syl}_p(G)$ and suppose $P \ntriangleleft G$. Assume that G has a normal p-complement, N. Show that $p + 1$ is a power of 2.

Hints Let G be a minimal counterexample and let q be an odd prime divisor of $|N : \mathbf{C}_N(P)|$. Let $Q \in \mathrm{Syl}_q(N)$ and show that $\mathbf{N}_N(Q)$ is abelian.

(14.8) Replace the hypothesis that G has a normal p-complement in Problem 14.7 by the weaker condition that G is p-solvable. Draw the same conclusion.

Hints Let $U = \mathbf{O}_p(G)$ and $M/U = \mathbf{O}_{p'}(G/U)$. It is no loss to assume that G/M is a p-group and that $\mathbf{O}^{p'}(G) = G$. If $U \not\subseteq \mathbf{Z}(G)$, let $T = I_G(\lambda)$, where λ is a linear constituent of χ_U and χ is the given faithful character of G. Show that χ_T has an irreducible constituent of degree p and that $M \cap T \lhd M$.

(14.9) Let G be a linear group which is generated by two elements, x and y. Suppose that each of x and y have only two distinct eigenvalues. Show that the irreducible constituents of G (that is, of the given faithful representation) have degree at most 2.

Hint If V is a nonzero vectorspace over $\mathbb{C}$ and V_1, V_2, V_3, and V_4 are subspaces such that $V = V_i \dotplus V_j$ whenever $i \neq j$, then there exists a subspace $U \subseteq V$ such that $\dim U = 2$ and $U \cap V_i \neq 0$ for all i.

(14.10) Let G be a solvable linear group of degree n. Show that G has a nilpotent normal subgroup of index $\leq n!$.

(14.11) Let p be a prime. Show that for every integer $n > 0$, there exists a p-solvable linear group G of degree n with $|P : \mathbf{O}_p(G)| \geq p^\alpha$, where $P \in \mathrm{Syl}_p(G)$ and

$$\alpha = \sum_{i=1}^{\infty} [n/p^i].$$

15 Changing the characteristic

Some of the deepest and most powerful results in group representation theory involve "modular" representations, that is, representations over fields of prime characteristic. These are important for at least two reasons. First, if $K, H \lhd G$, with $K \subseteq H$, and H/K is an elementary abelian p-group, then H/K may be viewed as an $F[G]$-module, where F is the field of order p. In this situation, the representation theory can give direct structural information about G. Perhaps an even more important way in which the characteristic p representations of G are relevant is that they can give new information about the characteristic zero situation. This is especially the case when $|G|$ is divisible by p since then it is possible to obtain results which relate the p-subgroups of G with the properties of $\mathrm{Irr}(G)$.

Our emphasis in this chapter will be on the relationship between the absolutely irreducible characteristic p representations of G and $\mathrm{Irr}(G)$. Following R. Brauer, who was the originator of this theory, we shall focus our attention on characters rather than on modules or representations. (In fact, we shall not even mention the indecomposable but not irreducible modules which seem to be crucial for many of the deeper results.)

The objective of this chapter is to familiarize the reader with the principal definitions and the most basic results of the theory. We do not attempt to give a comprehensive treatment of the subject and we shall not prove every fact that is mentioned.

We establish some notation which will remain fixed throughout this chapter. Let R be the full ring of algebraic integers in $\mathbb{C}$ and let p be a prime. We construct a particular field F of characteristic p by choosing a maximal ideal $M \supseteq pR$ of R and setting $F = R/M$. (Note that there is a certain amount of arbitrariness here since M is not uniquely determined.) Let $*$ denote the natural homomorphism $R \to F$. Since $p1^* = p^* = 0$, it follows that char$(F) = p$ as claimed.

(15.1) LEMMA Let $U = \{\varepsilon \in \mathbb{C} \mid \varepsilon^m = 1$ for some $m \in \mathbb{Z}$ with $p \nmid m\}$. Let R, F and $*$ be as above. Then

(a) $U \subseteq R$;
(b) $*$ maps U isomorphically onto $F^\times$;
(c) F is algebraically closed and algebraic over its prime field.

Proof Clearly $U \subseteq R$ and so $*$ is defined on U. If $\alpha \in U - \{1\}$, then α is a primitive nth root of 1 for some $n > 1$ with $p \nmid n$ and hence

$$1 + x + \cdots + x^{n-1} = (x^n - 1)/(x - 1) = \prod_{i=1}^{n-1} (x - \alpha^i).$$

Setting $x = 1$, we conclude that $1 - \alpha$ divides n in R. If $\alpha^* = 1$, then $(1 - \alpha)^* = 0$ and thus $n^* = 0$. Since $p^* = 0$ and $(p, n) = 1$, it follows that $1^* = 0$, a contradiction. Thus $*$ maps U isomorphically into $F^\times$.

If $\alpha \in F$, then $\alpha = a^*$ for some $a \in R$ and there exists monic $f \in \mathbb{Z}[x]$, with $f(a) = 0$. Let $K \subseteq F$ be the prime subfield. Then $0 \neq f^* \in K[x]$ and $f^*(\alpha) = f(a)^* = 0$. Thus F is algebraic over K.

To complete the proof, let E be an algebraic extension of F. Then $U^* \subseteq F^\times \subseteq E^\times$ and it suffices to show that $E^\times \subseteq U^*$ in order to conclude that $U^* = F^\times$ and that F is algebraically closed. Let $\beta \in E^\times$. Then β is algebraic over F and hence over K and thus $\beta^m = 1$, where $m = |K(\beta)| - 1$. Now $p \nmid m$ and so U^* contains m roots of $x^m - 1$. Thus $\beta \in U^*$ and the proof is complete. ∎

Continuing with the above notation, let $\mathfrak{X}$ be an F-representation of a group G. Let $\mathscr{S}$ be the set of *p-regular* elements of G, that is, elements of order not divisible by p. We define a function $\varphi\colon \mathscr{S} \to \mathbb{C}$ as follows. Let $x \in \mathscr{S}$ and let $\varepsilon_1, \varepsilon_2, \ldots, \varepsilon_f \in F^\times$ be the eigenvalues of $\mathfrak{X}(x)$, counting multiplicities. (Thus $f = \deg \mathfrak{X}$ and $\sum \varepsilon_i = \psi(x)$ where ψ is the F-character afforded by $\mathfrak{X}$.) For each i, there exists a unique $u_i \in U$ such that $(u_i)^* = \varepsilon_i$. Let $\varphi(x) = \sum u_i$.

The function $\varphi\colon \mathscr{S} \to R \subseteq \mathbb{C}$ is called the *Brauer character* of G afforded by $\mathfrak{X}$. Note that similar F-representations afford equal Brauer characters and that Brauer characters are constant on conjugacy classes. Both of these statements follow from the fact that $\mathfrak{X}(x)$ and $P^{-1}\mathfrak{X}(x)P$ have the same eigenvalues. Also, if $x \in \mathscr{S}$, then $\varphi(x^{-1}) = \overline{\varphi(x)}$ since the eigenvalues of $\mathfrak{X}(x^{-1})$ are the reciprocals of those of $\mathfrak{X}(x)$ and for $u \in U$ we have $(\bar{u})^* = (u^{-1})^* = (u^*)^{-1}$.

(15.2) LEMMA Let $\mathfrak{X}$ be an F-representation of G which affords the Brauer character φ and the F-character ψ. For $g \in G$, let $x = g_{p'} \in \mathscr{S}$. Then $\varphi(x)^* = \psi(g)$.

Proof We have $g = xy$, where $x, y \in \langle g \rangle$, $p \nmid o(x)$ and $o(y)$ is a power of p. Replace $\mathfrak{X}$ by a similar representation so as to assume that $\mathfrak{X}(g)$ is in upper triangular form. Since $\mathfrak{X}(g) = \mathfrak{X}(x)\mathfrak{X}(y)$ it follows that the eigenvalues α_i of $\mathfrak{X}(g)$ can be factored, $\alpha_i = \varepsilon_i \delta_i$, where $\varepsilon_1, \ldots, \varepsilon_f$ are the eigenvalues of $\mathfrak{X}(x)$ and $(\delta_i)^{o(y)} = 1$. Since $o(y)$ is a power of p, we have $\delta_i = 1$ and $\alpha_i = \varepsilon_i$. Thus $\psi(g) = \psi(x)$. That $\psi(x) = \varphi(x)^*$ is immediate from the definition of φ. ∎

Lemma 15.2 provides one reason why we only bother to define Brauer characters on p-regular elements: this is sufficient to reconstruct the full F-character afforded by $\mathfrak{X}$.

Some words of caution are appropriate here. Given a group G and a function, $\varphi: \mathscr{S} \to \mathbb{C}$, where $\mathscr{S}$ is the set of p-regular elements of G, it is *not* always meaningful to ask if φ is a Brauer character unless the ideal $M \subseteq R$ is specified or some other additional information is given. Examples exist where φ is a Brauer character with respect to some choice of M and is not one when some other maximal ideal is chosen. Also, if σ is an automorphism of the complex numbers and φ is a Brauer character of G, then φ^σ need not also be a Brauer character where φ^σ is defined by $\varphi^\sigma(x) = \varphi(x)^\sigma$ for $x \in \mathscr{S}$.

(15.3) LEMMA Let φ be a Brauer character of G. Then $\bar{\varphi}$, the complex conjugate function, is also a Brauer character.

Proof Let $\mathfrak{X}$ be an F-representation affording φ, where $F = R/M$, as usual. For $g \in G$, define $\mathfrak{Y}(g) = \mathfrak{X}(g^{-1})^{\mathrm{T}}$ and observe that $\mathfrak{Y}$ is an F-representation of G. If $\varepsilon_1, \ldots, \varepsilon_f$ are the eigenvalues of $\mathfrak{X}(g)$, then $\varepsilon_1^{-1}, \ldots, \varepsilon_f^{-1}$ are the eigenvalues of $\mathfrak{X}(g)^{-1} = \mathfrak{X}(g^{-1})$ and hence of $\mathfrak{Y}(g)$. The result now follows. ∎

Let $\mathfrak{X}_1, \ldots, \mathfrak{X}_r$ be a set of representatives for the similarity classes of irreducible F-representations of G and let φ_i be the Brauer character afforded by $\mathfrak{X}_i$. We say that the φ_i are *irreducible* Brauer characters and we write $\mathrm{IBr}(G) = \{\varphi_i\}$. When we use this notation, it is understood that a particular prime p and maximal ideal M have been fixed.

It is routine to prove that sums of Brauer characters are Brauer characters and that every Brauer character is of the form $\sum n_i \varphi_i$, where the $n_i \in \mathbb{Z}$ are nonnegative and not all zero. To prove that the φ_i are linearly independent we need the following fact from algebraic number theory. Although it is not terribly deep, we omit the proof.

(15.4) LEMMA Let $\alpha_1, \ldots, \alpha_m \in \mathbb{C}$ be algebraic over $\mathbb{Q}$ and let I be a proper ideal of R, the ring of algebraic integers. Suppose that not all $\alpha_i = 0$. Then there exists $\beta \in \mathbb{C}$ such that $\beta\alpha_i \in R$ for all i but not all $\beta\alpha_i \in I$.

Proof Omitted. ∎

(15.5) THEOREM The irreducible Brauer characters φ_i are distinct and linearly independent over $\mathbb{C}$. Also, if $\sum \alpha_i \varphi_i(x) \equiv 0 \bmod M$ for all $x \in \mathcal{S}$ with $\alpha_i \in R$, then all $\alpha_i \equiv 0 \bmod M$.

Proof We prove the second statement first. Since the $\mathfrak{X}_i$ are absolutely irreducible, we have $\mathfrak{X}_i(F[G]) = M_{f_i}(F)$, where $f_i = \deg \mathfrak{X}_i$. It follows that we can choose $b_i \in F[G]$, with $\psi_i(b_i) = 1$, where ψ_i is the character afforded by $\mathfrak{X}_i$. Also, by Theorem 9.6, we may suppose that $\psi_i(b_j) = 0$ if $i \neq j$.

We have $\sum \alpha_i \varphi_i(x) \equiv 0 \bmod M$ for all $x \in \mathcal{S}$ and thus $\sum \alpha_i{}^* \psi_i(g) = 0$ for all $g \in G$ by Lemma 15.2. It follows that $\sum_i \alpha_i{}^* \psi_i(b_j) = 0$ for all j, and thus $\alpha_i{}^* = 0$ for all i. This proves the assertion.

Now let E be the algebraic closure of $\mathbb{Q}$ in $\mathbb{C}$ and suppose that $\sum \alpha_i \varphi_i = 0$ with $\alpha_i \in E$. If not all $\alpha_i = 0$, we apply Lemma 15.4 and choose β with all $\beta\alpha_i \in R$ but not all $\beta\alpha_i \in M$. Since $\sum (\beta\alpha_i)\varphi_i = 0$, this contradicts the first part of the proof. Thus all $\alpha_i = 0$ and the φ_i are linearly independent over E. Since the φ_i have values in E, it follows by elementary linear algebra that they remain linearly independent over any extension field of E. ∎

The principal reason that Brauer characters are important is that they provide a link which connects $\text{Irr}(G)$ with the characteristic p representations of G.

(15.6) THEOREM Let χ be an ordinary character of G and let $\hat{\chi}$ denote the restriction of χ to $\mathcal{S}$, the set of p-regular elements of G. Then $\hat{\chi}$ is a Brauer character of G (for any choice of M).

In order to prove Theorem 15.6, we need to consider a somewhat larger ring than R. As always, we assume that we have fixed a particular maximal ideal, $M \supseteq pR$. Let $\tilde{R} = \{\alpha/\beta \mid \alpha, \beta \in R, \beta \notin M\} \subseteq \mathbb{C}$ and observe that $\tilde{R}$ is a ring and $\tilde{R} \supseteq R$. Let $\tilde{M} = \{\alpha/\beta \mid \alpha \in M, \beta \in R - M\}$. Then $\tilde{M}$ is an ideal of $\tilde{R}$ and every element of $\tilde{R} - \tilde{M}$ has an inverse in $\tilde{R}$. It follows that $\tilde{M}$ is the unique maximal ideal of $\tilde{R}$. We call $\tilde{R}$ a ring of *local integers* for the prime p.

We extend the homomorphism $*: R \to F$ to $\tilde{R}$ by defining $(\alpha/\beta)^* = \alpha^*/\beta^*$. Note that $\tilde{M}$ is the kernel of this extension and $M = \tilde{M} \cap R$.

(15.7) LEMMA (*Nakayama*) Let $\tilde{R}$ and $\tilde{M}$ be as in the preceding and let V be a finitely generated $\tilde{R}$-module. Suppose $V = V\tilde{M} + U$ for some submodule $U \subseteq V$. Then $U = V$.

Proof Since V is finitely generated, we can choose $v_1, v_2, \ldots, v_n \in V$ such that $V = \sum v_i \tilde{M} + U$. Do this with the minimal possible n. If $n = 0$, then $U = V$ and we are done. Suppose $n \geq 1$ and write

$$v_n = \sum_{i=1}^{n} v_i m_i + u,$$

where $m_i \in \tilde{M}$ and $u \in U$. Thus

$$v_n(1 - m_n) \in \sum_{i=1}^{n-1} v_i \tilde{M} + U.$$

Since $m_n \in \tilde{M}$, we have $1 - m_n \in \tilde{R} - \tilde{M}$ and thus $1 - m_n$ is invertible in $\tilde{R}$ and $v_n \in \sum_{i=1}^{n-1} v_i \tilde{M} + U$. It follows that $V = \sum_{i=1}^{n-1} v_i \tilde{M} + U$ and this contradicts the minimality of n and completes the proof. ∎

Suppose $\mathfrak{X}$ is a $\mathbb{C}$-representation of G with the property that all entries in the matrices $\mathfrak{X}(g)$ for $g \in G$ lie in $\tilde{R}$. We can construct an F-representation $\mathfrak{X}^*$ of G by setting $\mathfrak{X}^*(g) = \mathfrak{X}(g)^*$, that is, we apply $*$ to every entry of $\mathfrak{X}(g)$.

The following includes Theorem 15.6.

(15.8) THEOREM Let $\mathfrak{X}$ be a $\mathbb{C}$-representation of G. Then there exists a $\mathbb{C}$-representation $\mathfrak{Y}$ similar to $\mathfrak{X}$ such that all entries in $\mathfrak{Y}(g)$ lie in $\tilde{R}$ for all $g \in G$. If $\mathfrak{Y}$ is any such representation, then the F-representation $\mathfrak{Y}^*$ affords the Brauer character $\hat{\chi}$ where χ is the ordinary character afforded by $\mathfrak{X}$ (and by $\mathfrak{Y}$).

Proof Let E be the algebraic closure of $\mathbb{Q}$ in $\mathbb{C}$. By the results of Chapter 9, every $\mathbb{C}$-representation of G is similar to one of the form $\mathfrak{Z}^{\mathbb{C}}$ for some E-representation $\mathfrak{Z}$. It therefore suffices to assume that $\mathfrak{X}$ is an E-representation and to produce a similar E-representation $\mathfrak{Y}$ with entries in $\tilde{R}$.

Let V be an $E[G]$-module corresponding to $\mathfrak{X}$ and let $v_1, \ldots, v_n$ be an E-basis for V. Let W be the $\tilde{R}$-span of the finite set $\{v_i g \mid 1 \leq i \leq n, g \in G\}$ so that W is a finitely generated $\tilde{R}$-module which is G-invariant. Let

$$\tau\colon W \to W/W\tilde{M}$$

be the natural homomorphism and view $W/W\tilde{M}$ as an F-vector space via $(w\tau)\alpha^* = (w\alpha)\tau$ for $\alpha \in \tilde{R}$. Let $\{w_j\tau\}$ be an F-basis for $W/W\tilde{M}$ so that $(\sum w_j \tilde{R})\tau = W/W\tilde{M}$ and $W = W\tilde{M} + \sum w_j \tilde{R}$. By Nakayama's Lemma 15.7, we have $W = \sum w_j \tilde{R}$ and thus the w_j span V over E since W contains an E-basis for V.

We claim that the w_j are linearly independent over E. Suppose that $\sum w_j \alpha_j = 0$ with $\alpha_j \in E$ and not all $\alpha_j = 0$. By Lemma 15.4, we can multiply by a suitable $\beta \in E$ and assume that all $\alpha_j \in R \subseteq \tilde{R}$ but that not all $\alpha_j^* = 0$. Now $0 = (\sum w_j \alpha_j)\tau = \sum (w_j\tau)\alpha_j^*$ and this contradicts the linear independence of the $w_j\tau$ and proves the claim.

Now let $\mathfrak{Y}$ be the E-representation of G corresponding to V with respect to the basis $\{w_j\}$ so that $\mathfrak{Y}(g) = (a_{ij})$, when $w_i g = \sum w_j a_{ij}$. However, $w_i g \in W = \sum w_j \tilde{R}$ and so all $a_{ij} \in \tilde{R}$. This completes the proof of the first assertion.

Now for p-regular $g \in G$, we need to compute the eigenvalues of $\mathfrak{Y}^*(g) = \mathfrak{Y}(g)^*$. Let f be the characteristic polynomial of $\mathfrak{Y}(g)$ so that $f(x) = \det(xI - \mathfrak{Y}(g))$. Then $f \in \tilde{R}[x]$ and $f^* \in F[x]$ is the characteristic polynomial of $\mathfrak{Y}(g)^*$. Write $f(x) = \prod(x - \lambda_i)$ and note that the eigenvalues λ_i lie in $R \subseteq \tilde{R}$. Then $f^*(x) = \prod(x - \lambda_i^*)$ and hence the eigenvalues of $\mathfrak{Y}^*(g)$ are the λ_i^*. Thus $\hat{\chi}(g) = \sum \lambda_i$ is the value of the Brauer character afforded by $\mathfrak{Y}^*$ at the element g. The proof is complete. ∎

In the foregoing proof, it is not the case that the character χ uniquely determines the F-representation $\mathfrak{Y}^*$ up to similarity. It is possible that χ is afforded by another $\mathbb{C}$-representation $\mathfrak{Z}$ with entries in $\tilde{R}$ such that $\mathfrak{Y}^*$ and $\mathfrak{Z}^*$ are not similar over F. Of course, χ does uniquely determine the Brauer character $\hat{\chi}$ and we may write $\hat{\chi} = \sum n_\varphi \varphi$ where φ runs over $\mathrm{IBr}(G)$ and $0 \leq n \in \mathbb{Z}$. The coefficients n_φ are uniquely determined because of the linear independence of $\mathrm{IBr}(G)$ and since n_φ is the multiplicity of a particular irreducible F-representation as a constituent of $\mathfrak{Y}^*$, it follows that $\mathfrak{Y}^*$ and $\mathfrak{Z}^*$ have the same irreducible constituents with the same multiplicities. Since $\mathfrak{Y}^*$ and $\mathfrak{Z}^*$ need not be completely reducible, it does not follow that they are similar.

(15.9) DEFINITION Let $\chi \in \mathrm{Irr}(G)$ and let $\hat{\chi}$ be the restriction of χ to the p-regular elements of G. Write

$$\hat{\chi} = \sum_{\varphi \in \mathrm{IBr}(G)} d_{\chi\varphi} \varphi.$$

The uniquely defined nonnegative integers $d_{\chi\varphi}$ are the *decomposition numbers* of G for the prime p.

We view the decomposition numbers as forming a $|\mathrm{Irr}(G)| \times |\mathrm{IBr}(G)|$ matrix, called the *decomposition matrix*. Although the decomposition numbers are not even defined until the maximal ideal M is chosen, it is a fact (whose proof we omit) that the decomposition matrix of G for the prime p is uniquely determined up to permutations of the rows and columns.

(15.10) THEOREM The decomposition matrix $(d_{\chi\varphi})$ has linearly independent columns. Also, $\mathrm{IBr}(G)$ is a basis for the space of $\mathbb{C}$-valued functions defined on p-regular elements of G and constant on conjugacy classes.

Proof Let V be the space of p-regular class functions and let $W \subseteq V$ be the span of $\mathrm{IBr}(G)$. Let U be the span of the columns of $(d_{\chi\varphi})$ so that

$$\dim U \leq |\mathrm{IBr}(G)| = \dim W \leq \dim V,$$

where the equality follows from Theorem 15.5. The theorem will follow when we show that $\dim V \le \dim U$. It therefore suffices to find a one-to-one linear map from the dual space $\tilde{V}$ of V into U. The elements of U are columns indexed by $\chi \in \mathrm{Irr}(G)$. For $\alpha \in \tilde{V}$, define $u = \mathrm{col}(\alpha(\hat{\chi}))$. To show that $u \in U$, observe that $\alpha(\hat{\chi}) = \alpha(\sum_\varphi d_{\chi\varphi}\varphi) = \sum_\varphi d_{\chi\varphi}\alpha(\varphi)$. Thus u is a linear combination of the $\mathrm{col}(d_{\chi\varphi})$ for $\varphi \in \mathrm{IBr}(G)$. We map $\tilde{V} \to U$ by $\alpha \mapsto \mathrm{col}(\alpha(\hat{\chi}))$.

If $\alpha(\hat{\chi}) = 0$ for all $\chi \in \mathrm{Irr}(G)$, we claim that $\alpha = 0$. To see this, let $\vartheta \in V$ and extend ϑ to a class function ϑ_1 of G. Write $\vartheta_1 = \sum a_\chi \chi$ so that $\vartheta = \sum a_\chi \hat{\chi}$ and $\alpha(\vartheta) = 0$ as desired. The result now follows. ▮

(15.11) COROLLARY The number of conjugacy classes of p-regular elements of G is equal to $|\mathrm{IBr}(G)|$. This is also the number of similarity classes of irreducible K-representations of G for every splitting field K of characteristic p.

Proof The first assertion is immediate from Theorem 15.10. To prove the second statement, we may replace K by its algebraic closure by Corollary 9.8. Assuming that K is algebraically closed, it contains an isomorphic copy of F which is an algebraic closure for $\mathbb{Z}_p$. By Corollary 9.8 again, we may assume that $F = K$. The result now follows from the fact that nonsimilar irreducible F-representations afford distinct Brauer characters. ▮

(15.12) COROLLARY If $\varphi \in \mathrm{IBr}(G)$, then there exists $\chi \in \mathrm{Irr}(G)$ with $d_{\chi\varphi} \ne 0$.

Note that for each $\chi \in \mathrm{Irr}(G)$, it is trivial that there exists $\varphi \in \mathrm{IBr}(G)$ with $d_{\chi\varphi} \ne 0$ since $0 \ne \chi(1) = \sum d_{\chi\varphi}\varphi(1)$.

The interesting case of this theory is when $p \mid |G|$. The reason for this is given by the following.

(15.13) THEOREM Suppose $p \nmid |G|$. Then $\mathrm{IBr}(G) = \mathrm{Irr}(G)$.

Proof In this case, the group algebra $F[G]$ is completely reducible and thus by Corollary 1.17 we have $|G| = \dim F[G] = \sum (\deg \mathfrak{X}_i)^2$, where the $\mathfrak{X}_i$ are a set of representatives for the similarity classes of irreducible F-representations. If $\mathfrak{X}_i$ affords $\varphi_i \in \mathrm{IBr}(G)$, then $\deg \mathfrak{X}_i = \varphi_i(1)$ and hence

$$\sum_{\varphi \in \mathrm{IBr}(G)} \varphi(1)^2 = |G| = \sum_{\chi \in \mathrm{Irr}(G)} \chi(1)^2 = \sum_\chi \left(\sum_\varphi d_{\chi\varphi}\varphi(1) \right)^2$$

$$= \sum_{\varphi,\, \mu \in \mathrm{IBr}(G)} \varphi(1)\mu(1) \sum_\chi d_{\chi\varphi} d_{\chi\mu}.$$

If $\varphi \ne \mu$, then $\sum_\chi d_{\chi\varphi}d_{\chi\mu} \ge 0$ and if $\varphi = \mu$, then $\sum_\chi d_{\chi\varphi}d_{\chi\mu} \ge 1$ by Corollary 15.12. It follows that these inequalities are all equalities.

We now have $\sum_\chi (d_{\chi\varphi})^2 = 1$ and hence for each φ, there is a unique χ such that $d_{\chi\varphi} \ne 0$, and in fact $d_{\chi\varphi} = 1$. Since $\sum_\chi d_{\chi\varphi}d_{\chi\mu} = 0$ if $\varphi \ne \mu$, it follows that for each $\chi \in \mathrm{Irr}(G)$, there is a unique φ with $d_{\chi\varphi} \neq 0$. We conclude that $\chi = \hat{\chi} = \varphi$ and the result follows. ∎

A necessary condition that a function $\vartheta: \mathscr{S} \to \mathbb{C}$ be a Brauer character is that for each subgroup $H \subseteq G$ with $p \nmid |H|$, the restriction ϑ_H is a Brauer character and hence is an ordinary character of H. This is something which can be checked.

It is trivial that every $\varphi \in \mathrm{IBr}(G)$ is a $\mathbb{C}$-linear combination of the Brauer characters $\hat{\chi}$ for $\chi \in G$. This may be seen by extending φ to a class function of G.

(15.14) THEOREM Let $\varphi \in \mathrm{IBr}(G)$. Then φ is a $\mathbb{Z}$-linear combination of $\{\hat{\chi} \mid \chi \in \mathrm{Irr}(G)\}$.

Proof Extend φ to a class function ϑ of G by setting $\vartheta(g) = \varphi(g_{p'})$. It suffices to show that ϑ is a generalized character of G. We appeal to Brauer's Theorem 8.4(a).

Let $E \subseteq G$ be elementary and write $E = P \times Q$, where P is a p-group and $p \nmid |Q|$. If $g \in E$, write $g = xy$ with $x \in P$ and $y \in Q$. Then $\vartheta(g) = \varphi(y)$ and so $\vartheta_E = 1_P \times \varphi_Q$. By Theorem 15.13, φ_Q is a character of Q and hence ϑ_E is a character of E. It follows that ϑ is a generalized character and the proof is complete. ∎

We digress to show another way in which Theorem 15.13 can be used.

(15.15) LEMMA Let E be an algebraically closed field of characteristic not dividing $|N|$, where N is a group. Let H act on N and suppose that $\mathbf{C}_H(n) = 1$ for all $1 \neq n \in N$. Let $\mathfrak{Y}$ be a nonprincipal irreducible E-representation of N and write $\mathfrak{Y}^h(n^h) = \mathfrak{Y}(n)$ for $n \in N$ and $h \in H$. Then the representations $\mathfrak{Y}^h$ are pairwise nonsimilar for $h \in H$.

Proof Let K be the algebraic closure in E of the prime subfield of E. Then $\mathfrak{Y} = \mathfrak{Z}^E$ for some irreducible K-representation $\mathfrak{Z}$ of N. It suffices to show that the $\mathfrak{Z}^h$ are pairwise nonsimilar by Corollary 9.7 and hence we may assume that $K = E$.

If $\mathrm{char}(E) = 0$, then (up to isomorphism) $E \subseteq \mathbb{C}$ and the representations $(\mathfrak{Y}^{\mathbb{C}})^h$ are pairwise nonsimilar by Theorem 6.34 and Problem 7.1 which proves this case.

Suppose $\mathrm{char}(E) = p$. Choose a maximal ideal, $M \supseteq pR$ of R and let $F = R/M$ as usual. Then $F \cong E$ and we may assume $F = E$. Let $\mathfrak{Y}$ afford $\varphi \in \mathrm{IBr}(N) = \mathrm{Irr}(N)$. By Theorem 6.34, the characters φ^h for $h \in H$ are all distinct and the result follows. ∎

The following result is often quite useful.

(15.16) THEOREM Let G be a Frobenius group with kernel N and let K be a field with characteristic not dividing $|N|$. Let V be a $K[G]$-module and suppose that $\mathbf{C}_V(N) = 0$. Let H be a Frobenius complement for G. Then V has a basis which is permuted by H with orbits of size $|H|$. Also, if $H_0 \subseteq H$, then $\dim \mathbf{C}_V(H_0) = |H : H_0| \dim \mathbf{C}_V(H)$.

Proof The second statement follows from the first since if $\mathfrak{b}$ is a basis permuted by H, then $\dim \mathbf{C}_V(H_0)$ is equal to the number of orbits of the action of H_0 on $\mathfrak{b}$. Since each orbit of H_0 on $\mathfrak{b}$ has size $|H_0|$, the assertion follows.

To prove the first statement we argue that it suffices to assume that K is algebraically closed. Let $\mathfrak{X}$ be a K-representation of G corresponding to V and let $E \supseteq K$. The condition that $\mathbf{C}_V(N) = 0$ is equivalent to the restriction $\mathfrak{X}_N$ having no principal constituent and this property is inherited by $(\mathfrak{X}^E)_N$ by Theorem 9.6. The conclusion of the theorem is equivalent to $\mathfrak{X}_H$ being similar to a representation $\mathfrak{Y}$ in block diagonal form with each block being the regular representation of H. If we can prove that $(\mathfrak{X}_H)^E$ and $\mathfrak{Y}^E$ are similar, then $\mathfrak{X}_H$ and $\mathfrak{Y}$ are similar by Problem 9.5. It follows that we may replace K by any extension field and thus we assume that K is algebraically closed.

Now let $\mathcal{M}$ be a representative set of irreducible $K[N]$-modules. Since H acts on the set of similarity classes of K-representations of N, we can define a corresponding action of H on $\mathcal{M}$. Let $\mathcal{M}_0 \subseteq \mathcal{M}$ be a set of representatives for the H-orbits of nonprincipal $K[N]$-modules in $\mathcal{M}$. In the notation of Definition 1.12, let

$$W = \sum_{M \in \mathcal{M}_0} M(V) \subseteq V.$$

We claim that $V = \sum^{\cdot}_{h \in H} Wh$. This will suffice to prove the result since we obtain a basis for V by choosing any basis for W and taking all H-translates. To prove the claim, observe that $\mathcal{M} = \bigcup_{h \in H} (\mathcal{M}_0)^h \cup \{U\}$, where U is a principal module. By Lemma 15.15, this union is disjoint. Since V_N has no principal constituent, Lemma 1.13 yields

$$V = \sum^{\cdot}_{h \in H} \left(\sum_{M \in (\mathcal{M}_0)^h} M(V) \right).$$

The result now follows from Lemma 6.4. ∎

We now resume consideration of the general case where p can divide $|G|$. We introduce the concept of "blocks" which is at the heart of Brauer's theory.

For each $\chi \in \mathrm{Irr}(G)$, we have the algebra homomorphism ω_χ: $\mathbf{Z}(\mathbb{C}[G]) \to \mathbb{C}$ as in Chapter 3. The function ω_χ is determined by its values on the conjugacy class sums K_i which form a basis for $\mathbf{Z}(\mathbb{C}[G])$. Furthermore, the values $\omega_\chi(K_i)$ lie in R. For $\chi, \psi \in \mathrm{Irr}(G)$, write $\chi \sim \psi$ if $\omega_\chi(K_i)^* = \omega_\psi(K_i)^*$ for all i. This clearly establishes an equivalence relation on $\mathrm{Irr}(G)$.

(15.17) DEFINITION A *p-block* of G is a subset $B \subseteq \mathrm{Irr}(G) \cup \mathrm{IBr}(G)$ such that

(a) $B \cap \mathrm{Irr}(G)$ is an equivalence class under the relation $\sim$ defined above.

(b) $B \cap \mathrm{IBr}(G) = \{\varphi \in \mathrm{IBr}(G) \mid d_{\chi\varphi} \neq 0$ for some $\chi \in B \cap \mathrm{Irr}(G)\}$.

From its definition, it appears that the equivalence relation $\sim$ on $\mathrm{Irr}(G)$ depends on the choice of the maximal ideal M. In fact, this is not true.

(15.18) THEOREM Let $\chi, \psi \in \mathrm{Irr}(G)$. Then χ and ψ lie in the same p-block iff $\omega_\chi(K) - \omega_\psi(K)$ lies in every maximal ideal of R which contains pR for every class sum K.

Proof The "if" statement is clear. Assume $\chi \sim \psi$ and fix K. Let $\alpha = \omega_\chi(K) - \omega_\psi(K)$. We shall show that $\alpha^n \in pR$ for some integer n and the result will follow.

Let $\mathcal{G} = \mathcal{G}(\mathbb{Q}_{|G|}/\mathbb{Q})$, the Galois group, and let $\sigma \in \mathcal{G}$. Let ε be a primitive $|G|$th root of unity so that $\varepsilon^\sigma = \varepsilon^m$ for some m, with $(m, |G|) = 1$. Let $g \in G$ be in the class with sum equal to K and let L be the sum of the class containing g^m. We have $\chi(g)^\sigma = \chi(g^m)$. Also $|\mathbf{C}(g)| = |\mathbf{C}(g^m)|$ since $\langle g \rangle = \langle g^m \rangle$. It follows that $\omega_\chi(K)^\sigma = \omega_\chi(L)$ and similarly $\omega_\psi(K)^\sigma = \omega_\psi(L)$. It follows that $\alpha^\sigma = \omega_\chi(L) - \omega_\psi(L) \in M$ since $\chi \sim \psi$.

Let $f(x) = \prod_{\sigma \in \mathcal{G}} (x - \alpha^\sigma) \in (\mathbb{Q} \cap R)[x] = \mathbb{Z}[x]$. Since $\alpha^\sigma \in M$ for all $\sigma \in \mathcal{G}$, all of the coefficients of f except for the leading one lie in $M \cap \mathbb{Z} = p\mathbb{Z}$. Thus $0 = f(\alpha) \equiv \alpha^n \bmod pR$, where $n = |\mathcal{G}|$. The proof is now complete. ∎

Although it is clear from the definition that every $\chi \in \mathrm{Irr}(G)$ lies in a unique p-block, the analogous statement about $\mathrm{IBr}(G)$, though true, is not so obvious. To prove it, we relate the p-blocks of G to the F-algebra, $\mathbf{Z}(F[G])$.

We extend the map $*: \tilde{R} \to F$ to a ring homomorphism $*: \tilde{R}[G] \to F[G]$ by setting $g^* = g$ for $g \in G$. (Here, $\tilde{R}[G]$ is simply the $\tilde{R}$-span of the group elements in $\mathbb{C}[G]$.) Since the class sums K_i form a basis for $\mathbf{Z}(\mathbb{C}[G])$ and their images $K_i^* \in F[G]$ form a basis for $\mathbf{Z}(F[G])$, it follows that $\mathbf{Z}(\tilde{R}[G]) = \tilde{R}[G] \cap \mathbf{Z}(\mathbb{C}[G])$ maps onto $\mathbf{Z}(F[G])$ via $*$.

Now suppose $\chi \in \mathrm{Irr}(G)$. Then ω_χ maps $\mathbf{Z}(\tilde{R}[G])$ to $\tilde{R}$ and we can define $\omega_\chi^*: \mathbf{Z}(F[G]) \to F$ by setting $\omega_\chi^*(z^*) = \omega_\chi(z)^*$ for $z \in \mathbf{Z}(\tilde{R}[G])$. This is well defined since if $z_1, z_2 \in \mathbf{Z}(\tilde{R}[G])$, with $z_1^* = z_2^*$, then $z_1 - z_2$ has coefficients in $\tilde{M}$ and thus $(\omega_\chi(z_1 - z_2))^* = 0$. It is trivial to check that $\omega_\chi^*: \mathbf{Z}(F[G]) \to F$ is an algebra homomorphism and that if $\chi, \psi \in \mathrm{Irr}(G)$, then $\omega_\chi^* = \omega_\psi^*$ iff $\chi \sim \psi$; that is, iff χ and ψ lie in the same p-block.

Let $\mathrm{Bl}(G)$ be the set of p-blocks of G. If $B \in \mathrm{Bl}(G)$, let $\lambda_B = \omega_\chi{}^*$ for $\chi \in B \cap \mathrm{Irr}(G)$. Thus $B \mapsto \lambda_B$ is a one-to-one map from $\mathrm{Bl}(G)$ into the set of algebra homomorphisms $\mathbf{Z}(F[G]) \to F$.

(15.19) THEOREM Let $\varphi \in \mathrm{IBr}(G)$. Then φ lies in a unique p-block B. If $\mathfrak{Y}$ is an irreducible F-representation which affords φ, then $\mathfrak{Y}(u) = \lambda_B(u)I$ for all $u \in \mathbf{Z}(F[G])$.

Proof By Corollary 15.12, $d_{\chi\varphi} \neq 0$ for some $\chi \in \mathrm{Irr}(G)$. Let $B \in \mathrm{Bl}(G)$, with $\chi \in B$. Then $\varphi \in B$. Let $\mathfrak{Y}$ afford φ. We will be done when we show that $\mathfrak{Y}(u) = \lambda_B(u)I$ for all $u \in \mathbf{Z}(F[G])$ since this equation uniquely determines λ_B and thus uniquely determines B.

Let $\mathfrak{X}$ be a $\mathbb{C}$-representation which affords χ and which has entries in $\tilde{R}$ (Theorem 15.8). Thus $\mathfrak{X}^*$ affords the Brauer character $\hat{\chi}$ and by the linear independence of $\mathrm{IBr}(G)$, it follows that $\mathfrak{Y}$ has multiplicity $d_{\chi\varphi} > 0$ as a constituent of $\mathfrak{X}^*$. Now let $u \in \mathbf{Z}(F[G])$ and write $u = z^*$ for some $z \in \mathbf{Z}(\tilde{R}[G])$. Then

$$\mathfrak{X}^*(u) = \mathfrak{X}(z)^* = (\omega_\chi(z)I)^* = \omega_\chi{}^*(u)I = \lambda_B(u)I.$$

Since $\mathfrak{Y}$ is a constituent of $\mathfrak{X}^*$, the result follows. ∎

We define a graph with $\mathrm{Irr}(G)$ as its vertex set by linking $\chi, \psi \in \mathrm{Irr}(G)$ iff there exists $\varphi \in \mathrm{IBr}(G)$ such that $d_{\chi\varphi}$ and $d_{\psi\varphi}$ are both nonzero. This is called the *Brauer graph*. If χ and ψ are linked by φ, it follows that χ and ψ lie in the same p-block, namely the unique one which contains φ. This proves the following.

(15.20) COROLLARY Let B be a p-block of G. Then $B \cap \mathrm{Irr}(G)$ is a union of connected components of the Brauer graph.

We shall see that in fact $B \cap \mathrm{Irr}(G)$ is a single connected component.

One of the principal benefits of considering blocks in various applications of the theory is that in certain circumstances we can replace equations like

$$\sum_{\chi \in \mathrm{Irr}(G)} \chi(x)\overline{\chi(y)} = 0$$

(which arise from the second orthogonality relation) by equations like

$$\sum_{\chi \in \mathrm{Irr}(G) \cap B} \chi(x)\overline{\chi(y)} = 0,$$

where B is a p-block. In particular, this holds whenever x_p and y_p are not conjugate in G. We shall prove a weak form of this "block orthogonality."

For each $\varphi \in \mathrm{IBr}(G)$ we define

$$\Phi_\varphi = \sum_{\chi \in \mathrm{Irr}(G)} d_{\chi\varphi}\chi.$$

The Φ's are called *projective characters* of G. (There is no connection with projective representations in the sense of Chapter 11, but there is a connection with projective modules in ring theory. The Φ's are also called "principal indecomposable characters" in the literature.)

(15.21) LEMMA Let $\mathscr{A} \subseteq \mathrm{Irr}(G)$ be a union of connected components of the Brauer graph and let $\mathscr{B} = \{\varphi \in \mathrm{IBr}(G) \mid d_{\chi\varphi} \neq 0 \text{ for some } \chi \in \mathscr{A}\}$. Let $x, y \in G$ with $p \nmid o(x)$. Then

$$\sum_{\chi \in \mathscr{A}} \chi(x)\overline{\chi(y)} = \sum_{\varphi \in \mathscr{B}} \varphi(x)\overline{\Phi_\varphi(y)}.$$

Proof For $\chi \in \mathscr{A}$, we have

$$\chi(x) = \sum_{\varphi \in \mathscr{B}} d_{\chi\varphi}\varphi(x)$$

since $d_{\chi\mu} = 0$ for all $\mu \in \mathrm{IBr}(G) - \mathscr{B}$. Also, if $\varphi \in \mathscr{B}$, then

$$\Phi_\varphi(y) = \sum_{\chi \in \mathscr{A}} d_{\chi\varphi}\chi(y)$$

since $d_{\xi\varphi} = 0$ for all $\xi \in \mathrm{Irr}(G) - \mathscr{A}$.

Now

$$\sum_{\chi \in \mathscr{A}} \chi(x)\overline{\chi(y)} = \sum_{\chi \in \mathscr{A};\, \varphi \in \mathscr{B}} d_{\chi\varphi}\varphi(x)\overline{\chi(y)} = \sum_{\varphi \in \mathscr{B}} \varphi(x)\overline{\Phi_\varphi(y)}$$

and the proof is complete. ∎

(15.22) COROLLARY For each $\varphi \in \mathrm{IBr}(G)$ we have $\Phi_\varphi(y) = 0$ if $p \mid o(y)$. Furthermore, $|P|$ divides $\Phi_\varphi(1)$ where $P \in \mathrm{Syl}_p(G)$.

Proof Let $x \in G$ be p-regular and let $p \mid o(y)$. By the second orthogonality relation we have

$$\sum_{\chi \in \mathrm{Irr}(G)} \chi(x)\overline{\chi(y)} = 0.$$

By Lemma 15.21 with $\mathscr{A} = \mathrm{Irr}(G)$ and $\mathscr{B} = \mathrm{IBr}(G)$, we conclude that

$$\sum_{\varphi \in \mathrm{IBr}(G)} \varphi(x)\overline{\Phi_\varphi(y)} = 0.$$

Since this holds for all p-regular $x \in G$, the linear independence of $\mathrm{IBr}(G)$ yields $\Phi_\varphi(y) = 0$ for all φ.

The second statement follows since $|P|[(\Phi_\varphi)_P, 1_P] = \Phi_\varphi(1)$. ∎

(15.23) COROLLARY (*Weak Block Orthogonality*) Let $x, y \in G$ with $p \nmid o(x)$ and $p \mid o(y)$. Let B be a p-block of G. Then

$$\sum_{\chi \in B \cap \mathrm{Irr}(G)} \chi(x)\overline{\chi(y)} = 0.$$

Proof Apply Lemma 15.21 with $\mathscr{A} = \mathrm{Irr}(G) \cap B$ and $\mathscr{B} = \mathrm{IBr}(G) \cap B$. Since $\Phi_\varphi(y) = 0$ for all φ, the result is immediate. ▮

Before going on to develop more of the theory of blocks we digress to give an application of what we have already done.

(15.24) THEOREM (*Brauer*) Let G be a simple group of order $p^a q^b r$ where p, q, r are distinct primes. Let $S \in \mathrm{Syl}_r(G)$. Then $S = \mathbf{C}_G(S)$.

The p-block containing 1_G is called the *principal* block.

(15.25) LEMMA Let G be simple and let B be the principal p-block of G. Suppose $\chi \in B \cap \mathrm{Irr}(G)$ and $\chi(1)$ is a power of p. Then $\chi = 1_G$.

Proof Let $g \in G$ and let $K \in \mathbb{C}[G]$ be the sum of the elements in $\mathrm{Cl}(g)$, the conjugacy class containing g. Then $\omega_1(K) = |\mathrm{Cl}(g)|$ and since χ and 1_G lie in the same p-block, we have

$$(*) \qquad \frac{\chi(g)|\mathrm{Cl}(g)|}{\chi(1)} = \omega_\chi(K) \equiv |\mathrm{Cl}(g)| \bmod M.$$

Now let $P \in \mathrm{Syl}_p(G)$. If $P = 1$, then since $\chi(1)$ divides $|G|$, we have $\chi(1) = 1$ and by simplicity $\chi = 1_G$. Assume then, that $P > 1$ and take $g \in \mathbf{Z}(P) - \{1\}$. Then $p \nmid |\mathrm{Cl}(g)|$ and so $|\mathrm{Cl}(g)| \not\equiv 0 \bmod M$ since $M \cap \mathbb{Z} = p\mathbb{Z}$. Thus $(*)$ yields $\chi(g) \neq 0$. However $(\chi(1), |\mathrm{Cl}(g)|) = 1$ and Burnside's Theorem 3.8 yields that $g \in \mathbf{Z}(\chi)$. Since $g \neq 1$ and G is simple, it follows that $\chi = 1_G$. ▮

Proof of Theorem 15.24 Suppose $\mathbf{C}_G(S) > S$. Then there exists $x \in G$ of order pr or qr. Say $o(x) = pr$ and let B be the principal p-block of G. By Corollary 15.23, we have

$$-1 = \sum_{\substack{\chi \in B \cap \mathrm{Irr}(G) \\ q | \chi(1)}} \chi(1)\overline{\chi(x)} + \sum_{\substack{\chi \in B \cap \mathrm{Irr}(G) \\ \chi \neq 1_G;\, q \nmid \chi(1)}} \chi(1)\overline{\chi(x)}.$$

We claim that the second sum is zero. If $\chi \in B \cap \mathrm{Irr}(G)$ and $\chi \neq 1_G$ and $q \nmid \chi(1)$, then $r | \chi(1)$ by Lemma 15.25. Since $r | o(x)$, we have $\chi(x) = 0$ by Theorem 8.17 and the claim follows. We conclude that $-1/q$ is an algebraic integer and this contradiction completes the proof. ▮

For $\chi \in \mathrm{Irr}(G)$, let $e_\chi \in \mathbf{Z}(\mathbb{C}[G])$ be the idempotent corresponding to χ. By Theorem 2.12 we have

$$e_\chi = \frac{\chi(1)}{|G|} \sum_{g \in G} \overline{\chi(g)} g.$$

Note that in general, $e_\chi \notin \tilde{R}[G]$ (where $\tilde{R}$ is a ring of local integers for p) and so we cannot find idempotents in $F[G]$ simply by applying $*$ to e_χ. Since $e_\chi e_\psi = 0$ if $\chi \neq \psi$, sums of the various e_χ's are also idempotents and we shall consider when such a sum lies in $\tilde{R}[G]$. Note that if $\mathscr{A} \subseteq \mathrm{Irr}(G)$, then the set $\mathscr{A}$ can be recovered from $s = \sum_{\chi \in \mathscr{A}} e_\chi$ since $\mathscr{A} = \{\chi \in \mathrm{Irr}(G) \mid se_\chi \neq 0\}$.

(15.26) THEOREM (*Osima*) Let $\mathscr{A}$ be a connected component of the Brauer graph and let $f = \sum_{\chi \in \mathscr{A}} e_\chi$. Write $f = \sum a_g g$. Then

(a) $a_g = (1/|G|) \sum_{\chi \in \mathscr{A}} \chi(1)\overline{\chi(g)}$;
(b) $a_g \in \tilde{R}$ for all $g \in G$;
(c) $a_g = 0$ if $p \mid o(g)$.

Proof Statement (a) is immediate from the formula for e_χ in Theorem 2.12. By Lemma 15.21, we have $a_g = (1/|G|) \sum_{\psi \in \mathscr{B}} \varphi(1)\overline{\Phi_\varphi(g)}$, where $\mathscr{B} = \{\varphi \in \mathrm{IBr}(G) \mid d_{\chi\varphi} \neq 0$ for some $\chi \in \mathscr{A}\}$. If $p \mid o(g)$, then $\Phi_\varphi(g) = 0$ by Corollary 15.22 and (c) follows.

Now assume that $p \nmid o(g)$. Lemma 15.21 then yields

$$a_g = (1/|G|) \sum_{\varphi \in \mathscr{B}} \Phi_\varphi(1)\overline{\varphi(g)}.$$

However, $|P|$ divides $\Phi_\varphi(1)$ where $P \in \mathrm{Syl}_p(G)$ and thus $\Phi_\varphi(1)/|G| \in \tilde{R}$. Since $\overline{\varphi(g)} \in R$, it follows that $a_g \in \tilde{R}$ and (b) is proved. ∎

(15.27) THEOREM The connected components of the Brauer graph are exactly the sets $\mathrm{Irr}(G) \cap B$ for p-blocks B. Furthermore, every set $\mathscr{A} \subseteq \mathrm{Irr}(G)$ such that $\sum_{\chi \in \mathscr{A}} e_\chi \in \tilde{R}[G]$ is a union of sets of the form $\mathrm{Irr}(G) \cap B$.

Proof We prove the second statement first. If $\chi \in \mathrm{Irr}(G)$ is afforded by $\mathfrak{X}$, then $\mathfrak{X}(e_\chi)$ is the identity matrix and $\mathfrak{X}(e_\psi) = 0$ if $\psi \neq \chi$. Thus $\omega_\chi(e_\chi) = 1$ and $\omega_\chi(e_\psi) = 0$ for $\psi \neq \chi$. Write $f = \sum_{\mathscr{A}} e_\chi$. It follows that $\chi \in \mathscr{A}$ iff $\omega_\chi(f) = 1$ and otherwise $\omega_\chi(f) = 0$. Thus $\chi \in \mathscr{A}$ iff $\omega_\chi(f)^* \neq 0$. Now if $f \in \tilde{R}[G]$, then all $\omega_\chi(f)^*$ are equal as χ runs over $\mathrm{Irr}(G) \cap B$ for a block B. The assertion follows.

Now let $\mathscr{A}$ be a connected component of the Brauer graph. By Theorem 15.26, $\sum_{\mathscr{A}} e_\chi \in \tilde{R}[G]$ and thus $\mathscr{A}$ is a union of sets of the form $\mathrm{Irr}(G) \cap B$. Since each $\mathrm{Irr}(G) \cap B$ is a union of connected components of the Brauer graph, the result follows. ∎

We now have three different characterizations of the sets $B \cap \mathrm{Irr}(G)$. In addition to their definition as equivalence classes under $\sim$, they are also the connected components of the Brauer graph and they are the minimal nonempty subsets $\mathscr{A} \subseteq \mathrm{Irr}(G)$ such that $\sum_{\chi \in \mathscr{A}} e_\chi \in \tilde{R}[G]$.

The following is a strengthening of Theorem 15.14.

(15.28) LEMMA Let B be a block of G and let $\varphi \in \mathrm{IBr}(G) \cap B$. Then φ is a $\mathbb{Z}$-linear combination of Brauer characters of the form $\hat{\chi}$ for $\chi \in \mathrm{Irr}(G) \cap B$.

Proof By Theorem 15.14, there exist $b_\chi \in \mathbb{Z}$ such that $\varphi = \sum_{\chi \in \mathrm{Irr}(G)} b_\chi \hat{\chi}$. Thus $\varphi = \vartheta_B + \vartheta_0$, where

$$\vartheta_B = \sum_{\chi \in \mathrm{Irr}(G) \cap B} b_\chi \hat{\chi} \qquad \text{and} \qquad \vartheta_0 = \sum_{\chi \in \mathrm{Irr}(G) - B} b_\chi \hat{\chi}.$$

Now we express ϑ_B and ϑ_0 in terms of $\mathrm{IBr}(G)$ using the decomposition numbers $d_{\chi\mu}$ for $\mu \in \mathrm{IBr}(G)$. If $\chi \in B$, then $d_{\chi\mu} = 0$ when $\mu \notin B$ and hence ϑ_B is a linear combination of $\mu \in B$. Similarly, if $\chi \notin B$, then $d_{\chi\mu} = 0$ when $\mu \in B$ and hence ϑ_0 is a linear combination of $\mu \notin B$. The equation $\varphi = \vartheta_B + \vartheta_0$ and the linear independence of $\mathrm{IBr}(G)$ now yield $\varphi = \vartheta_B$ and the proof is complete. ∎

(15.29) THEOREM Let B be a p-block of G. Then

$$|B \cap \mathrm{Irr}(G)| \geq |B \cap \mathrm{IBr}(G)|.$$

Let $\chi \in B \cap \mathrm{Irr}(G)$. Then the following are equivalent.

(a) $|B \cap \mathrm{Irr}(G)| = |B \cap \mathrm{IBr}(G)|$.
(b) $p \nmid (|G|/\chi(1))$.
(c) $B \cap \mathrm{Irr}(G) = \{\chi\}$.

Also in this case, $B \cap \mathrm{IBr}(G) = \{\hat{\chi}\}$.

Proof Let $D = (d_{\chi\varphi})$ be the decomposition matrix and let D_B be the submatrix corresponding to the rows and columns indexed by elements of B. For each $\varphi \in B$, the part of the corresponding column of D outside of D_B consists of zeros. Since the columns of D are linearly independent by Theorem 15.10, it follows that the columns of D_B are linearly independent and thus D_B has at least as many rows as columns.

Now assume (a). Then D_B is a square nonsingular matrix and we let $D_B^{-1} = (a_{\varphi\chi})$. For fixed $\chi \in \mathrm{Irr}(G) \cap B$, we have

$$\sum_{\varphi \in B} a_{\varphi\chi} \Phi_\varphi = \sum_{\xi \in B \cap \mathrm{Irr}(G)} \left(\sum_{\varphi \in B} d_{\xi\varphi} a_{\varphi\chi} \right) \xi = \chi$$

and so χ is a linear combination of the Φ_φ and hence vanishes on elements of order divisible by p (Corollary 15.22). If $P \in \mathrm{Syl}_p(G)$, then $|P|[\chi_P, 1_P] = \chi(1)$ and (b) follows.

Assuming (b) we have

$$e_\chi = (\chi(1)/|G|) \sum_{g \in G} \overline{\chi(g)} g \in \tilde{R}[G]$$

and so $B \cap \mathrm{Irr}(G) = \{\chi\}$ by Theorem 15.27.

Finally assume (c). Then

$$0 < |\mathrm{IBr}(G) \cap B| \leq |\mathrm{Irr}(G) \cap B| = 1$$

and (a) follows. Also, if $\mathrm{IBr}(G) \cap B = \{\varphi\}$, then $\varphi = b\hat{\chi}$ for some $b \in \mathbb{Z}$ by Lemma 15.28. Thus $\hat{\chi} = d_{\chi\varphi}\varphi = d_{\chi\varphi}b\hat{\chi}$ and $d_{\chi\varphi}b = 1$. It follows that $d_{\chi\varphi} = 1$ and the proof is complete. ∎

Note that Theorem 15.29 provides an alternate proof of Theorem 8.17. Also observe that if $p \nmid |G|$, then each p-block contains a unique irreducible character.

We use Theorem 15.26 to obtain further connections between the p-blocks of G and $\mathbf{Z}(F[G])$. For each $B \in \mathrm{Bl}(G)$, write $f_B = \sum_{\chi \in B \cap \mathrm{Irr}(G)} e_\chi$. This *Osima idempotent* lies in $\mathbf{Z}(\tilde{R}[G])$ and we let $e_B = f_B{}^* \in \mathbf{Z}(F[G])$.

(15.30) THEOREM We have the following.

(a) $\lambda_B(e_B) = 1$ and $\lambda_{B'}(e_B) = 0$ for blocks $B \neq B'$.
(b) The e_B are idempotents and $e_B e_{B'} = 0$ for $B \neq B'$.
(c) e_B is an F-linear combination of class sums of p-regular classes.
(d) $\sum e_B = 1$.
(e) If $\lambda_B(z) = 0$ for all $B \in \mathrm{Bl}(G)$, then z is nilpotent.
(f) The λ_B are all of the algebra homomorphisms $\mathbf{Z}(F[G]) \to F$.
(g) Every idempotent of $\mathbf{Z}(F[G])$ is a sum of some of the e_B.

Proof (a) For $\chi \in \mathrm{Irr}(G)$, we have $\omega_\chi(f_B) = 1$ if $\chi \in B$ and otherwise $\omega_\chi(f_B) = 0$. If $\chi \in B'$, then $\omega_\chi{}^* = \lambda_{B'}$ and (a) follows.

(b) Since $f_B f_{B'} = \delta_{BB'} f_B$ we have $e_B e_{B'} = \delta_{BB'} e_B$. Since $\lambda_B(e_B) = 1$, we have $e_B \neq 0$.

(c) Immediate from Osima's Theorem 15.26(c).

(d) $\sum f_B = \sum e_\chi = 1$ and thus $\sum e_B = 1^* = 1$.

(e) Let $\mathfrak{Y}$ be any irreducible F-representation of G and let $\mathfrak{Y}$ afford $\varphi \in \mathrm{IBr}(G) \cap B$. Then $\mathfrak{Y}(z) = \lambda_B(z)I = 0$ by Theorem 15.19. Thus z is in the Jacobson radical, $J(F[G])$ and hence is nilpotent (Problem 1.4).

(f) Let λ: $\mathbf{Z}(F[G]) \to F$ be an algebra homomorphism. Then $\ker\lambda$ has codimension 1 and so $\mathbf{Z}(F[G]) = \ker\lambda + F \cdot 1$ and λ is determined by its kernel. If $\lambda \neq \lambda_B$, then $\ker\lambda_B \nsubseteq \ker\lambda$ and we can choose $z_B \in \ker\lambda_B$ with $\lambda(z_B) \neq 0$. Assuming $\lambda \notin \{\lambda_B\}$, let $z = \prod z_B$. Then $\lambda_B(z) = 0$ for all B and hence z is nilpotent by (e). However, $\lambda(z) = \prod \lambda(z_B) \neq 0$. This is a contradiction.

(g) Let $e \in \mathbf{Z}(F[G])$ be an idempotent. Then $e = e\sum e_B = \sum ee_B$ and it suffices to show that either $ee_B = 0$ or $ee_B = e_B$. Suppose $ee_B \neq 0$. Then since ee_B is not nilpotent and $\lambda_{B'}(ee_B) = 0$ for $B' \neq B$, we have $\lambda_B(ee_B) \neq 0$ and thus $\lambda_B(ee_B) = 1 = \lambda_B(e_B)$. Thus $\lambda_{B'}(e_B(1 - e)) = 0$ for all $B' \in \mathrm{Bl}(G)$ and since $e_B(1 - e) = (e_B(1 - e))^2$, we have $e_B(1 - e) = 0$ and the result follows. ∎

We now discuss some of the connections between block theory and the collection of p-subgroups of G. We work in the situation of Theorem 15.30. If $\mathscr{K}$ is a conjugacy class of G we consider the Sylow p-subgroups of $\mathbf{C}_G(x)$ for $x \in \mathscr{K}$. These are called the *p-defect groups* for $\mathscr{K}$. They constitute a conjugacy class of p-subgroups of G. The collection of p-defect groups for $\mathscr{K}$ is denoted $\delta(\mathscr{K})$.

We use the notation $\hat{\mathscr{K}}$ for the sum in $F[G]$ of the elements of the class $\mathscr{K}$ so that the $\hat{\mathscr{K}}$ form a basis for $\mathbf{Z}(F[G])$. If $B \in \mathrm{Bl}(G)$, write $e_B = \sum a_B(\mathscr{K})\hat{\mathscr{K}}$ so that a_B is a uniquely determined function from the set of classes of G into F. In fact, $a_B(\mathscr{K}) = ((1/|G|)\sum_{\chi \in \mathrm{Irr}(G) \cap B} \chi(1)\overline{\chi(g)})^*$ for $g \in \mathscr{K}$ by Theorem 15.26(a). By 15.30(c) we have $a_B(\mathscr{K}) = 0$ if $\mathscr{K}$ does not consist of p-regular elements.

Since $1 = \lambda_B(e_B) = \sum a_B(\mathscr{K})\lambda_B(\hat{\mathscr{K}})$, it follows that for each $B \in \mathrm{Bl}(G)$, there exists at least one class $\mathscr{K}$ such that $a_B(\mathscr{K}) \neq 0$ and $\lambda_B(\hat{\mathscr{K}}) \neq 0$. We call such a class a *defect class* for B.

(15.31) THEOREM (*Min–Max*) Let $\mathscr{K}$ be a defect class for $B \in \mathrm{Bl}(G)$ and let $D \in \delta(\mathscr{K})$. Let $\mathscr{L}$ be any class of G.

(a) If $a_B(\mathscr{L}) \neq 0$, then D contains a defect group for $\mathscr{L}$.
(b) If $\lambda_B(\hat{\mathscr{L}}) \neq 0$, then D is contained in a defect group for $\mathscr{L}$.

To prove the min–max theorem, we define the *Brauer homomorphism* β_P. Let $P \subseteq G$ be a p-subgroup. Let $N = \mathbf{N}_G(P)$ and $C = \mathbf{C}_G(P)$. We map β_P: $\mathbf{Z}(F[G]) \to \mathbf{Z}(F[N])$ by

$$\beta_P(\hat{\mathscr{K}}) = \sum_{x \in \mathscr{K} \cap C} x,$$

and extend by linearity. Since $\mathscr{K} \cap C$ is a union of classes of N, we do have $\beta_P(\hat{\mathscr{K}}) \in \mathbf{Z}(F[N])$.

(15.32) LEMMA The map β_P: $\mathbf{Z}(F[G]) \to \mathbf{Z}(F[N])$ is an algebra homomorphism.

Proof If suffices to check that $\beta_P(\hat{\mathscr{K}}\hat{\mathscr{L}}) = \beta_P(\hat{\mathscr{K}})\beta_P(\hat{\mathscr{L}})$ for classes $\mathscr{K}, \mathscr{L}$. For $c \in C$, let $\mathscr{A} = \{(x, y) | x \in \mathscr{K}, y \in \mathscr{L}, xy = c\}$ and

$$\mathscr{A}_0 = \{(x, y) | x \in \mathscr{K} \cap C, y \in \mathscr{L} \cap C, xy = c\}.$$

Then $|\mathscr{A}|^*$ is the coefficient of c in $\beta_P(\hat{\mathscr{K}}\hat{\mathscr{L}})$ and $|\mathscr{A}_0|^*$ is the coefficient of c in $\beta_P(\hat{\mathscr{K}})\beta_P(\hat{\mathscr{L}})$. It thus suffices to check that $|\mathscr{A}| \equiv |\mathscr{A}_0| \bmod p$.

Since $P \subseteq \mathbf{C}(c)$, it follows that P acts on $\mathscr{A}$ by $(x, y)^u = (x^u, y^u)$ for $u \in P$. Then $\mathscr{A}_0$ is exactly the set of fixed points of $\mathscr{A}$ under the action of P. The result follows. ∎

(15.33) LEMMA Let P be a p-subgroup of G and let β_P be the corresponding Brauer homomorphism. Let $z = \sum a_{\mathscr{K}}\hat{\mathscr{K}}$. Then $\beta_P(z) \neq 0$ iff P is contained in some $D \in \delta(\mathscr{K})$ with $a_{\mathscr{K}} \neq 0$.

Proof The sets $\mathscr{K} \cap \mathbf{C}(P)$ are disjoint for distinct classes $\mathscr{K}$ and thus the nonzero elements of the form $\beta_P(\hat{\mathscr{K}})$ are linearly independent. It follows that $\beta_P(z) \neq 0$ iff $\beta_P(\hat{\mathscr{K}}) \neq 0$ for some $\mathscr{K}$ with $a_{\mathscr{K}} \neq 0$.

Now $\beta_P(\hat{\mathscr{K}}) \neq 0$ iff $\mathscr{K} \cap \mathbf{C}(P) \neq \varnothing$ and this happens iff $P \subseteq \mathbf{C}(x)$ for some $x \in \mathscr{K}$. Since $P \subseteq \mathbf{C}(x)$ iff P is contained in some Sylow p-subgroup of $\mathbf{C}(x)$, the result follows. ▮

(15.34) LEMMA Let $B \in \mathrm{Bl}(G)$ and let $\mathscr{L}$ be a class of G with $a_B(\mathscr{L}) \neq 0$. Let $P \in \delta(\mathscr{L})$ and $N = \mathbf{N}_G(P)$. Then there exists $b \in \mathrm{Bl}(N)$ such that

$$\lambda_B = \beta_P \lambda_b \colon \mathbf{Z}(F[G]) \to \mathbf{Z}(F[N]) \to F.$$

Proof By Lemma 15.33, $\beta_P(e_B) \neq 0$. Since β_P is a homomorphism, $e = \beta_P(e_B)$ is a nonzero idempotent in $\mathbf{Z}(F[N])$ and so is not nilpotent. Thus there exists $b \in \mathrm{Bl}(N)$ with $\lambda_b(e) \neq 0$. Let $\mu = \beta_P \lambda_b$. Then $\mu(e_B) = \lambda_b(e) \neq 0$ and μ is an algebra homomorphism $\mathbf{Z}(F[G]) \to F$. Since $\mu(e_B) \neq 0$, we have $\mu = \lambda_B$ and the proof is complete. ▮

Proof of Theorem 15.31 If $a_B(\mathscr{L}) \neq 0$, let $P \in \delta(\mathscr{L})$. Then $\lambda_B = \beta_P \lambda_b$ for some $b \in \mathrm{Bl}(N(P))$ by Lemma 15.34. Since $\mathscr{K}$ is a defect class, we have $0 \neq \lambda_B(\hat{\mathscr{K}}) = \lambda_b(\beta_P(\hat{\mathscr{K}}))$ and so $\beta_P(\hat{\mathscr{K}}) \neq 0$. Thus P is contained in some defect group of $\mathscr{K}$ by Lemma 15.33. Now (a) follows.

Now apply Lemma 15.34 to $\mathscr{K}$. Since $a_B(\mathscr{K}) \neq 0$ we have $\lambda_B = \beta_D \lambda_b$ for some $b \in \mathrm{Bl}(\mathbf{N}(D))$. Suppose $\lambda_B(\hat{\mathscr{L}}) \neq 0$. Then $\beta_D(\hat{\mathscr{L}}) \neq 0$ and hence D is contained in a defect group of $\mathscr{L}$.

(15.35) DEFINITION Let B be a p-block of G. Then the p-defect groups of the defect classes of B are called *defect groups* of B. The set of these is denoted $\delta(B)$.

(15.36) COROLLARY Let $B \in \mathrm{Bl}(G)$. Then $\delta(B)$ is a single conjugacy class of subgroups.

Proof Let $\mathscr{K}_1$ and $\mathscr{K}_2$ be defect classes for G. It suffices to show that $\delta(\mathscr{K}_1) = \delta(\mathscr{K}_2)$. Let $D_i \in \delta(\mathscr{K}_i)$. By Theorem 15.31, each of the D_i contains a conjugate of the other. The result follows. ▮

Which p-subgroups of G can be defect groups for blocks? It is a fact (which we shall not prove) that a defect group for a block is necessarily of the form $P \cap Q$ for some $P, Q \in \mathrm{Syl}_p(G)$. We prove the weaker assertion that $\mathbf{O}_p(G)$ is contained in every defect group of a block.

(15.37) LEMMA Let $P = \mathbf{O}_p(G)$. Then $P \subseteq \ker \mathfrak{Y}$ for every irreducible F-representation $\mathfrak{Y}$ of G.

Proof Since P has a unique p-regular class, the principal representation of P is its unique irreducible F-representation by Corollary 15.11. If $\mathfrak{Y}$ is an irreducible F-representation of G, then the restriction $\mathfrak{Y}_P$ is completely reducible by Corollary 6.6. The result follows from these two facts. ∎

(15.38) THEOREM Let $\mathscr{K}$ be a class of G and assume that $\mathscr{K} \cap \mathbf{C}(\mathbf{O}_p(G)) = \varnothing$. Then $\hat{\mathscr{K}}$ is nilpotent.

Proof Let $P = \mathbf{O}_p(G)$ act on $\mathscr{K}$ by conjugation and let $\mathscr{O}$ be an orbit of this action. Then $|\mathscr{O}| > 1$ and so $p \mid |\mathscr{O}|$. Let $x \in \mathscr{O}$. If $y \in \mathscr{O}$, then $y = x^u = x[x, u]$ for some $u \in P$. Since $P \lhd G$, we have $[x, u] \in P$ and so $\mathscr{O} \subseteq xP$.

Now let $\mathfrak{Y}$ be an irreducible F-representation of G so that $P \subseteq \ker \mathfrak{Y}$ by Lemma 15.37 and thus $\mathfrak{Y}$ has the constant value $\mathfrak{Y}(x)$ on the coset xP. Therefore $\sum_{y \in \mathscr{O}} \mathfrak{Y}(y) = |\mathscr{O}|\mathfrak{Y}(x) = 0$ since $p \mid |\mathscr{O}|$. We thus have $\mathfrak{Y}(\hat{\mathscr{K}}) = 0$ for all irreducible $\mathfrak{Y}$ and hence $\hat{\mathscr{K}} \in J(F[G])$, the Jacobson radical. It follows that $\hat{\mathscr{K}}$ is nilpotent, ∎

(15.39) COROLLARY Every defect group for a p-block of G contains $\mathbf{O}_p(G)$.

Proof Let $B \in \mathrm{Bl}(G)$ and let $\mathscr{K}$ be a defect class for B. Then $\lambda_B(\hat{\mathscr{K}}) \neq 0$ and hence $\hat{\mathscr{K}}$ is not nilpotent. Thus $\mathscr{K} \cap \mathbf{C}(\mathbf{O}_p(G)) \neq \varnothing$ and it follows that $\mathbf{O}_p(G) \subseteq D$ for every $D \in \delta(\mathscr{K})$. ∎

(15.40) COROLLARY Let G be p-solvable with $\mathbf{O}_{p'}(G) = 1$. Then G has a unique p-block.

Proof Let $B, B' \in \mathrm{Bl}(G)$. We claim that $\lambda_B(e_{B'}) = a_{B'}(\{1\})$. Since this is independent of B, the result will follow. We have $\lambda_B(e_{B'}) = \sum a_{B'}(\mathscr{K})\lambda_B(\hat{\mathscr{K}})$, where the sum runs over classes $\mathscr{K}$. Since $\lambda_B(1) = 1$, it suffices to show that if $\mathscr{K} \neq \{1\}$, then either $a_{B'}(\mathscr{K}) = 0$ or $\lambda_B(\hat{\mathscr{K}}) = 0$.

Let $P = \mathbf{O}_p(G)$. If $\lambda_B(\hat{\mathscr{K}}) \neq 0$, then $\mathscr{K} \cap \mathbf{C}(P) \neq \varnothing$. However, the Hall–Higman "Lemma 1.2.3" yields $\mathbf{C}(P) \subseteq P$ and thus $\mathscr{K} \subseteq P$. Since $\mathscr{K} \neq \{1\}$, the elements of $\mathscr{K}$ are not p-regular and thus $a_{B'}(\mathscr{K}) = 0$. The proof is complete. ∎

If $B \in \mathrm{Bl}(G)$, let D be a defect group for B and write $|D| = p^d$. We call d the *defect* of B and write $d = d(B)$. (This is well defined by Corollary 15.36.) We show how to compute $d(B)$ from a knowledge of $B \cap \mathrm{Irr}(G)$ or $B \cap \mathrm{IBr}(G)$.

We mention that if $m, n \in \mathbb{Z}$ with $p \nmid n$, then $m/n \in \tilde{R}$. Thus if $p \mid m$, we have $m/n \in p\tilde{R} \subseteq \tilde{M}$.

(15.41) THEOREM Let $|G| = p^a m$ with $p \nmid m$ and let $B \in \mathrm{Bl}(G)$ with $d(B) = d$. Then p^{a-d} is the largest power of p which divides all $\chi(1)$ for $\chi \in \mathrm{Irr}(G) \cap B$.

Proof Let $\mathscr{K}$ be a defect class for B and let $g \in \mathscr{K}$. Then $|\mathscr{K}| = |G|/|\mathbf{C}(g)|$ and so p^{a-d} is the p-part of $|\mathscr{K}|$ since $|D| = p^d$ for $D \in \mathrm{Syl}_p(\mathbf{C}(g))$. Let $\chi \in B \cap \mathrm{Irr}(G)$. We have

$$0 \neq \lambda_B(\hat{\mathscr{K}}) = (\omega_\chi)^*(\hat{\mathscr{K}}) = (\chi(g)|\mathscr{K}|/\chi(1))^*.$$

Since $\chi(g) \in R$ and $\chi(g)|\mathscr{K}|/\chi(1) \in \tilde{R} - \tilde{M}$, we conclude that $|\mathscr{K}|/\chi(1) \notin \tilde{M}$ and hence the p-part of $|\mathscr{K}|$ cannot exceed that of $\chi(1)$. Thus p^{a-d} divides $\chi(1)$.

The coefficient of g in the Osima idempotent f_B is given by

$$a_g = (1/|G|) \sum_{\chi \in B \cap \mathrm{Irr}(G)} \chi(1)\overline{\chi(g)}$$

and thus

$$a_g = (1/|G|) \sum_{\varphi \in B \cap \mathrm{IBr}(G)} \Phi_\varphi(1)\overline{\varphi(g)}$$

by Lemma 15.21 (since g is p-regular).

We have $0 \neq a_B(\mathscr{K}) = a_g{}^*$ and $a_g \notin \tilde{M}$. However, $p^a | \Phi_\varphi(1)$ by Corollary 15.22 and thus $\Phi_\varphi(1)/|G| \in \tilde{R}$ for all $\varphi \in \mathrm{IBr}(G)$. We conclude that $\overline{\varphi(g)} \notin \tilde{M}$ for some $\varphi \in \mathrm{IBr}(G) \cap B$.

Now $\varphi(g)$ is a $\mathbb{Z}$-linear combination of $\chi(g)$ for $\chi \in \mathrm{Irr}(G) \cap B$ by Lemma 15.28 and thus $\overline{\chi(g)} \notin \tilde{M}$ for some $\chi \in \mathrm{Irr}(G) \cap B$. Now $\overline{\chi(g)}|\mathscr{K}|/\chi(1) = \alpha \in R$ and $\overline{\chi(g)} = \alpha\chi(1)/|\mathscr{K}|$. It follows that $\chi(1)/|\mathscr{K}| \notin \tilde{M}$ and so the p-part of $\chi(1)$ cannot exceed p^{a-d}. The proof is complete. ∎

(15.42) COROLLARY In the situation of Theorem 15.41, p^{a-d} is the largest power of p which divides all $\varphi(1)$ for $\varphi \in \mathrm{IBr}(G) \cap B$.

Proof In fact $\{\chi(1) | \chi \in \mathrm{Irr}(G) \cap B\}$ and $\{\varphi(1) | \varphi \in \mathrm{IBr}(G) \cap B\}$ have the same greatest common divisor. This follows since each $\chi(1)$ is a $\mathbb{Z}$-linear combination of the $\varphi(1)$ using decomposition numbers and each $\varphi(1)$ is a $\mathbb{Z}$-linear combination of the $\chi(1)$ by Lemma 15.28. ∎

In connection with Corollary 15.42, we mention that $\varphi(1)$ need not divide $|G|$ for $\varphi \in \mathrm{IBr}(G)$.

In the situation of Theorem 15.41, if $\chi \in B \cap \mathrm{Irr}(G)$, then the p-part of $\chi(1)$ can be written in the form p^{a-d+h}, where $h \geq 0$. The integer h is called the *height* of χ. Brauer has conjectured that all $\chi \in \mathrm{Irr}(G) \cap B$ have height zero iff a defect group of B is abelian. Note that if $d(B) = 1$ and $\chi \in B$ has positive height, then p^a divides $\chi(1)$ and thus $B \cap \mathrm{Irr}(G) = \{\chi\}$ by Theorem 15.29. This forces $d(B) = 0$, a contradiction. Thus if $d(B) = 1$, then all characters in B have height zero.

Suppose $H \subseteq G$ and $b \in \mathrm{Bl}(H)$. Let $\lambda = \lambda_b$: $\mathbf{Z}(F[H]) \to F$ be the corresponding central homomorphism. We construct a linear map λ^G: $\mathbf{Z}(F[G]) \to F$

by setting

$$\lambda^G(\hat{\mathscr{K}}) = \lambda\left(\sum_{x \in \mathscr{K} \cap H} x\right).$$

It may be that λ^G is an algebra homomorphism, in which case $\lambda^G = \lambda_B$ for some unique $B \in \mathrm{Bl}(G)$. When this happens we say that b^G is *defined* and we write $b^G = B$. The block b^G is called the *induced* block.

(15.43) LEMMA Let $b \in \mathrm{Bl}(H)$ for $H \subseteq G$ and suppose b^G is defined. Then every defect group for b is contained in a defect group for b^G.

Proof Let $\mathscr{K}$ be a defect class for b^G and let $D \in \delta(b)$. We have $0 \neq (\lambda_b)^G(\hat{\mathscr{K}}) = \lambda_b(\sum \hat{\mathscr{L}})$, where $\mathscr{L}$ runs over the classes of H contained in $\mathscr{K}$. In particular, $\mathscr{K} \cap H \neq \varnothing$ and there exists $\mathscr{L} \subseteq \mathscr{K} \cap H$ such that $\lambda_b(\hat{\mathscr{L}}) \neq 0$. By the Min–Max Theorem 15.31, there exists $P \in \delta(\mathscr{L})$ with $D \subseteq P$. Now $P \in \mathrm{Syl}_p(\mathbf{C}_H(x))$ for some $x \in \mathscr{L} \subseteq \mathscr{K}$ and thus there exists $S \in \mathrm{Syl}_p(\mathbf{C}_G(x))$, with $S \supseteq P$. Thus $D \subseteq S \in \delta(\mathscr{K}) = \delta(b^G)$. ∎

The Brauer homomorphism can be used to give a sufficient condition for induced blocks to be defined.

(15.44) LEMMA Let $P \subseteq G$ be a p-subgroup and let $\mathbf{C}(P) \subseteq H \subseteq \mathbf{N}(P)$. Then b^G is defined for all $b \in \mathrm{Bl}(H)$. If $b \in \mathrm{Bl}(H)$ and $B \in \mathrm{Bl}(G)$, then $b^G = B$ iff $\lambda_B = \beta_P \lambda_b$, where β_P is the Brauer homomorphism.

Proof The image of β_P: $\mathbf{Z}(F[G]) \to \mathbf{Z}(F[\mathbf{N}(P)])$ actually lies in $\mathbf{Z}(F[H])$. Let λ: $\mathbf{Z}(F[H]) \to F$ be an algebra homomorphism and let

$$\mu = \beta_P \lambda\colon \mathbf{Z}(F[G]) \to \mathbf{Z}(F[H]) \to F$$

so that μ is an algebra homomorphism. We claim that $\mu = \lambda^G$.

Let $C = \mathbf{C}(P) \subseteq H$ and let $\mathscr{K}$ be a class of G. Write $\sum_{x \in \mathscr{K} \cap H} x = u + v$, where $u = \sum_{x \in \mathscr{K} \cap C} x$. Then $\lambda^G(\hat{\mathscr{K}}) = \lambda(u + v)$ and $\mu(\hat{\mathscr{K}}) = \lambda(\beta_P(\hat{\mathscr{K}})) = \lambda(u)$. We must therefore show that $\lambda(v) = 0$. Now v is a sum of elements of the form $\hat{\mathscr{L}}$, where $\mathscr{L}$ is a class of H such that $\mathscr{L} \cap C = \varnothing$. Since $P \triangleleft H$ we have $\mathbf{C}(\mathbf{O}_p(H)) \subseteq C$ and it follows that $\hat{\mathscr{L}}$ is nilpotent by Theorem 15.38. Thus $\lambda(\hat{\mathscr{L}}) = 0$ and hence $\lambda(v) = 0$ and $\lambda^G = \mu$ as claimed.

If $\lambda = \lambda_b$, then $\lambda^G = \beta_P \lambda$ is an algebra homomorphism and $b^G = B$ is defined, where B is the unique block such that $\lambda_B = \beta_P \lambda$. ∎

Brauer's first and second "main theorems" concern induced blocks.

(15.45) THEOREM (*First Main*) Let $D \subseteq G$ be a p-subgroup and let $N = \mathbf{N}(D)$. Then $b \mapsto b^G$ is a bijection of

$$\{b \in \mathrm{Bl}(N) \mid D \in \delta(b)\} \qquad \text{onto} \qquad \{B \in \mathrm{Bl}(G) \mid D \in \delta(B)\}.$$

(15.46) LEMMA Let P be a p-subgroup of G. Let $C = \mathbf{C}(P)$ and $N = \mathbf{N}(P)$. Then $\mathscr{K} \mapsto \mathscr{K} \cap C$ is a bijection of the set of classes of G with p-defect group P onto the set of classes of N with p-defect group P.

Proof Suppose $P \in \delta(\mathscr{K})$ and $x, y \in \mathscr{K} \cap C$. Then $P \in \mathrm{Syl}_p(\mathbf{C}(x))$ and $P \in \mathrm{Syl}_p(\mathbf{C}(y))$. Write $y = x^g$. Then $P, P^g \in \mathrm{Syl}_p(\mathbf{C}(y))$ and hence $P = P^{gc}$ for some $c \in \mathbf{C}(y)$. Then $gc \in N$ and $y = x^{gc}$. It follows that $\mathscr{L} = \mathscr{K} \cap C$ is a class of N. Clearly $P \in \delta(\mathscr{L})$ and the map $\mathscr{K} \mapsto \mathscr{K} \cap C$ is one-to-one.

If $\mathscr{L}$ is a class of N with $P \in \delta(\mathscr{L})$, let $\mathscr{K}$ be the unique class of G which contains $\mathscr{L}$. Now $\mathscr{L} \subseteq C$ since $\mathscr{L} \cap C \neq \varnothing$ and $C \lhd N$. If $x \in \mathscr{L}$, let $P \subseteq S \in \mathrm{Syl}_p(\mathbf{C}_G(x))$. If $S > P$, then $P < \mathbf{N}_S(P) = S \cap N \subseteq \mathbf{C}_N(x)$. Since $S \cap N$ is a p-group, this contradicts $P \in \delta(\mathscr{L})$. We conclude that $P = S \in \delta(\mathscr{K})$ and $\mathscr{K} \mapsto \mathscr{L}$. The proof is complete. ∎

Proof of Theorem 15.45 Write $\mathscr{B} = \{b \in \mathrm{Bl}(N) \mid D \in \delta(b)\}$. If $b \in \mathscr{B}$, then b^G is defined by Lemma 15.44. Let $b^G = B \in \mathrm{Bl}(G)$. Since $D \in \delta(b)$, we have $D \subseteq P$ for some $P \in \delta(B)$ by Lemma 15.43. We claim that $D = P$.

Let $\mathscr{L}$ be a defect class of b and let $\mathscr{K} \supseteq \mathscr{L}$ be a class of G. Then $\mathscr{K} \cap C = \mathscr{L}$ and $D \in \delta(\mathscr{K})$ by Lemma 15.46. By Lemma 15.44 we have

$$\lambda_B(\hat{\mathscr{K}}) = \lambda_b(\beta_D(\hat{\mathscr{K}})) = \lambda_b(\hat{\mathscr{L}}) \neq 0.$$

By the Min–Max Theorem 15.31, it follows that D contains some defect group of B. Thus $P^g \subseteq D \subseteq P$ for some $g \in G$ and hence $D = P \in \delta(B)$ as desired. Thus block induction maps $\mathscr{B}$ into $\{B \in \mathrm{Bl}(G) \mid D \in \delta(B)\}$.

Now let $B \in \mathrm{Bl}(G)$ with $D \in \delta(B)$. Let $\mathscr{K}$ be a defect class for B. Then $a_B(\mathscr{K}) \neq 0$ and $D \in \delta(\mathscr{K})$. Thus $\lambda_B = \beta_D \lambda_b$ for some $b \in \mathrm{Bl}(N)$ by Lemma 15.34 and hence $b^G = B$ by Lemma 15.44. We must show that $b \in \mathscr{B}$. Let $P \in \delta(b)$ so that $D \subseteq P$ by Corollary 15.39 since $D \lhd N$. By Lemma 15.43, P is contained in some defect group for B and we have $D \subseteq P \subseteq D^g$ for some $g \in G$. Thus $D = P \in \delta(b)$ and $b \in \mathscr{B}$.

Finally, let $b_1, b_2 \in \mathscr{B}$ with $b_1{}^G = B = b_2{}^G$. Let $\mathscr{L}$ be a class of N with $D \supseteq P \in \delta(\mathscr{L})$. If $D > P$, then by the min–max theorem, $\lambda_{b_1}(\hat{\mathscr{L}}) = 0 = \lambda_{b_2}(\hat{\mathscr{L}})$. If $D = P$, then by Lemma 15.46, $\mathscr{L} = C \cap \mathscr{K}$ for some class $\mathscr{K}$ of G. Then $\lambda_{b_i}(\hat{\mathscr{L}}) = \lambda_{b_i}(\beta_D(\hat{\mathscr{K}})) = \lambda_B(\hat{\mathscr{K}})$ for $i = 1, 2$ and hence the λ_{b_i} agree on all $\hat{\mathscr{L}}$ for classes $\mathscr{L}$ with defect group contained in D. By the min–max theorem, it follows that $\lambda_{b_2}(e_{b_1}) = \lambda_{b_1}(e_{b_1}) = 1$ and thus $b_1 = b_2$. ∎

To state Brauer's "second main theorem" we need to introduce "generalized decomposition numbers."

(15.47) LEMMA Let $\pi \in G$ with $o(\pi) = p^e$ and let $C = \mathbf{C}_G(\pi)$. For $\chi \in \mathrm{Irr}(G)$ and $\varphi \in \mathrm{IBr}(C)$, there exist unique $d^{\pi}_{\chi\varphi} \in R \cap \mathbb{Q}_{p^e}$ such that

$$\chi(x\pi) = \sum_{\varphi \in \mathrm{IBr}(C)} d^{\pi}_{\chi\varphi}\,\varphi(x)$$

for all p-regular $x \in C$.

Proof Write $\chi_C = \sum a_\psi \psi$ with $a_\psi \in \mathbb{Z}$ and $\psi \in \mathrm{Irr}(C)$. We have $\psi_{\mathbf{Z}(C)} = \psi(1)\mu_\psi$ for some linear $\mu_\psi \in \mathrm{Irr}(\mathbf{Z}(C))$. Then $\psi(x\pi) = \psi(x)\mu_\psi(\pi)$ and therefore

$$\chi(x\pi) = \sum_{\psi \in \mathrm{Irr}(C);\, \varphi \in \mathrm{IBr}(C)} a_\psi \mu_\psi(\pi) d_{\psi\varphi} \varphi(x)$$

and so we take $d^\pi_{\chi\varphi} = \sum_{\psi \in \mathrm{Irr}(C)} a_\psi \mu_\psi(\pi) d_{\psi\varphi}$. Uniqueness follows from the linear independence of $\mathrm{IBr}(C)$. ∎

The algebraic integers $d^\pi_{\chi\varphi}$ are called *generalized decomposition numbers.* Note that if $\pi = 1$, then $C = G$ and $d^\pi_{\chi\varphi} = d_{\chi\varphi}$. Also, if $g \in G$ is arbitrary, we can take $\pi = g_p$ and $x = g_{p'}$ and thus express $\chi(g)$ in terms of generalized decomposition numbers and Brauer characters.

Note that if b is a block of $C = \mathbf{C}(\pi)$ for a p-element π then b^G is defined by Lemma 15.44.

(15.48) THEOREM (*Second Main*) Let $\chi \in \mathrm{Irr}(G)$ and $\varphi \in \mathrm{IBr}(C)$, where $C = \mathbf{C}_G(\pi)$ for some p-element $\pi \in G$. Let $\chi \in B \in \mathrm{Bl}(G)$ and $\varphi \in b \in \mathrm{Bl}(C)$. Then $d^\pi_{\chi\varphi} = 0$ if $b^G \neq B$.

Proof Omitted. ∎

Theorem 15.48 is extremely powerful and useful. We give a few consequences.

(15.49) COROLLARY Let $\chi \in \mathrm{Irr}(G)$ and $g \in G$. Suppose g_p is not contained in any defect group for the p-block containing χ. Then $\chi(g) = 0$.

Proof Let $\pi = g_p$ and write $g = \pi x$, where $x \in \mathbf{C}(\pi)$ is p-regular. Then $\chi(g) = \sum_\varphi d^\pi_{\chi\varphi}\varphi(x)$. If $d^\pi_{\chi\varphi} \neq 0$ then $\varphi \in b \in \mathrm{Bl}(\mathbf{C}(\pi))$ and $\chi \in b^G$. Let $Q \in \delta(b)$. Then $\pi \in \mathbf{O}_p(\mathbf{C}(\pi)) \subseteq Q$ by Corollary 15.39. Also, Lemma 15.43 yields $P \in \delta(b^G)$ with $Q \subseteq P$ and thus $\pi \in P$, a contradiction. Thus all $d^\pi_{\chi\varphi} = 0$ and the result follows. ∎

Note that Corollary 15.49 generalizes Theorem 8.17 since if χ has p-defect zero, then the subgroup 1 is the unique defect group for the p-block containing χ.

If ϑ is a class function of G and $B \in \mathrm{Bl}(G)$, write $\vartheta_B = \sum_{\chi \in \mathrm{Irr}(G) \cap B} [\vartheta, \chi]\chi$ so that $\vartheta = \sum_{B \in \mathrm{Bl}(G)} \vartheta_B$.

(15.50) THEOREM Let ϑ be a class function of G and let $\pi \in G$ be a p-element. Suppose $\vartheta(\pi x) = 0$ for all p-regular $x \in \mathbf{C}(\pi)$. Then $\vartheta_B(\pi x) = 0$ for all such x and all $B \in \mathrm{Bl}(G)$.

Proof We have

$$0 = \vartheta(\pi x) = \sum_{\chi \in \mathrm{Irr}(G)} [\vartheta, \chi]\chi(\pi x) = \sum_{\chi \in \mathrm{Irr}(G);\, \varphi \in \mathrm{IBr}(C)} [\vartheta, \chi] d^\pi_{\chi\varphi} \varphi(x)$$

for all p-regular $x \in \mathbf{C}(\pi) = C$. The linear independence of $\mathrm{IBr}(C)$ yields $\sum_{\chi \in \mathrm{Irr}(G)} [\vartheta, \chi] d^{\pi}_{\chi\varphi} = 0$ for each $\varphi \in \mathrm{IBr}(C)$. If $\varphi \in \mathrm{IBr}(C) \cap b$ with $b \in \mathrm{Bl}(C)$, Theorem 15.48 yields $d^{\pi}_{\chi\varphi} = 0$ for $\chi \notin b^G$ and thus $\sum_{\chi \in \mathrm{Irr}(G) \cap b^G} [\vartheta, \chi] d^{\pi}_{\chi\varphi} = 0$.

Now let $\mathscr{A} = \bigcup \{b \cap \mathrm{IBr}(C) \mid b \in \mathrm{Bl}(C), b^G = B\}$. Then for each $\varphi \in \mathscr{A}$ we have $\sum_{\chi \in B} [\vartheta, \chi] d^{\pi}_{\chi\varphi} = 0$ and thus

$$\vartheta_B(\pi x) = \sum_{\chi \in B; \varphi \in \mathscr{A}} [\vartheta, \chi] d^{\pi}_{\chi\varphi} \varphi(x) = 0. \qquad \blacksquare$$

(15.51) COROLLARY (*Block Orthogonality*) Let $g, h \in G$ be such that g_p and h_p are not conjugate in G. Then

$$\sum_{\chi \in \mathrm{Irr}(G) \cap B} \chi(g)\overline{\chi(h)} = 0$$

for every p-block B.

Proof Write $\vartheta = \sum_{\chi \in \mathrm{Irr}(G)} \overline{\chi(h)}\chi$ and let $\pi = g_p$. If $x \in \mathbf{C}(\pi)$ is p-regular, then $\pi = (\pi x)_p$ is not conjugate to h_p in G and hence $\vartheta(\pi x) = 0$ by the second orthgonality relation. Thus $0 = \vartheta_B(g) = \sum_{\chi \in B} \chi(g)\overline{\chi(h)}$ by Theorem 15.50. $\blacksquare$

Problems

(15.1) Let D be the decomposition matrix for G. The matrix $D^{\mathsf{T}} D = C$ is the *Cartan matrix*. (The rows and columns of C are indexed by $\mathrm{IBr}(G)$.) For φ, $\vartheta \in \mathrm{IBr}(G)$, let $\gamma_{\varphi\vartheta} = (1/|G|) \sum_{x \in \mathscr{S}} \varphi(x)\overline{\vartheta(x)}$, where $\mathscr{S}$ is the set of p-regular elements of G. Define the matrix $\Gamma = (\gamma_{\varphi\vartheta})$. Show that $\Gamma = C^{-1}$.

Hint Let X_0 be the part of the character table corresponding to the p-regular classes and let Y be the Brauer character table. Then $X_0 = DY$.

(15.2) (a) Let $\varphi, \vartheta \in \mathrm{IBr}(G)$ lie in different p-blocks. Show that

$$\sum_{x \in \mathscr{S}} \varphi(x)\overline{\vartheta(x)} = 0.$$

(b) Let $\chi, \psi \in \mathrm{Irr}(G)$ lie in different p-blocks. Show that $\sum_{x \in \mathscr{S}} \chi(x)\overline{\psi(x)} = 0$.

(15.3) Let $|G| = p^a m$ with $p \nmid m$ and let $\psi \in \mathrm{Irr}(G) \cup \mathrm{IBr}(G)$. Define the class function, Ξ_ψ by $\Xi_\psi(x) = p^a \psi(x)$ if $p \nmid o(x)$ and $\Xi_\psi(x) = 0$ if $p \mid o(x)$. Show that Ξ_ψ is a generalized character of G. Conclude that $\det(C)$ is a power of p, where C is the Cartan matrix as in Problem 15.1.

(15.4) (a) Show that $\{\Phi_\varphi \mid \varphi \in \mathrm{IBr}(G)\}$ is a basis for the space of class functions on G which vanish on $G - \mathscr{S}$.

(b) If $\varphi, \mu \in \mathrm{IBr}(G)$, show that $(1/|G|) \sum_{x \in \mathscr{S}} \Phi_\varphi(x)\overline{\mu(x)} = \delta_{\varphi\mu}$.

(15.5) (a) Show that the product of two Brauer characters is a Brauer character.

(b) If φ, $\mu \in \mathrm{IBr}(G)$, define Ξ on G by $\Xi(x) = \Phi_\varphi(x)\mu(x)$ for p-regular x; $\Xi(x) = 0$ when $p \mid o(x)$. Show that Ξ is a nonnegative integer linear combination of $\{\Phi_\nu \mid \nu \in \mathrm{IBr}(G)\}$.

(c) If φ, $\mu \in \mathrm{IBr}(G)$, show that $\Phi_\varphi \Phi_\mu$ is a nonnegative integer combination of $\{\Phi_\nu \mid \nu \in \mathrm{IBr}(G)\}$.

(15.6) Let $\mathscr{C}$ be the collection of classes of G which are defect classes for blocks of defect d. Show that $|\{B \in \mathrm{Bl}(G) \mid d(B) = d\}| \leq |\mathscr{C}|$.

Hint Let $A \subseteq \mathbf{Z}(F[G])$ be the span of the $\hat{\mathscr{K}}$ for $\mathscr{K} \in \mathscr{C}$. Then the restrictions of the algebra homomorphisms λ_B to A are linearly independent for those B with defect d.

(15.7) Let $N = \mathbf{O}_{p'}(G)$. Show that if χ, $\psi \in \mathrm{Irr}(G)$ lie in the same p-block, then χ_N and ψ_N have the same irreducible constituents. If n denotes the number of classes of G contained in N, conclude that $|\mathrm{Bl}(G)| \geq n$.

(15.8) Let $P \in \mathrm{Syl}_p(G)$. Show that the number of p-blocks of G with defect group P is equal to the number of p-regular classes of $\mathbf{N}(P)$ contained in $\mathbf{C}(P)$.

Hint Use Problems 15.6 and 15.7.

(15.9) Let N be a normal p-complement for G and let $\mathscr{T}$ be the set of orbits of the action of G on $\mathrm{Irr}(N)$. For each $B \in \mathrm{Bl}(G)$, let $\mathscr{O}(B)$ denote $\{\vartheta \in \mathrm{Irr}(N) \mid [\chi_N, \vartheta] \neq 0$ for some $\chi \in B \cap \mathrm{Irr}(G)\}$. Show that $B \mapsto \mathscr{O}(B)$ is a bijection of $\mathrm{Bl}(G)$ onto $\mathscr{T}$.

(15.10) In the situation and notation of Problem 15.9, show that $\delta(B) = \bigcup_{\vartheta \in \mathscr{O}(B)} \mathrm{Syl}_p(I_G(\vartheta))$.

Hints If $\chi \in \mathrm{Irr}(G)$, $\vartheta \in \mathrm{Irr}(N)$, and $[\chi_N, \vartheta] \neq 0$, then $\omega_\chi(K) = \omega_\vartheta(K)$, where K is any conjugacy class sum of G for a class contained in N. If $\mathscr{K}$ is a defect class for B and $P \in \mathrm{Syl}_p(I_G(\vartheta))$ for some $\vartheta \in \mathscr{O}(B)$, show that P fixes one of the classes of N contained in $\mathscr{K}$ and conclude that P is contained in a defect group for B.

Appendix
Some character tables

In the following tables, each conjugacy class is denoted by the order of its elements. If there are more than one class of elements of a given order, they will be distinguished by subscripts.

1. $G = \Sigma_4$
   $|G| = 24 = 2^3 \times 3$

Class:	1	2_1	2_2	3	4
$\lvert \mathbf{C}(g) \rvert$:	24	4	8	3	4
$\lvert \mathrm{Cl}(g) \rvert$:	1	6	3	8	6
χ_1:	1	1	1	1	1
χ_2:	1	-1	1	1	-1
χ_3:	2	0	2	-1	0
χ_4:	3	1	-1	0	-1
χ_5:	3	-1	-1	0	1

Note Class 2_1 is the class of transpositions.

2. $G = \mathrm{SL}(2, 3)$
$|G| = 24 = 2^3 \times 3$

Class:	1	2	4	3_1	3_2	6_1	6_2
$\lvert\mathbf{C}(g)\rvert$:	24	24	4	6	6	6	6
$\lvert\mathrm{Cl}(g)\rvert$:	1	1	6	4	4	4	4
χ_1:	1	1	1	1	1	1	1
χ_2:	1	1	1	ω	ω^2	ω	ω^2
χ_3:	1	1	1	ω^2	ω	ω^2	ω
χ_4:	3	3	-1	0	0	0	0
χ_5:	2	-2	0	-1	-1	1	1
χ_6:	2	-2	0	$-\omega$	$-\omega^2$	ω	ω^2
χ_7:	2	-2	0	$-\omega^2$	$-\omega$	ω^2	ω

Irrational Entries $\omega = e^{2\pi i/3}$.

3. $G = A_5 \cong \mathrm{PSL}(2, 5) \cong \mathrm{SL}(2, 4)$
$|G| = 60 = 2^2 \times 3 \times 5$

Class:	1	2	3	5_1	5_2
$\lvert\mathbf{C}(g)\rvert$:	60	4	3	5	5
$\lvert\mathrm{Cl}(g)\rvert$:	1	15	20	12	12
χ_1:	1	1	1	1	1
χ_2:	4	0	1	-1	-1
χ_3:	5	1	-1	0	0
χ_4:	3	-1	0	α_1	α_2
χ_5:	3	-1	0	α_2	α_1

Irrational Entries $\alpha_1 = (1 + \sqrt{5})/2 = 1 + \varepsilon + \varepsilon^4$ and $\alpha_2 = (1 - \sqrt{5})/2 = 1 + \varepsilon^2 + \varepsilon^3$, where $\varepsilon = e^{2\pi i/5}$.

4. $G = \mathrm{PSL}(2, 7) \cong \mathrm{GL}(3, 2)$
$|G| = 168 = 2^3 \times 3 \times 7$

Class:	1	2	4	3	7_1	7_2
$\lvert \mathbf{C}(g) \rvert$:	168	8	4	3	7	7
$\lvert \mathrm{Cl}(g) \rvert$:	1	21	42	56	24	24
χ_1:	1	1	1	1	1	1
χ_2:	6	2	0	0	-1	-1
χ_3:	7	-1	-1	1	0	0
χ_4:	8	0	0	-1	1	1
χ_5:	3	-1	1	0	α	$\bar{\alpha}$
χ_6:	3	-1	1	0	$\bar{\alpha}$	α

Irrational Entries $\alpha = (-1 + i\sqrt{7})/2 = \varepsilon + \varepsilon^2 + \varepsilon^4$, where $\varepsilon = e^{2\pi i/7}$.

5. $G = A_6 \cong \mathrm{PSL}(2, 9)$
$|G| = 360 = 2^3 \times 3^2 \times 5$

Class:	1	2	4	3_1	3_2	5_1	5_2
$\lvert \mathbf{C}(g) \rvert$:	360	8	4	9	9	5	5
$\lvert \mathrm{Cl}(g) \rvert$:	1	45	90	40	40	72	72
χ_1:	1	1	1	1	1	1	1
χ_2:	5	1	-1	2	-1	0	0
χ_3:	5	1	-1	-1	2	0	0
χ_4:	9	1	1	0	0	-1	-1
χ_5:	10	-2	0	1	1	0	0
χ_6:	8	0	0	-1	-1	α_1	α_2
χ_7:	8	0	0	-1	-1	α_2	α_1

Irrational Entries $\alpha_1 = (1 + \sqrt{5})/2 = 1 + \varepsilon + \varepsilon^4$ and $\alpha_2 = (1 - \sqrt{5})/2 = 1 + \varepsilon^2 + \varepsilon^3$, where $\varepsilon = e^{2\pi i/5}$.

6. $G = \mathrm{SL}(2, 8)$
$|G| = 504 = 2^3 \times 3^2 \times 7$

Class:	1	2	3	9_1	9_2	9_3	7_1	7_2	7_3
$\|\mathbf{C}(g)\|$:	504	8	9	9	9	9	7	7	7
$\|\mathrm{Cl}(g)\|$:	1	63	56	56	56	56	72	72	72
χ_1:	1	1	1	1	1	1	1	1	1
χ_2:	8	0	-1	-1	-1	-1	1	1	1
χ_3:	7	-1	-2	1	1	1	0	0	0
χ_4:	7	-1	1	β_1	β_2	β_3	0	0	0
χ_5:	7	-1	1	β_2	β_3	β_1	0	0	0
χ_6:	7	-1	1	β_3	β_1	β_2	0	0	0
χ_7:	9	1	0	0	0	0	α_1	α_2	α_3
χ_8:	9	1	0	0	0	0	α_2	α_3	α_1
χ_9:	9	1	0	0	0	0	α_3	α_1	α_2

Irrational Entries $\alpha_1 = \varepsilon + \varepsilon^6$, $\alpha_2 = \varepsilon^2 + \varepsilon^5$ and $\alpha_3 = \varepsilon^3 + \varepsilon^4$, where $\varepsilon = e^{2\pi i/7}$. $\beta_1 = -(\delta + \delta^8)$, $\beta_2 = -(\delta^2 + \delta^7)$, and $\beta_3 = -(\delta^4 + \delta^5)$, where $\delta = e^{2\pi i/9}$.

7. $G = \mathrm{PSL}(2, 11)$
$|G| = 660 = 2^2 \times 3 \times 5 \times 11$

Class:	1	2	3	6	5_1	5_2	11_1	11_2
$\|\mathbf{C}(g)\|$:	660	12	6	6	5	5	11	11
$\|\mathrm{Cl}(g)\|$:	1	55	110	110	132	132	60	60
χ_1:	1	1	1	1	1	1	1	1
χ_2:	10	2	1	-1	0	0	-1	-1
χ_3:	10	-2	1	1	0	0	-1	-1
χ_4:	11	-1	-1	-1	1	1	0	0
χ_5:	12	0	0	0	α_1	α_2	1	1
χ_6:	12	0	0	0	α_2	α_1	1	1
χ_7:	5	1	-1	1	0	0	β	$\bar{\beta}$
χ_8:	5	1	-1	1	0	0	$\bar{\beta}$	β

Irrational Entries $\alpha_1 = (-1 + \sqrt{5})/2 = \varepsilon + \varepsilon^4$, $\alpha_2 = (-1 - \sqrt{5})/2 = \varepsilon^2 + \varepsilon^3$, where $\varepsilon = e^{2\pi i/5}$. $\beta = (-1 + i\sqrt{11})/2 = \delta + \delta^3 + \delta^4 + \delta^5 + \delta^9$, where $\delta = e^{2\pi i/11}$.

8. $G = M_{11}$
$|G| = 7920 = 2^4 \times 3^2 \times 5 \times 11$

Class:	1	2	4	8_1	8_2	3	6	5	11_1	11_2
$\lvert \mathbf{C}(g) \rvert$:	7920	48	8	8	8	18	6	5	11	11
$\lvert \mathrm{Cl}(g) \rvert$:	1	165	990	990	990	440	1320	1584	720	720
χ_1:	1	1	1	1	1	1	1	1	1	1
χ_2:	10	2	2	0	0	1	-1	0	-1	-1
χ_3:	11	3	-1	-1	-1	2	0	1	0	0
χ_4:	10	-2	0	α	$\bar{\alpha}$	1	1	0	-1	-1
χ_5:	10	-2	0	$\bar{\alpha}$	α	1	1	0	-1	-1
χ_6:	16	0	0	0	0	-2	0	1	β	$\bar{\beta}$
χ_7:	16	0	0	0	0	-2	0	1	$\bar{\beta}$	β
χ_8:	44	4	0	0	0	-1	1	-1	0	0
χ_9:	45	-3	1	-1	-1	0	0	0	1	1
χ_{10}:	55	-1	-1	1	1	1	-1	0	0	0

Irrational Entries $\alpha = i\sqrt{2}$. $\beta = (-1 + i\sqrt{11})/2 = \varepsilon + \varepsilon^3 + \varepsilon^4 + \varepsilon^5 + \varepsilon^9$, where $\varepsilon = e^{2\pi i/11}$.

Bibliographic notes

There exist several excellent bibliographies in group theory and character theory (for instance in the books by Huppert [26], Dornhoff [15], and Curtis and Reiner [12]) and so, instead of giving a comprehensive list of publications, we shall only mention some of the items which are directly relevant to the various chapters of this book.

General Some other books on characters and representations are Dornhoff [15], Feit [16], and Curtis and Reiner [12]. Each of these has a point of view somewhat different from the others and from this book. As a reference on group theory we mention Huppert [26] which also has an extensive chapter on characters. Finally, we come to Burnside [10]. This classic, although somewhat difficult for the modern reader, contains a wealth of material.

Chapter 1 Further relevant information on rings and algebras can be found in Curtis and Reiner [12] and Herstein [25].

Chapter 2 Other methods exist for obtaining the basic results about characters such as the orthogonality relations. For instance, instead of using the central idempotents of $\mathbb{C}[G]$, Feit [16] and Dornhoff [15] use a matrix approach which results in additional information, namely the Schur relations which appear here as Problem 2.20.

Chapter 3 There is, of course, a large literature in algebraic number theory. A reference for those parts of the subject most relevant to group theory is the appropriate chapter in Curtis and Reiner [12]. The proof of

Theorem 3.12 given here was discovered (independently) by G. Glauberman and the author. For another proof, see (4.2) of [16]. Whitcomb's result on isomorphism of integer group rings occurs in Whitcomb [37].

Chapter 4 The Brauer–Fowler proof [8] of Theorem 4.11(a) does not depend on characters. They obtain a slightly better bound which is of the same order of magnitude.

Chapter 5 There is a great deal more to be said about the relationship between permutation groups and representation theory. For instance, see Chapter V of Wielandt [38] and Sections V.20 and V.21 of Huppert [26].

Chapter 6 Clifford's results appear in Ref. [11]. This paper includes significantly more than Theorem 6.5; it also has part of Theorem 6.11 and bears heavily on Chapter 11. The "going down" Theorem 6.18 and its dual Problem 6.12 appears in Isaacs [28], however Corollary 6.19 goes back to Burnside's book [10]. A result on relative M-groups which is more general than Theorem 6.22 occurs in Price [35]. Theorems 6.22 and 6.23 give sufficient conditions for a group to be an M-group which generalize a result of Huppert (Satz V.18.4 of Huppert [26]). The connections between the extendibility of ϑ and $\det(\vartheta)$ are due to Gallagher [19]. A proof of a version of Tate's Theorem 6.31 without characters appears as Satz IV.4.7 of Huppert [26] and Thompson's proof (including Problem 6.20) is found in Ref. [36].

Chapter 7 A proof of Theorem 7.8 for the case that $|P| = 8$ that does not depend on "modular characters" has recently been discovered by Glauberman [22]. For a proof of Theorem 7.10 without characters, see Bender [2]. A large fraction of the known applications of character theory to "pure" group theory are either directly or indirectly related to the content of this chapter. We mention as examples Sections 28 and 32 of Feit [16].

Chapter 8 Brauer's Theorem 8.4 occurs in Brauer and Tate [9] and in some of Brauer's earlier papers. Banaschewski's Lemma 8.5 appears (in a somewhat more complicated form) in Ref. [1]. Dade's Theorem 8.24 and its consequence Theorem 8.26 appears in more general form in Ref. [13].

Chapter 9 An alternate source for much of this material is Curtis and Reiner [12].

Chapter 10 The more standard version of Theorem 10.7 is due to Brauer and Witt (Theorem 70.28 of [12] or Yamada's notes [41]). Theorem 10.12 appears in Goldschmidt and Isaacs [24]. What amounts to a special case of Theorem 10.16 occurs in Burnside [10] as Exercise 8 on page 319. That every integer can occur as a Schur index was proved by Brauer [5] using groups similar to those of Theorem 10.16. A great deal of further information can be found in Yamada [41].

Chapter 11 Much of the theory relating projective representations with the properties of a character triple was originated by Clifford [11]. Our more character theoretic setting of Clifford's work occurs in Isaacs [31]. Berger's Theorem 11.33 appears in more general form in Ref. [3].

Chapter 12 Much of this material is the work of Passman and the author and appears in Isaacs and Passman [33, 34], and Isaacs [29]. Also relevant are Isaacs [32], Berger [4], and Garrison [20].

Chapter 13 This chapter is taken almost entirely from Glauberman's paper [21]. Parts of Theorems 13.6 and 13.14 also appear in Isaacs [27]. A proof of Theorem 13.25 can be found in Isaacs [31].

Chapter 14 The literature on linear groups is very extensive and we mention just a sample. Dixon's book [14] is a good reference. For information on solvable and p-solvable linear groups, see Winter [39, 40] and Isaacs [30]. We also mention the lecture notes by Feit and Sibley [18] for results without solvability hypotheses.

Chapter 15 Brauer's papers [6] and [7] are good sources for further reading on blocks and Brauer characters. There is also a chapter on the subject in Curtis and Reiner [12]. The material is treated from a different point of view in Part B of Dornhoff [15] and in Feit's notes [17]. Goldschmidt's notes [23] provide a development of the subject along lines similar to those used here.

References

1. B. Banaschewski, On the character rings of finite groups, *Canad. J. Math.* **15** (1963), 605–612.
2. H. Bender, Finite groups with large subgroups, *Illinois J. Math.* **18** (1974), 223–228.
3. T. R. Berger, Primitive solvable groups, *J. Algebra* **33** (1975), 9–21.
4. T. R. Berger, Characters and derived length in groups of odd order, *J. Algebra* (to be published).
5. R. Brauer, Gruppen linearer Substitutionen II, *Math. Z.* **31** (1930), 733–747.
6. R. Brauer, Zur Darstellungstheorie der Gruppen endlicher Ordnung I, II, *Math. Z.* **63** (1956) 406–444; **72** (1959), 25–46.
7. R. Brauer, Some applications of the theory of blocks of characters I–V, *J. Algebra* **1** (1964), 152–167; **1** (1964), 307–334; **3** (1966), 225–255; **17** (1971), 489–521; **28** (1974), 433–460.
8. R. Brauer and K. A. Fowler, Groups of even order, *Ann. Math.* (2) **62** (1955), 565–583.
9. R. Brauer and J. Tate, On the characters of finite groups, *Ann. Math.* (2) **62** (1955), 1–7.
10. W. Burnside, *Theory of groups of finite order* (2nd ed. 1911), reprinted by Dover, New York, 1955.
11. A. H. Clifford, Representations induced in an invariant subgroup, *Ann. Math.* (2) **38** (1937), 533–550.
12. C. W. Curtis and I. Reiner, *Representation theory of finite groups and associative algebras*, Wiley (Interscience), New York, 1962.
13. E. C. Dade, Isomorphisms of Clifford extensions, *Ann. Math.* (2) **92** (1970), 375–433.
14. J. D. Dixon, *The structure of linear groups*, Van Nostrand-Reinhold, Princeton, New Jersey, 1971.
15. L. Dornhoff, *Group representation theory*, Dekker, New York, 1971.
16. W. Feit, *Characters of finite groups*, Benjamin, New York, 1967.
17. W. Feit, *Representations of finite groups*, Mimeographed notes, Yale Univ., 1969.
18. W. Feit and D. A. Sibley, *Finite linear groups of relatively small degree*, Mimeographed notes, Yale Univ., 1974.
19. P. X. Gallagher, Group characters and normal Hall subgroups, *Nagoya Math. J.* **21** (1962), 223–230.

20. S. C. Garrison, *On groups with a small number of character degrees*, Ph.D. Thesis, Univ. of Wisconsin, Madison, 1973.
21. G. Glauberman, Correspondences of characters for relatively prime operator groups, *Canad. J. Math.* **20** (1968), 1465–1488.
22. G. Glauberman, On groups with a quaternion Sylow 2-subgroup, *Illinois J. Math.* **18** (1974), 60–65.
23. D. M. Goldschmidt (to be published).
24. D. M. Goldschmidt and I. M. Isaacs, Schur indices in finite groups, *J. Algebra* **33** (1975), 191–199.
25. I. N. Herstein, *Noncommutative rings*, Carus monograph **15**, Math. Assoc. Amer., 1968.
26. B. Huppert, *Endliche Gruppen I*, Springer–Verlag, Berlin, 1967.
27. I. M. Isaacs, Extensions of certain linear groups, *J. Algebra* **4** (1966), 3–12.
28. I. M. Isaacs, Fixed points and characters . . . , *Canad. J. Math.* **20** (1968), 1315–1320.
29. I. M. Isaacs, Groups having at most three irreducible character degrees, *Proc. Amer. Math. Soc.* **21** (1969), 185–188.
30. I. M. Isaacs, Complex p-solvable linear groups, *J. Algebra* **24** (1973), 513–530.
31. I. M. Isaacs, Characters of solvable and symplectic groups, *Amer. J. Math.* **95** (1973), 594–635.
32. I. M. Isaacs, Character degrees and derived length of a solvable group, *Canad. J. Math.* **27** (1975), 146–151.
33. I. M. Isaacs and D. S. Passman, A characterization of groups in terms of the degrees of their characters I, II, *Pacific J. Math.* **15** (1965), 877–903; **24** (1968), 467–510.
34. I. M. Isaacs and D. S. Passman, Finite groups with small character degrees and large prime divisors II, *Pacific J. Math.* **29** (1969), 311–324.
35. D. Price, *A generalization of M-groups*, Ph.D. Thesis, Univ. of Chicago, 1971.
36. J. G. Thompson, Normal p-complements and irreducible characters, *J. Algebra* **14** (1970), 129–134.
37. A. Whitcomb, *The group ring problem*, Ph.D. Thesis, Univ. of Chicago, 1971.
38. H. Wielandt, *Finite permutation groups*, Academic Press, New York, 1964.
39. D. L. Winter, p-solvable linear groups of finite order, *Trans. Amer. Math. Soc.* **157** (1971), 155–160.
40. D. L. Winter, On the structure of certain p-solvable linear groups II, *J. Algebra* **33** (1975), 170–190.
41. T. Yamada, *The Schur subgroup of the Brauer group*, Springer–Verlag, Berlin, 1974.

Index

D

E

F

G

H

I

J

K

L

M

N

O

P

Q

R

S

A 6
B 7
C 8
D 9
E 0
F 1
G 2
H 3
I 4
J 5